AF775123

Proceedings
of the
STEKLOV INSTITUTE
OF MATHEMATICS

1983, ISSUE 1

Spectral Theory of Functions and Operators. II

Edited by

N. K. Nikol′skiĭ

Translation of
ТРУДЫ
ордена Ленина
МАТЕМАТИЧЕСКОГО ИНСТИТУТА
имени В. А. СТЕКЛОВА

Том 155 (1981)

АКАДЕМИЯ НАУК

СОЮЗА СОВЕТСКИХ СОЦИАЛИСТИЧЕСКИХ РЕСПУБЛИК

ТРУДЫ

ордена Ленина

МАТЕМАТИЧЕСКОГО ИНСТИТУТА

имени В. А. СТЕКЛОВА

CLV

СПЕКТРАЛЬНАЯ ТЕОРИЯ ФУНКЦИЙ И ОПЕРАТОРОВ. II

Ответственный редактор (Editor-in-chief)
академик С. М. НИКОЛЬСКИЙ (S. M. Nikol'skiĭ)

Заместитель ответственного редактора (Assistant to the editor-in-chief)
профессор Е. А. ВОЛКОВ (E. A. Volkov)

издательство "наука"
Ленинград 1981

ABSTRACT. The papers comprising this collection are devoted to the connections of spectral theory with the theory of functions of a complex variable. The classical problem of interpolation by uniformly convergent Fourier or Taylor series is illuminated, and imbedding theorems dual to interpolation problems are proved; a constructive characterization of Besov and Sobolev classes, well-known and widely used in analysis, is given; an analogue of von Neumann's inequality is obtained by a combination of spectral theory methods, combinatorial constructions and complex analysis; positive projections in Banach lattices are described.

1980 *Mathematics Subject Classification.* Primary 30G20, 42A16, 42A20, 42A28, 42A50, 42B05, 42B30, 42B99, 46B30, 46E35, 47A20, 47A60, 47B38, 47B55, 60G15, 60G17; Secondary 22C05, 22D40, 26A15, 26B15, 30B30, 30E05, 41A10, 41A17, 42A24, 42A55, 42C10, 43A22, 46E15, 46E30, 47A35, 47B37, 76F99.

Library of Congress Cataloging in Publication Data
(Revised for volume 4)
Main entry under title:

Spectral theory of functions and operators.

(Proceedings of the Steklov Institute of Mathematics; 1978, issue 4–)
Translation of Spektral' nai͡a teorii͡a funkt͡sii i operatov.
Numbers in Russian series statement: tom 130 (1978), tom 155 (1981).
Bibliography: p.
1. Analytic functions–Addresses, essays, lectures. 2. Spectral theory (Mathematics)–Addresses, essays, lectures. 3. Linear operators–Addresses, essays, lectures. I. Nikolskiĭ, Sergeĭ Mikhaĭlovich. II. Trudy Ordena Lenina Matematicheskogo instituta imeni V. A. Steklova. English.; 1978, issue 4, etc.
QA1.A413 no. 130 [QA331] 510s [515.9] 80-11102
ISBN 0-8212-3072-4 (v. 2)

March 1983

Translation authorized by the All-Union Agency for Author's Rights, Moscow

Information on Copying and Reprinting can be found at the back of this journal.

Copyright © 1983, by the American Mathematical Society

PROCEEDINGS OF THE STEKLOV INSTITUTE OF MATHEMATICS IN THE ACADEMY OF SCIENCES OF THE USSR

TABLE OF CONTENTS

1983 subscription consists of 4 issues, published quarterly by American Mathematical Society. Subscription price of 1983 is list $226.00, member $113.00. Previously translated Russian Numbers are sold as separate books. Write to the Society for prices.

Subscription and orders should be addressed to American Mathematical Society, P. O. Box 1571, Annex Station, Providence, R. I. 02901. *All orders must be accompanied by payment.* Other correspondence should be addressed to P. O. Box 6248, Providence, R. I. 02940.

Proceedings of the Steklov Institute of Mathematics (ISSN 0081-5438) is published quarterly by the American Mathematical Society at 201 Charles Street, Providence, Rhode Island 02904. Second Class postage paid at Providence, Rhode Island 02940. Postmaster: Send address changes to Proceedings of the Steklov Institute of Mathematics, American Mathematical Society, P. O. Box 6248, Providence, RI 02940.

RUSSIAN TABLE OF CONTENTS*

* The American Mathematical Society scheme for transliteration of Cyrillic may be found at the end of index issues of *Mathematical Reviews.*

Correspondence between *Trudy Mat. Inst. Steklov.* and *Proc. Steklov Inst. Math.*

Russian imprint	Vol.	English imprint
1966	74	1967
1965	75	1967
1965	76	1967
1965	77	1967
1965	78	1967
1965	79	1966
1965	80	1968
1966	81	1968
1966	82	1967
1965	83	1967
1965	84	1968
1966	85	1967
1965	86	1967
1966	87	1967
1967	88	1969
1967	89	1968
1967	90	1969
1967	91	1969
1966	92	1968
1967	93	1970
1968	94	1969
1968	95	1971
1968	96	1970
1968	97	1969
1968	98	1971
1967	99	1968
1971	100	1974
1972	101	1975
1967	102	1970
1968	103	1970
1968	104	1971
1969	105	1971
1969	106	1972

Russian imprint	Vol.	English imprint
1971	107	1976
1968	108	1971
1971	109	1974
1970	110	1972
1970	111	1972
1971	112	1973
1970	113	1972
1970	114	1974
1971	115	1974
1971	116	1973
1972	117	1974
1972	118	1976
1973	119	1976
1974	120	1976
1972	121	1974
1973	122	1975
1973	123	1975
1976	124	1978, Issue 3
1973	125	1975
1973	126	1975
1975	127	1977
1972	128	1974
1973	129	1976
1978	130	1979, Issue 4
1974	131	1975
1973	132	1975
1973	133	1977
1975	134	1977
1975	135	1978, Issue 1
1975	136	1978, Issue 2
1976	137	1978, Issue 4
1975	138	1977

Russian imprint	Vol.	English imprint Year	Issue
1980	139	1982	1
1976	140	1979	1
1976	141	1979	2
1976	142	1979	3
1977	143	1980	1
1979	144	1980	2
1980	145	1981	1
1978	146	1980	3
1980	147	1981	2
1978	148	1980	4
1978	149	1981	3
1979	150	1981	4
1980	151	1982	2
1980	152	1982	3
1981	153	1982	4
1981	155	1983	1
1980	156	1983	2

EDITOR'S PREFACE

The papers in this collection are in the mainstream of the ideas outlined in the preface to the preceding issue under the same title (Trudy Mat. Inst. Steklov. **130** (1978) = Proc. Steklov Inst. Math. **1979**, Issue 4). These papers are again at the crossroads of spectral theory and theory of functions of complex variables with respect to both formulation of problems and methods of investigation. In three papers (those by S. A. Vinogradov, S. V. Kislyakova and S. V. Khrushchev) the classical problem of interpolation by uniformly convergent Fourier or Taylor series is considered from different viewpoints. The traces of such functions are studied in detail on subsets of a circle (of tori, of compact abelian groups); the behavior of Taylor coefficients is studied (their moduli turned out to be indistinguishable, in a certain sense, from the moduli of Taylor coefficients for an arbitrary function in the Hardy space H^2); embedding theorems dual to interpolation problems are proved (for the Cauchy transform of functionals over the space under investigation).

In two more papers (those of E. M. Dyn'kin and V. V. Peller) the theory of functions of a complex variable is interwoven with purely real methods and methods of harmonic analysis. In the former paper a constructive characterization is given of well-known Besov and Sobolev spaces that are widely used in analysis, and here it has turned out to be essential both conceptually and technically to regard a line segment as a degenerate case of a domain in the complex plane $\mathbf{C}$. In the latter paper a purely quantitative result (an analog of von Neumann's inequality for L^p-spaces and general spaces of measurable functions) is obtained by combining methods of spectral theory, combinatorial constructions and complex analysis. In the paper preceding Peller's V. G. Kulakova describes positive projections in KB-spaces.

Returning to the idea of a close relationship between complex analysis and spectral theory stated in the preface to the first issue of the same title, I should like again to note with satisfaction the spectral nature of the main problems and results published here.

N. K. NIKOL'SKIĬ

Труды Матем. Инст. Стеклов.
155(1981)

Proceedings of the Steklov
Institute of Mathematics
1983, Issue 1

A STRENGTHENED FORM OF KOLMOGOROV'S THEOREM ON THE CONJUGATE FUNCTION AND INTERPOLATION PROPERTIES OF UNIFORMLY CONVERGENT POWER SERIES

UDC 517.948: 513.8 + 519.4

S. A. VINOGRADOV

ABSTRACT. The first part of the paper (§2) is devoted to the proof of the following stronger form of a theorem of Kolmogorov on conjugate functions:

$$\operatorname{meas}\{\theta \in [-\pi, \pi]: |(K\mu)(re^{i\theta})| > t\} \leqslant \frac{a_0}{t} \sum_{n\geqslant 0} \operatorname{var} \mu_n, \quad t > 0, r \in (0, 1), \quad (*)$$

where

$$(K\mu)(z) = \sum_{n\geqslant 0} z^n \int_{\mathbf{T}} \frac{d\mu_n(\zeta)}{1 - \bar{\zeta} z}, \qquad |z| < 1,$$

a_0 is an absolute constant, $(\mu_n)_{n\geqslant 0}$ is a sequence of complex Borel measures on the unit circle $\mathbf{T} = \{\zeta \in \mathbf{C}: |\zeta| = 1\}$, for which $\sum_{n\geqslant 0} \operatorname{var} \mu_n < +\infty$, where $\operatorname{var} \mu_n$ is the total variation of the measure μ_n and meas is Lebesgue measure on the line. Functions of the form $K\mu$ unavoidably appear in the analysis of the space U_A of power series that converge uniformly in the unit disk, and correspond in a natural way to the bounded linear functionals on that space. In the second part (§§3 and 4) the inequality (*) is applied to the study of the behavior of Taylor coefficients and the values of functions of class U_A and other closely related classes of functions on "sparse" sets.

Bibliography: 29 titles.

Contents

§0. Introduction

In the first part of the paper (§§1 and 2) we show that the function $K\mu$,

$$(K\mu)(\eta) \stackrel{\text{def}}{=} \sum_{n\geqslant 0} \eta^n \int_{|\zeta|=1} \frac{d\mu_n(\zeta)}{1 - \bar{\zeta}\eta}, \qquad |\eta| < 1,$$

belongs to every Hardy class H^p with $p \in (0, 1)$, for an arbitrary sequence of complex measures $\mu = (\mu_n)_{n\geqslant 0}$ on the unit circle $\mathbf{T}$, provided that $\sum_{n\geqslant 0} \operatorname{var} \mu_n < +\infty$. If $\mu_n \equiv 0$ for all sufficiently large n, this becomes the well-known theorem of

1980 *Mathematics Subject Classification*. Primary 42A50; Secondary 42A55, 30B30, 30E05, 46E15.

©1983 American Mathematical Society
0081-5438/83 $1.00 + $.25 per page

Smirnov [1] on integrals of Cauchy-Stieltjes type, which in turn follows from Kolmogorov's [2] inequality for the conjugate function. We deduce the relation $K\mu \in \bigcap_{0<p<1} H^p$ from the following strengthened form of Kolmogorov's theorem (see §2):

$$\operatorname{meas}\left\{\theta \in [-\pi, \pi]: \left|(K\mu)(re^{i\theta})\right| > t\right\} \leqslant \frac{a_0}{t} \sum_{n\geqslant 0} \operatorname{var} \mu_n, \qquad t > 0, r \in (0, 1), \quad (0.1)$$

where a_0 is an absolute constant, $a_0 \in (0, +\infty)$, and meas denotes Lebesgue measure on the line. This is based in an essential way on Carleson's theorem on the almost everywhere convergence of the Fourier series of elements of the space $L^2(\mathbb{T})$.

Functions of the form $K\mu$ inevitably appear in the study of the space U_A of power series that converge uniformly in the unit disk $\mathbb{D}$, and correspond in a natural way to the continuous linear functionals on that space (see §3). The inequality (0.1) will be applied in the second part of the paper (§§3 and 4) to the study of the Taylor coefficients of functions of class U_A (and of other closely related classes). In addition, this second part contains stronger forms of the author's earlier results [3]–[5]; some of these will be stated presently. We also answer (see §4, Theorem 4.17) a question asked by Stechkin in [6].

However, inequality (0.1) seems to have independent interest. We outline the main features of its proof, which follows the conventional plan of obtaining estimates for singular integral operators in L^p spaces (see, for example, [7]). In this plan a decisive role is played, as a rule, first by the boundedness of the operator in L^{p_0} for some $p_0 > 1$ (often $p_0 = 2$), and second, by some smoothness property of the kernel. The first of these is easily established for many singular operators. In our case, however, it is equivalent to having a function of the form

$$(Kf)(\eta) \overset{\text{def}}{=} \sum_{n\geqslant 0} \eta^n \int_{|\zeta|=1} \frac{f_n(\zeta)\, d\zeta}{\zeta - \eta}, \qquad |\eta| < 1, f = (f_n)_{n\geqslant 0},$$

belonging to H^{p_0}, where $\sum_{n\geqslant 0} |f_n| \in L^{p_0}(\mathbb{T})$, i.e. to the inequality

$$\|Kf\|_{H^{p_0}} \leqslant \text{const} \left\| \sum_{n\geqslant 0} |f_n| \right\|_{L^{p_0}}.$$

This last inequality—a strengthened form of M. Riesz's theorem—is nontrivial in the highest degree, even for $p_0 = 2$. It is easily shown (this is done in §2) that when $p_0 = 2$ it is equivalent to the Carleson-Hunt theorem [8] on the behavior of the partial sums of the Fourier series of an element of $L^2(\mathbb{T})$.

We note that for the proof of the main theorem of the first part (Theorem 2.4) we use only Carleson's result [9] on the almost everywhere convergence of the Fourier series of an element of $L^2(\mathbb{T})$ (without Hunt's estimate of the supremum of the partial sums). We do this in order to indicate one of the simplest methods for deriving Hunt's estimates for $p > 2$ from Carleson's result. In addition, if instead of using Carleson's difficult theorem we use only Kolmogorov's result [10] on the partial sums of Fourier series whose indices are chosen from a lacunary sequence, we can, by following our proof of Theorem 2.4, easily obtain a different strengthened form of Kolmogorov's theorem on the conjugate function (of course, weaker than (0.1)):

$$\operatorname{meas}\left\{\theta \in [-\pi, \pi]: \left|(K_0\mu)(re^{i\theta})\right| > t\right\} \leqslant \frac{a_1}{t} \sum_{n\geqslant 0} \operatorname{var} \mu_n, \qquad t > 0, r \in (0, 1),$$

where

$$(K_0\mu)(\eta) = \sum_{n\geqslant 0} \eta^{k_n} \int \frac{d\mu_n(\zeta)}{1-\bar{\zeta}\eta}, \qquad \eta \in \mathbb{D},$$

$\mu = (\mu_n)_{n\geqslant 0}$ is a sequence of complex measures concentrated on the circle $\mathbb{T}$, such that $\sum_{n\geqslant 0} \operatorname{var} \mu_n < +\infty$, $(k_n)_{n\geqslant 0}$ is a lacunary sequence of positive integers ($\sup\{k_n/k_{n+1}: n = 0, 1, 2, \ldots\} < 1$), and a_1 is an absolute constant, $a_1 \in (0, +\infty)$.

We now give a short outline of the paper and at the same time mention the most representative results. §1 contains basic notation and some preliminary remarks. §2 is devoted to the proof of the strengthened Kolmogorov theorem on the conjugate function. In §3 we present the argument used to obtain (sometimes with the use of the propositions in §2) results of the following kind.

0.1. THEOREM. *There is an absolute constant $a_2 > 0$ such that, for every positive integer n and every sequence $(d_k)_{k\geqslant 0}$ of nonnegative numbers,*

$$a_2\left(\sum_{k=0}^{n} d_k^2\right)^{1/2} \leqslant \sup\left\{\sum_{k=0}^{n} d_k|\hat{f}(k)| : f \in U_A, \|f\|_{U_A} \leqslant 1\right\} \leqslant \left(\sum_{k=0}^{n} d_k^2\right)^{1/2},$$

where

$$f(\eta) = \sum_{k\geqslant 0} \hat{f}(k)\eta^k, \qquad \eta \in \overline{\mathbb{D}};$$

$$\|f\|_{U_A} = \sup\left\{\left|\sum_{k=n}^{\infty} \hat{f}(k)\eta^k\right| : |\eta| < 1, n = 0, 1, 2, \ldots\right\}$$

(*the definition of U_A has been given above*).

This theorem strengthens one of Paley's results [11]. Recently S. V. Kislyakov, using a method from papers of de Leeuw, Katznelson, and Kahane, and inequality (0.1), has obtained an even stronger result than Theorem 0.1 (see his article in this issue).

0.2. THEOREM. *Let E be a closed countable subset of the unit circle $\mathbb{T}$, and $(n_k)_{k\geqslant 0}$ a lacunary sequence of nonnegative integers. Then for every function g that is continuous on E, and every sequence $x = (x_k)_{k\geqslant 0}$, $x \in l^2$, there is a function $f \in U_A$ such that*

$$f(\eta) = g(\eta), \quad \eta \in E; \qquad \hat{f}(n_k) = x_k, \quad k = 0, 1, 2, \ldots$$

$$\left(f(\eta) \overset{\text{def}}{=} \sum_{n\geqslant 0} \hat{f}(n)\eta^n, |\eta| \leqslant 1\right).$$

In §4 the majority of the propositions of §3 are extended to spaces whose elements are functions that are analytic except on a "sparse" set. Here is one of the results.

0.3. THEOREM. *Let E be a closed subset of* $\mathbb{T}$, $\operatorname{meas}\{t \in [-\pi, \pi]: e^{it} \in E\} > 0$, *and let $(n_j)_{j\geqslant 0}$ and $(k_j)_{j\geqslant 0}$ be two lacunary sequences of positive integers. Then for any two sequences $x = (x_j)_{j\geqslant 0}$ and $y = (y_j)_{j\geqslant 0}$ ($x, y \in l^2$) there is a function f, analytic and bounded in $\overline{\mathbb{C}} \setminus E$ ($\overline{\mathbb{C}}$ is the extended complex plane), such that*

$$\hat{f}(n_j) = x_j, \quad \hat{f}(-k_j) = y_j \qquad (j = 0, 1, 2, \ldots),$$

$$f(\eta) = \sum_{j\geqslant 0} \hat{f}(j)\eta^j \quad (|\eta| < 1), \qquad f(\eta) = \sum_{j<0} \hat{f}(j)\eta^j \quad (|\eta| > 1),$$

$$\sup\left\{\left|\sum_{j=0}^{n}\hat{f}(j)\eta^j\right|:|\eta|<1, n\geqslant 0\right\}+\sup\left\{\left|\sum_{j=n}^{-1}\hat{f}(j)\eta^j\right|:|\eta|>1, n<0\right\}<+\infty.$$

The paper concludes with an appendix (§5) containing the proofs of some technical lemmas. Some of the results of this paper were announced in [5].

The author is indebted to S. V. Kislyakov, M. Z. Solomyak, V. P. Khavin and S. V. Hrushchev for helpful discussions and advice.

§1. Notation and preliminary remarks

1.1. Sets. $\overline{\mathbb{C}}$ is the extended complex plane, $\mathbb{D}=\{\eta\in\overline{\mathbb{C}}: |\eta|<1\}$, $\mathbb{T}=\{\eta\in\overline{\mathbb{C}}: |\eta|=1\}$, $\mathbb{N}$ is the set of nonnegative integers, and $\mathbb{Z}$ is the set of integers.

ζ is the identity map of the unit circle $\mathbb{T}$, and z is the identity map of the unit disk $\mathbb{D}$ or of its exterior.(1)

$f \mid E$ is the restriction of f to E.

1.2. Spaces L^p and M. If Λ is a countable set, $s_0(\Lambda)$ denotes the collection of all finitely supported systems of numbers defined on Λ (a system $x=(x_\lambda)_{\lambda\in\Lambda}$ is said to be *finitely supported* if the set $\{\lambda\in\Lambda: x_\lambda\neq 0\}$ is finite).

$l^p(\Lambda)$ is the space of systems of complex numbers $x=(x_\lambda)_{\lambda\in\Lambda}$ for which the norm

$$\|x\|_{l^p}\overset{\text{def}}{=}\left(\sum_{\lambda\in\Lambda}|x_\lambda|^p\right)^{1/p},\quad p\in[1,+\infty);\qquad \|x\|_{l^\infty}\overset{\text{def}}{=}\sup\{|x_\lambda|:\lambda\in\Lambda\}$$

is finite. If $1/p+1/p'=1$, the duality between $l^p(\Lambda)$ and $l^{p'}(\Lambda)$ is exhibited by the formula

$$\langle x,y\rangle=\sum_{\lambda\in\Lambda}x_\lambda y_\lambda,\qquad x\in l^p(\Lambda), y\in l^{p'}(\Lambda).$$

Let m be normalized Lebesgue measure on the circle $\mathbb{T}$, and let $L^p(\mathbb{T})$ $(0<p\leqslant+\infty)$ be the usual L^p-space with respect to m:

$$\|f\|_p\overset{\text{def}}{=}\left(\int_{\mathbb{T}}|f|^p\,dm\right)^{1/p},\qquad f\in L^p(\mathbb{T}), p\in(0,+\infty);$$

$$\|f\|_\infty\overset{\text{def}}{=}\operatorname*{ess\,sup}_{\eta\in\mathbb{T}}|f(\eta)|,\qquad f\in L^\infty(\mathbb{T}).$$

We call $\|f\|_p$ the "norm" even when it is not actually a norm, i.e. for $p\in(0,1)$.

$\hat{f}=(\hat{f}(k))_{k\in\mathbb{Z}}$ is the sequence of Fourier coefficients of $f\in L^1(\mathbb{T})$; $\hat{f}(k)=\int_{\mathbb{T}}\bar{\zeta}^k f\,dm$, $k\in\mathbb{Z}$.

$M(E)$ is the space of complex Borel measures supported on $E\subset\mathbb{T}$, $\|\mu\|_{M(E)}$ is the total variation of μ, and $\hat{\mu}(k)=\int_{\mathbb{T}}\bar{\zeta}^k\,d\mu$, $k\in\mathbb{Z}$.

$L^p(l^1)$, $1\leqslant p\leqslant+\infty$, is the space of sequences of functions $f=(f_n)_{n\in\mathbb{N}}$ such that

$$f_n\in L^p(\mathbb{T}),\quad n\in\mathbb{N};\qquad \|f\|_{L^p(l^1)}\overset{\text{def}}{=}\left\|\sum_{n\in\mathbb{N}}|f_n|\right\|_p<+\infty.$$

(1)We believe that this ambiguity will not lead to any misunderstanding.

$L^p(l^\infty)$, $1 \leqslant p \leqslant +\infty$, is the space of sequences of functions $f = (f_n)_{n\in\mathbb{N}}$ such that

$$f_n \in L^p(\mathbb{T}), \quad n \in \mathbb{N}; \qquad \|f\|_{L^p(l^\infty)} \stackrel{\text{def}}{=} \Big\| \sup_{n\in\mathbb{N}} |f_n| \Big\|_p < +\infty.$$

If X is a Banach space, $c_0(X)$ denotes the space of sequences $x = (x_n)_{n\in\mathbb{N}}$ ($x_n \in X$, $n \in \mathbb{N}$) such that $\lim_{n\to\infty} \|x_n\|_X = 0$:

$$\|x\|_{c_0(X)} \stackrel{\text{def}}{=} \max\{\|x_n\|_X : n \in \mathbb{N}\}, \qquad x \in c_0(X);$$

$l^1(X)$ is the space of the sequences $x = (x_n)_{n\in\mathbb{N}}$ ($x_n \in X$, $n \in \mathbb{N}$) for which the norm

$$\|x\|_{l^1(X)} \stackrel{\text{def}}{=} \sum_{n\in\mathbb{N}} \|x_n\|_X$$

is finite.

It is clear that the space $(c_0(X))^*$ dual to $c_0(X)$ is linearly isometric to $l^1(X^*)$.

1.3. The duality between $L^p(\mathbb{T})$ **and** $L^{p'}(\mathbb{T})$, $1/p + 1/p' = 1$, $1 \leqslant p < +\infty$, is conveniently exhibited by the formula

$$\langle f, g \rangle \stackrel{\text{def}}{=} \int_{\mathbb{T}} f(\zeta) g(\bar{\zeta})\, dm = (f * g)(1), \qquad f \in L^p(\mathbb{T}), g \in L^{p'}(\mathbb{T}); \quad (1.1)$$

and similarly

$$\langle f, \mu \rangle \stackrel{\text{def}}{=} \int_{\mathbb{T}} f(\bar{\zeta})\, d\mu, \qquad f \in C(\mathbb{T}), \mu \in M(\mathbb{T}).$$

1.4. The Lorentz-Marcinkiewicz space $L_{pq}(\mathbb{T})$ ($1 \leqslant p, q \leqslant +\infty$). $L_{pq}(\mathbb{T})$ is the set of functions f that are measurable with respect to Lebesgue measure m and are such that

$$\|f\|^*_{pq} = \left(\frac{q}{p} \int_0^\infty (t^{1/p} f^*(t))^q \frac{dt}{t} \right)^{1/q} < +\infty, \quad \text{if } p, q \in [1, +\infty),$$

$$\|f\|^*_{pq} = \sup\{t^{1/p} f^*(t) : t > 0\} < +\infty, \quad \text{if } q = +\infty, p \in [1, +\infty),$$

where f^* is the nonincreasing equimeasurable rearrangement of f. Evidently $L_{pp}(\mathbb{T}) = L^p(\mathbb{T})$, $1 \leqslant p < +\infty$.

If $p \in (1, +\infty)$ and $q \in [1, +\infty]$, one can introduce a norm $\|\cdot\|_{pq}$ into $L_{pq}(\mathbb{T})$ (see [12], Chapter V, §3) so that it becomes a Banach space, and

$$\|f\|^*_{pq} \leqslant \|f\|_{pq} \leqslant p(p-1)^{-1} \|f\|^*_{pq}, \qquad f \in L_{pq}(\mathbb{T}).$$

Moreover, if $p \in (1, +\infty)$, $q \in [1, +\infty)$, $1/p + 1/p' = 1$ and $1/q + 1/q' = 1$, then (see [13]) the space $L_{p'q'}(\mathbb{T})$ is linearly isomorphic to $(L_{pq}(\mathbb{T}))^*$, the dual space of $L_{pq}(\mathbb{T})$ (this isomorphism can be expressed by (1.1)).

1.5. Spaces of analytic functions. If G is an open subset of $\overline{\mathbb{C}}$, then $A(G)$ is the space of functions defined and analytic in G, provided with the topology of uniform convergence on compact subsets of G. We define

$$H^p = \Big\{ f \in A(\mathbb{D}) : \|f\|_{H^p} = \sup_{0<r<1} \|f(r\zeta)\|_p < +\infty \Big\}, \qquad 0 < p < +\infty;$$

these are the Hardy spaces (see [14] or [15]).

C_A is the space of functions analytic in $\mathbb{D}$ and continuous in its closure $\overline{\mathbb{D}}$:

$$\|f\|_{C_A} = \|f\|_{H^\infty} = \sup\{|f(\eta)| : \eta \in \mathbb{D}\}, \qquad f \in C_A,$$

$$U_A = \left\{f \in C_A \colon \lim_n \left\| \sum_{k \geqslant n} \hat{f}(k) z^k \right\|_{C_A} = 0\right\},$$

$$\|f\|_{U_A} = \sup\left\{\left\| \sum_{k \geqslant n} \hat{f}(k) z^k \right\|_{C_A} : n \in \mathbb{N}\right\}, \qquad f \in U_A.$$

C^∞ is the space of infinitely differentiable functions on $\mathbb{T}$, and $C_A^\infty = \{f \in C_A\colon f|\mathbb{T} \in C^\infty\}$.

1.6. Reflection. We sometimes make use of the following construction. Let $\mathcal{E}$ be a Banach space of functions analytic in $\mathbb{D}$ that satisfy

$$C_A^\infty \subset \mathcal{E} \subset H^1, \tag{1.2}$$

where the embeddings are continuous. Let

$$(\tau f)(\eta) = \eta^{-1} f(\eta^{-1}), \qquad |\eta| > 1, f \in H^1.$$

Then $\tau\mathcal{E}$ denotes the image of $\mathcal{E}$ under τ; $\tau\mathcal{E}$ is isometric to $\mathcal{E}$ if we introduce the norm $\|\tau f\|_{\tau\mathcal{E}} = \|f\|_{\mathcal{E}}$ in $\tau\mathcal{E}$. We denote by $\mathcal{E}_{\pm}(\mathbb{T})$ the space of functions $f \in L^1(\mathbb{T})$ that have finite norm,

$$\|f\|_{\mathcal{E}_\pm} \stackrel{\text{def}}{=} \max\{\|\mathcal{K}_+ f\|_{\mathcal{E}}, \|\mathcal{K}_- f\|_{\tau\mathcal{E}}\},$$

where

$$(\mathcal{K}_+ f)(\eta) = \sum_{n>0} \hat{f}(n)\eta^n \qquad (\eta \in \mathbb{D}),$$

$$(\mathcal{K}_- f)(\eta) = \sum_{n<0} \hat{f}(n)\eta^n \qquad (\eta \in \overline{\mathbb{C}} \backslash \overline{\mathbb{D}})$$

are the Cauchy transforms of f. The radial boundary values of $\mathcal{K}_+ f$ and $\mathcal{K}_- f$, $f \in L^1(\mathbb{T})$, are denoted by f_+ and f_- respectively.

In addition, we shall usually write $\mathcal{E}_+$, $\mathcal{E}_-$ and $\mathcal{E}_\pm$ for $\mathcal{E}$, $\tau\mathcal{E}$ and $\mathcal{E}_\pm(\mathbb{T})$.

1.7. The Cauchy transform. Let $(C^\infty)^*$ be the space of continuous functionals on C^∞, and let $\Phi \in (C^\infty)^*$.

We denote by $\hat{\Phi}$ the sequence of Fourier coefficients of Φ:

$$\hat{\Phi}(n) = \Phi(\zeta^n), \qquad n \in \mathbb{Z}.$$

Moreover,

$$\langle f, \Phi \rangle \stackrel{\text{def}}{=} \Phi(f) = \sum_{n \in \mathbb{Z}} \hat{\Phi}(n)\hat{f}(n), \qquad f \in C^\infty$$

(note that $\sum_{n \in \mathbb{Z}} |\hat{f}(n)\hat{\Phi}(n)| < +\infty$); and

$$\Phi_+(f) \stackrel{\text{def}}{=} \Phi(f_+), \qquad \Phi_-(f) \stackrel{\text{def}}{=} \Phi(f_-),$$

where $f_+ = \sum_{n \geqslant 0} \hat{f}(n)\zeta^n$ and $f_- = \sum_{n<0} \hat{f}(n)\zeta^n$, $f \in C^\infty$. Evidently $\Phi_+, \Phi_- \in (C^\infty)^*$, $\Phi = \Phi_+ + \Phi_-$, and

$$\hat{\Phi}_+(n) = \begin{cases} \hat{\Phi}(n), & n \in \mathbb{N}, \\ 0, & n \in \mathbb{Z}\backslash\mathbb{N}, \end{cases} \qquad \hat{\Phi}_-(n) = \begin{cases} 0, & n \in \mathbb{N}, \\ \hat{\Phi}(n), & n \in \mathbb{Z}\backslash\mathbb{N}. \end{cases}$$

With each functional $\Phi \in (C^\infty)^*$ we associate the two analytic functions $\mathcal{K}_+ \Phi$ and $\mathcal{K}_- \Phi$:

$$(\mathcal{K}_+ \Phi)(\eta) = \sum_{n\geqslant 0} \hat{\Phi}(n)\eta^n, \qquad \eta \in \mathbb{D},$$
$$(\mathcal{K}_- \Phi)(\eta) = \sum_{n<0} \hat{\Phi}(n)\eta^n, \qquad \eta \in \overline{\mathbb{C}}\backslash\overline{\mathbb{D}}.$$

In accordance with [16], $\mathcal{K}_+ \Phi$ and $\mathcal{K}_- \Phi$ are called the *Cauchy transforms* of Φ in $\mathbb{D}$ or $\overline{\mathbb{C}}\backslash\overline{\mathbb{D}}$, respectively.

1.8. Functionals and operators. Let $\Phi \in (C_A^\infty)^*$ and $(j\Phi)(f) \stackrel{\text{def}}{=} \Phi(f_+)$, $f \in C^\infty$. Evidently j establishes an isomorphism of $(C_A^\infty)^*$ on the subspace $\{\Psi \in (C^\infty)^*: \Psi = \Psi_+\}$ of $(C^\infty)^*$, and consequently we shall identify the functionals Φ and $j\Phi$, $\Phi \in (C_A)^*$. Similarly we shall identify the elements of $(\tau C_A^\infty)^*$ and $\{\Psi \in (C^\infty)^*: \Psi = \Psi_-\}$.

If $\mathcal{E}$ satisfies (1.2) and also

$$C_A^\infty \text{ is dense in } \mathcal{E}, \tag{1.3}$$

then $\mathcal{E}^*$ is embedded in $(C_A)^*$ in a natural way.

It is easy to see that $\Phi \in \mathcal{E}_\pm^*$ if and only if $\Phi_+ \in \mathcal{E}^*$ and $\Phi_- \in (\tau\mathcal{E})^*$. Furthermore, if $\Phi \in \mathcal{E}_\pm^*$ then

$$\Phi(f) = \Phi_+(f) + \Phi_-(f), \quad f \in \mathcal{E}_\pm; \qquad \|\Phi\|_{\mathcal{E}_\pm^*} = \|\Phi_+\|_{\mathcal{E}^*} + \|\Phi_-\|_{(\tau\mathcal{E})^*}.$$

The space $\mathcal{E}^*$ is linearly isometric to $(\tau\mathcal{E})^*$.

Let X and Y be normed spaces. By $\|T \mid X \to Y\|$ we denote the norm of a bounded linear operator T from X to Y; if $X = Y$, we write $\|T \mid X\|$ instead of $\|T \mid X \to X\|$.

1.9. Kolmogorov-Smirnov inequality (see [15]):

$$\left\| \int_{\mathbb{T}} \frac{d\mu(\zeta)}{1-\bar{\zeta}z} \right\|_{H^p} \leqslant \left(\frac{4}{1-p}\right)^{1/p} \|\mu\|_{M(\mathbb{T})}, \qquad \mu \in M(\mathbb{T}), 0 < p < 1. \tag{1.4}$$

§2. Strengthened form of Kolmogorov's theorem on conjugate functions

LEMMA 2.1. *Let $u \in L^1(l^1)$ and $t > 0$. Then there are a sequence of pairwise disjoint arcs $(I_n)_{n\in\mathbb{N}\backslash\{0\}}$, $I_n \subset \mathbb{T}$, and a sequence of vector-valued functions $(w^n)_{n\in\mathbb{N}}$, $w^n \in L^1(l^1)$, such that*

$$u = \sum_{n\geqslant 0} w^n; \tag{2.1}$$

$$\sum_{n\geqslant 0} \|w^n\|_{L^1(l^1)} \leqslant 3\|u\|_{L^1(l^1)}; \tag{2.2}$$

$$\|w^0(\eta)\|_{l^1} \leqslant 2t, \quad a.e.\, \eta \in T; \tag{2.3}$$

$$\int_{\mathbb{T}} w^n\, dm = \mathbf{0}, \quad n \in \mathbb{N}\backslash\{0\}; \qquad w^n(\eta) = \mathbf{0}, \quad \eta \notin I_n, n \in \mathbb{N}\backslash\{0\}; \tag{2.4}$$

$$\sum_{n\geqslant 1} mI_n \leqslant \frac{1}{t}\|u\|_{L^1(l^1)}. \tag{2.5}$$

This was proved for scalar functions by Hörmander [7]. The proof for functions of class $L^1(l^1)$ is practically the same as for the scalar case.

We are going to establish a modification of a proposition of Hörmander's (see [7]). For the sake of completeness, we feel compelled to present the proof in detail, although it almost entirely follows Hörmander's discussion.

Let $\mathfrak{k} \in L^1(T)$, $(h_n)_{n\in\mathbb{N}} \in L^\infty(l^\infty)$, $\|h_n\|_\infty \leqslant 1$ ($n \in \mathbb{N}$), and define

$$Tu \overset{\text{def}}{=} \sum_{n\geqslant 0} h_n(\mathfrak{k} * u_n), \qquad u = (u_n)_{n\in\mathbb{N}},\ u \in L^1(l^1);$$

$$T^\varphi f \overset{\text{def}}{=} T((f\cdot\varphi_n)_{n\in\mathbb{N}}), \qquad f \in L^1(\mathbb{T}), \qquad \varphi = (\varphi_n)_{n\in\mathbb{N}} \in L^\infty(l^1),\ \|\varphi\|_{L^\infty(l^1)} \leqslant 1.$$

LEMMA 2.2. *Let $p \in (1, +\infty)$*

$$B_0 \overset{\text{def}}{=} \sup\left\{\int_{e_y} |\mathfrak{k}(\bar\eta\zeta) - \mathfrak{k}(\zeta)|\, dm \colon y > 0,\ \eta \in \mathbb{T},\ |\arg\eta| \leqslant y\right\},$$

where

$$e_y \overset{\text{def}}{=} \{\xi \in \mathbb{T} \colon 2y \leqslant |\arg\xi| \leqslant \pi\};$$

and

$$B_1 \overset{\text{def}}{=} \sup\{\|T^\varphi \mid L_{p1} \to L^p\| \colon \|\varphi\|_{L^\infty(l^1)} \leqslant 1\} < +\infty.$$

There is a constant $B_2 \in (0, +\infty)$, depending only on p, B_0 and B_1, such that

$$m\{\eta \in \mathbb{T} \colon |(Tu)(\eta)| > t\} \leqslant (B_2/t)\|u\|_{L^1(l^1)}, \qquad t > 0,$$

for every function $u \in L^1(l^1)$.

PROOF. First we introduce the notation I and I^* for arcs of the unit circle $\mathbb{T}$ that are related as follows: if $I = \{\eta \in \mathbb{T} \colon |\arg\eta\bar\xi_0| < y\}$ ($\xi_0 \in \mathbb{T}$, $y > 0$), then $I^* = \{\eta \in \mathbb{T} \colon |\arg\eta\bar\xi_0| < 2y\}$. Note that $mI^* \leqslant 2mI$.

Let $g = (g_n)_{n\in\mathbb{N}} \in L^1(l^1)$, $\int_{\mathbb{T}} g\, dm = \mathbf{0}$, $g(\eta) = \mathbf{0}$ outside an arc I, where $I \subset \mathbb{T}$. Then

$$\int_{\mathbb{T}\setminus I^*} |Tg|\, dm \leqslant B_0 \int_I \|g(\eta)\|_{l^1}\, dm(\eta). \tag{2.6}$$

In fact, since $\int_{\mathbb{T}} g\, dm = \mathbf{0}$ (i.e., $\int_{\mathbb{T}} g_n\, dm = 0$, $n \in \mathbb{N}$), we have

$$(Tg)(\xi) = \sum_{n\geqslant 0} h_n \int_{\mathbb{T}} (\mathfrak{k}(\xi\bar\eta) - \mathfrak{k}(\xi\bar\xi_0)) g_n(\eta)\, dm(\eta), \qquad \xi \in \mathbb{T}$$

(ξ_0 is the center of I),

$$|(Tg)(\xi)| \leqslant \sum_{n\geqslant 0} \int_{\mathbb{T}} |\mathfrak{k}(\xi\bar\eta) - \mathfrak{k}(\xi\bar\xi_0)|\, |g_n(\eta)|\, dm(\eta)$$

$$= \int_I |\mathfrak{k}(\xi\bar\eta) - \mathfrak{k}(\xi\bar\xi_0)|\, \|g(\eta)\|_{l^1}\, dm(\eta), \qquad \xi \in \mathbb{T}.$$

Moreover,

$$\int_{\mathbb{T}\setminus I^*} |(Tg)(\xi)|\, dm(\xi) \leqslant \int_{\mathbb{T}\setminus I^*} \left(\int_I |\mathfrak{k}(\xi\bar\eta) - \mathfrak{k}(\xi\bar\xi_0)|\, \|g(\eta)\|_{l^1}\, dm(\eta)\right) dm(\xi)$$

$$= \int_{\mathbb{T}\setminus\bar\xi_0 I^*} \left(\int_{\bar\xi_0 I} |\mathfrak{k}(\xi\bar\eta) - \mathfrak{k}(\xi)|\, \|g(\xi_0\eta)\|_{l^1}\, dm(\eta)\right) dm(\xi)$$

(we have made the change of variables $\xi \mapsto \xi\bar{\xi}_0$, $\eta \mapsto \eta\bar{\xi}$). Since $\bar{\xi}_0 I$ and $\bar{\xi}_0 I^*$ are arcs centered at unity and $\bar{\xi}_0 I^* = (\bar{\xi}_0 I)^*$, we have

$$\int_{\mathbb{T}\setminus I^*} |Tg|\,dm \leqslant \int_{\bar{\xi}_0 I} \|g(\xi_0\eta)\|_{l^1}\left(\int_{\mathbb{T}\setminus(\bar{\xi}_0 I)^*} |\mathfrak{k}(\xi\bar{\eta}) - \mathfrak{k}(\xi)|\,dm(\xi)\right) dm(\eta)$$

$$\leqslant B_0\int_I \|g\|_{l^1}\,dm.$$

We can now prove the lemma. Let $u = (u_n)_{n\in\mathbb{N}} \in L^1(l^1)$, $\|u\|_{L^1(l^1)} \leqslant 1$ and $t > 0$. Then by Lemma 2.1 there are a sequence of pairwise disjoint arcs $(I_n)_{n\in\mathbb{N}\setminus\{0\}}$ and a sequence of functions $(w^n)_{n\in\mathbb{N}}$ satisfying (2.1)–(2.5). It is clear that

$$|(Tu)(\xi)| \leqslant \sum_{n\geqslant 0} |(Tw^n)(\xi)|, \qquad \xi \in \mathbb{T}.$$

By (2.6),

$$\int_{\mathbb{T}\setminus I_n^*} |Tw^n|\,dm \leqslant B_0\int_{I_n} \|w^n(\eta)\|_{l^1}\,dm(\eta), \qquad n \in \mathbb{N}\setminus\{0\}. \tag{2.7}$$

Let $E = \bigcup_{n\geqslant 1} I_n^*$ and $W = \sum_{n\geqslant 1} |Tw^n|$. Applying (2.7), we obtain

$$\int_{\mathbb{T}\setminus E} W\,dm \leqslant \sum_{n\geqslant 1}\int_{\mathbb{T}\setminus E} |Tw^n|\,dm \leqslant \sum_{n\geqslant 1}\int_{\mathbb{T}\setminus I_n^*} |Tw^n|\,dm$$

$$\leqslant B_0 \sum_{n\geqslant 1} \|w^n\|_{L^1(l^1)} \leqslant 3B_0\|u\|_{L^1(l^1)} \leqslant 3B_0.$$

Consequently

$$m\{\eta \in \mathbb{T}\setminus E\colon W(\eta) > t/2\} \leqslant 6B_0/t.$$

Since

$$mE \leqslant \sum_{n\geqslant 1} mI_n^* \leqslant 2\sum_{n\geqslant 1} mI_n \leqslant \frac{2}{t}$$

(see (2.5)), we have

$$m\{\eta \in \mathbb{T}\colon W(\eta) > t/2\} \leqslant (6B_0 + 2)t^{-1}. \tag{2.8}$$

In addition, let

$$f(\xi) = \|w^0(\xi)\|_{l^1}, \quad \psi(\xi) = w^0(\xi)/f(\xi), \qquad \xi \in \mathbb{T}.$$

Since $\|T^\psi \mid L_{p1} \to L^p\| \leqslant B_1$ and $T^\psi f = Tw^0$, we obtain

$$\|Tw^0\|_p \leqslant B_1\|f\|_{p1} \leqslant B_1\cdot 2\left(\frac{p+1}{p-1}\right)t^{1-1/p}\|f\|_1^{1/p}$$

$$\leqslant 6B_1\frac{p+1}{p-1}\|u\|_{L^1(l^1)}^{1/p}t^{1-1/p} \leqslant 6B_1\frac{p+1}{p-1}t^{1-1/p}\,\mathrm{def}\,\tilde{B}_1 t^{1-1/p}$$

(with reference to the inequality for $\|f\|_{p1}$ see §5, Lemma 5.1).

Therefore

$$m\left\{\eta \in \mathbb{T}\colon |(Tw^0)(\eta)| > \frac{t}{2}\right\} \leqslant \tilde{B}_1^p t^{p-1}\left(\frac{2}{t}\right)^p = \frac{(2\tilde{B}_1)^p}{t}.$$

This yields, together with (2.8), the inequality

$$m\{\eta\in\mathbb{T}:|(Tu)(\eta)|>t\}\leqslant m\{\eta\in\mathbb{T}:|(Tw^0)(\eta)|+W(\eta)>t\}$$
$$\leqslant m\{\eta\in\mathbb{T}:|(Tw^0)(\eta)|>t/2\}+m\{\eta\in\mathbb{T}:W(\eta)>t/2\}$$
$$\leqslant\left((2\tilde{B}_1)^p+6B_0+2\right)t^{-1}. \quad\bullet$$

COROLLARY 2.3. *Let the hypotheses of the preceding lemma be satisfied. Then*

$$\|T^\varphi\,|\,L^s\|\leqslant 2\left(\frac{sB_2}{s-1}+\frac{pB_1}{p-s}\right),\qquad s\in(1,p),\ \varphi\in L^\infty(l^1),\ \|\varphi\|_{L^\infty(l^1)}\leqslant 1.$$

PROOF. It follows from the lemma just proved that

$$m\{\eta\in\mathbb{T}:|(T^\varphi f)(\eta)|>t\}\leqslant(B_2/t)\|f\|_1,\qquad f\in L^1(\mathbb{T}).$$

Therefore we can apply the theorem on interpolation of linear operators in L_{pq} spaces (see [12], Chapter V, §3). ●

THEOREM 2.4. *Let $r\in(0,1)$ and*

$$(K_ru)(\eta)\overset{\text{def}}{=}\sum_{n\geqslant 0}r^n\eta^n\int_{\mathbb{T}}\frac{u_n\,dm}{1-r\eta\bar{\zeta}},\qquad \eta\in\mathbb{T},\ u=(u_n)_{n\in\mathbb{N}}\in L^1(l^1).$$

Then there is an absolute constant $a_0\in(0,+\infty)$ such that

$$m\{\eta\in\mathbb{T}:|(K_ru)(\eta)|>t\}\leqslant(a_0/t)\|u\|_{L^1(l^1)},$$

for $t>0$ and $r\in(0,1)$.

PROOF. Let $r\in(0,1)$, $\varphi=(\varphi_n)_{n\in\mathbb{N}}\in L^\infty(l^1)$, and $\|\varphi\|_{L^\infty(l^1)}\leqslant 1$. Put

$$(R_r^\varphi f)(\eta)=\sum_{n\geqslant 0}\varphi_n(\eta)\sum_{k\geqslant n}\hat{f}(k)r^k\eta^{k-n},\qquad \eta\in\mathbb{T},\ f\in L^1(\mathbb{T}).$$

It follows from Carleson's theorem [9] that

$$c_1\overset{\text{def}}{=}\sup\left\{\|R_r^\varphi\,|\,L^2\to L_{2\infty}\|:r\in(0,1),\ \|\varphi\|_{L^\infty(l^1)}\leqslant 1\right\}<+\infty.$$

Moreover, let

$$(K_r^\varphi g)(\eta)=\sum_{n\geqslant 0}r^n\eta^n\int_{\mathbb{T}}\frac{\varphi_n g}{1-r\eta\bar{\zeta}}\,dm,\qquad \eta\in\mathbb{T},\ g\in L^1(\mathbb{T}).$$

It is easily verified that

$$\langle K_r^\varphi g,f\rangle=\langle g,R_r^\varphi f\rangle,\qquad f\in L^2(\mathbb{T}),\ g\in L_{21}(\mathbb{T}),$$

whence

$$\|K_r^\varphi\,|\,L_{21}\to L^2\|=\|R_r^\varphi\,|\,L^2\to L_{2\infty}\|$$

and

$$\sup\left\{\|K_r^\varphi\,|\,L_{21}\to L^2\|:r\in(0,1),\ \|\varphi\|_{L^\infty(l^1)}\leqslant 1\right\}=c_1<+\infty.$$

Since in addition

$$\int_{e_y}\left|\frac{1}{1-r\bar{\eta}\zeta}-\frac{1}{1-r\zeta}\right|dm\leqslant 1,\qquad e_y\overset{\text{def}}{=}\{\xi\in\mathbb{T}:2y\leqslant|\arg\xi|\leqslant\pi\},$$

if $y > 0$, $\eta \in \mathbb{T}$, $|\arg \eta| \leqslant y$ and $r \in (0, 1)$, we obtain what was required by applying Lemma 2.2 (with $p = 2$) to the kernel $\mathfrak{k} = 1/(1 - r\zeta)$ and the operators $T = K_r$ and $T^\varphi = K_r^\varphi$. ●

The following proposition follows at once from the preceding theorem.

COROLLARY 2.5. *For every sequence* $\mu = (\mu_n)_{n\in\mathbb{N}}$, $\mu \in l^1(M(\mathbb{T}))$, *the function* F *defined by*

$$F(\eta) \overset{\text{def}}{=} \sum_{n\geqslant 0} \eta^n \int_{\mathbb{T}} \frac{d\mu_n}{1 - \eta\bar{\zeta}}, \qquad \eta \in \mathbb{D},$$

belongs to the space $\bigcap_{0<p<1} H^p$. *Moreover, there are absolute constants* c_2 *and* c_3 ($c_2, c_3 \in (0, +\infty)$) *such that*

$$1)\quad m\{\eta \in \mathbb{T}: |F(\eta)| > t\} \leqslant \frac{c_2}{t} \sum_{n\geqslant 0} \|\mu_n\|_{M(\mathbb{T})}, \qquad t > 0,$$

($F(\eta)$ *is the angular boundary value of* F *at* $\eta \in \mathbb{T}$; *in the present case* $F(\eta)$ *exists almost everywhere on* $\mathbb{T}$, *since* $F \in H^p$, $0 < p < 1$); *and*

$$2)\quad \|F\|_{H^p} \leqslant \frac{c_3}{1 - p} \sum_{n\geqslant 0} \|\mu_n\|_{M(\mathbb{T})}, \qquad 0 < p < 1. \tag{2.9}$$

PROOF. We need only apply the preceding theorem and the equalities

$$\int_{\mathbb{T}} |h|^p \, dm = \int_0^{+\infty} m\{\eta \in \mathbb{T}: |h(\eta)|^p > t\}\, dt,$$

$$F(r\eta) = \sum_{n\geqslant 0} \eta^n \int_{\mathbb{T}} \frac{r^n u_n(r\zeta)}{1 - \eta\bar{\zeta}}\, dm, \qquad \eta \in \mathbb{D}, r \in (0, 1),$$

where

$$u_n(r\zeta) \overset{\text{def}}{=} \sum_{k\in\mathbb{Z}} r^{|k|} \hat{\mu}_n(k)\zeta^k, \qquad n \in \mathbb{N}. \quad ●$$

REMARK 2.6. Theorem 2.4 makes it possible to obtain the following inequalities (see [8]) for $S_* f$:

$$(S_* f)(\xi) \overset{\text{def}}{=} \sup\left\{\left|\sum_{k=0}^{n} \hat{f}(k)\xi^k\right| : n \in \mathbb{N}\right\}, \qquad \xi \in \mathbb{T}, f \in L^1(\mathbb{T});$$

$$\|S_* f\|_q \leqslant 4\left(a_0 q + 2c_1 \frac{q - 1}{q - 2}\right)\|f\|_q, \qquad f \in L^q(\mathbb{T}), 2 < q < +\infty; \tag{2.10}$$

$$m\left\{\eta \in \mathbb{T}: \left|(S_* f)(\eta)\right| > t\right\} \leqslant 2\exp\frac{-t}{50(a_0 + c_1)\|f\|_\infty}, \qquad t > 0. \tag{2.11}$$

PROOF. It follows from Theorem 2.4 that

$$m\{\eta \in \mathbb{T}: |(K_r^\varphi f)(\eta)| > t\} \leqslant (a_0/t)\|f\|_1, \qquad t > 0, r \in (0, 1),$$
$$\varphi = (\varphi_n)_{n\in\mathbb{N}} \in L^\infty(l^1), \qquad \|\varphi\|_{L^\infty(l^1)} \leqslant 1.$$

Moreover,

$$\sup\left\{\|K_r^\varphi \mid L_{21} \to L^2\| : r \in (0, 1), \|\varphi\|_{L^\infty(l^1)} \leqslant 1\right\} = c_1 < +\infty$$

(see the proof of Theorem 2.4). Applying Corollary 2.3 (with $p=2$) to K_r^φ, we obtain

$$\| K_r^\varphi \mid L^s \| \leqslant 2\left(\frac{a_0 s}{s-1} + \frac{2c_1}{2-s} \right), \qquad s \in (1,2).$$

It is easily seen that

$$\| K_r^\varphi \mid L^s \| = \| R_r^\varphi \mid L^q \|, \qquad 1/s + 1/q = 1, s \in (1,2), q \in (2,+\infty)$$

(see the proof of Theorem 2.4) and

$$\| S_* f \|_q \leqslant 2 \sup\left\{ \| R_r^\varphi f \|_q : r \in (0,1), \|\varphi\|_{L^\infty(l^1)} \leqslant 1 \right\}.$$

Hence

$$\| S_* f \|_q \leqslant 4\left(a_0 q + 2c_1 \frac{q-1}{q-2} \right) \| f \|_q,$$

if $2 < q < +\infty$.

Furthermore, if we use (2.10) we obtain

$$\begin{aligned} \int_{\mathbb{T}} | S_* f |^n \, dm &\leqslant \left(8(a_0 + c_1) n \| f \|_\infty \right)^n \\ &\leqslant \left(8e(a_0 + c_1) \| f \|_\infty \right)^n n!, \qquad n \in \{3,4,5,\dots\}, \end{aligned}$$

$$\begin{aligned} \int_{\mathbb{T}} | S^* f |^n \, dm &\leqslant \left(\int_{\mathbb{T}} | S_* f |^3 \, dm \right)^{n/3} \\ &\leqslant \left(8(a_0 + c_1) \cdot 3 \| f \|_\infty \right)^n, \qquad n \in \{0,1,2\}, f \in L^\infty(\mathbb{T}); \end{aligned}$$

$$\begin{aligned} t^n m\left\{ \eta \in \mathbb{T} : \left| \left(S_* f \right)(\eta) \right| > t \right\} &\leqslant \int_{\mathbb{T}} | S_* f |^n \, dm \\ &\leqslant \left(24(a_0 + c_1) \| f \|_\infty \right)^n n!, \qquad n \in \mathbb{N}, t > 0; \end{aligned}$$

$$\frac{1}{n!} \left(\frac{t}{50(a_0 + c_1) \| f \|_\infty} \right)^n m\left\{ \eta \in \mathbb{T} : \left| \left(S_* f \right)(\eta) \right| > t \right\} \leqslant \frac{1}{2^n}, \qquad n \in \mathbb{N}, t > 0;$$

$$m\left\{ \eta \in \mathbb{T} : \left| \left(S_* f \right)(\eta) \right| > t \right\} \cdot \exp \frac{t}{50(a_0 + c_1) \| f \|_\infty} \leqslant 2, \qquad t > 0. \quad \bullet$$

§3. Inequalities for Taylor coefficients. Simultaneous interpolation

LEMMA 3.1. *The set $\mathcal{K}_+(U_A^*)$ consists of the functions F that are analytic in $\mathbb{D}$ and representable in the form*

$$F(\eta) = \sum_{n \geqslant 0} \eta^n \int_{\mathbb{T}} \frac{\bar{\xi}^n \, d\mu_n}{1 - \bar{\xi}\eta}, \qquad \eta \in \mathbb{D},$$

$$\mu_n \in M(\mathbb{T}), \qquad \sum_{n \geqslant 0} \| \mu_n \|_{M(\mathbb{T})} < +\infty. \tag{3.1}$$

Moreover,

$$\| \Phi \|_{U_A^*} = \inf\left\{ \sum_{n \geqslant 0} \| \mu_n \|_{M(\mathbb{T})} : (\mu_n)_{n \in \mathbb{N}} \in \mathfrak{M}_{\mathcal{K}_+\Phi} \right\},$$

where $\mathfrak{M}_F$ is the set of sequences $(\mu_n)_{n\in\mathbb{N}} \in l^1(M(\mathbb{T}))$ that represent F in the sense of (3.1).

PROOF. It is well known that the space $c_0(C(\mathbb{T}))^*$ dual to $c_0(C(\mathbb{T}))$ is, in a natural way, linearly isometric to $l^1(M(\mathbb{T}))$ (for the definitions of $c_0(X)$ and $l^1(X)$ see §1.9). Let $j\colon U_A \to c_0(C(\mathbb{T}))$ be the natural embedding of U_A in $c_0(C(\mathbb{T}))$:

$$jf = \Bigl(\sum_{k\geqslant n} \hat{f}(k)\zeta^k\Bigr)_{n\in\mathbb{N}}, \qquad f \in U_A,$$

and $\Phi \in U_A^*$. Since jU_A is a closed subspace of $c_0(C(\mathbb{T}))$, it follows by the Hahn-Banach theorem that the functional $\Phi \circ j_l^{-1}$ ($j_l^{-1}\colon jU_A \to U_A$, $j_l^{-1} \circ j = \mathrm{id}$) can be extended, with preservation of the norm, to a linear functional on $c_0(C(\mathbb{T}))$. Therefore there is a sequence $(\mu_n)_{n\in\mathbb{N}} \in l^1(M(\mathbb{T}))$ such that $\|\Phi\|_{U_A^*} = \sum_{n\geqslant 0}\|\mu_n\|_{M(\mathbb{T})}$ and

$$\Phi(f) = \sum_{n\geqslant 0}\Bigl(\mu_n * \sum_{k\geqslant n}\hat{f}(k)\zeta^k\Bigr)(1), \qquad f\in U_A.$$

If $\eta \in \mathbb{D}$, we have

$$F(\eta) \overset{\text{def}}{=} (\mathcal{K}_+\Phi)(\eta) = \Phi\Bigl(\frac{1}{1-\eta\zeta}\Bigr) = \sum_{n\geqslant 0}\Bigl(\mu_n * \sum_{k\geqslant n}\eta^k\zeta^k\Bigr)(1)$$

$$= \sum_{n\geqslant 0}\Bigl(\mu_n * \frac{\eta^n\zeta^n}{1-\eta\zeta}\Bigr)(1) = \sum_{n\geqslant 0}\eta^n\int_{\mathbb{T}}\frac{\bar{\zeta}^n\,d\mu_n}{1-\eta\bar{\zeta}}.$$

Consequently

$$\|\Phi\|_{U_A^*} \geqslant \inf\Bigl\{\sum_{n\geqslant 0}\|\nu_n\|_{M(\mathbb{T})}\colon (\nu_n)_{n\in\mathbb{N}} \in \mathfrak{M}_F\Bigr\}. \tag{3.2}$$

Now suppose that $(\mu_n)_{n\in\mathbb{N}} \in l^1(M(\mathbb{T}))$; then, when $f\in U_A$,

$$\sum_{n\geqslant 0}\Bigl|\Bigl(\mu_n * \sum_{k\geqslant n}\hat{f}(k)\zeta^k\Bigr)(1)\Bigr| \leqslant \sum_{n\geqslant 0}\|\mu_n\|\Bigl\|\sum_{k\geqslant n}\hat{f}(k)\zeta^k\Bigr\|_\infty$$

$$\leqslant \Bigl(\sum_{n\geqslant 0}\|\mu_n\|_{M(\mathbb{T})}\Bigr)\|f\|_{U_A} < +\infty.$$

Hence the functional Φ,

$$\Phi(f) \overset{\text{def}}{=} \sum_{n\geqslant 0}\Bigl(\mu_n * \sum_{k\geqslant n}\hat{f}(k)\zeta^k\Bigr)(1), \qquad f\in U_A,$$

is bounded on U_A (the linearity of Φ is evident), its norm does not exceed $\sum_{n\geqslant 0}\|\mu_n\|_{M(\mathbb{T})}$, and

$$\mathcal{K}_+\Phi = \sum_{n\geqslant 0} z^n\int_{\mathbb{T}}\frac{\bar{\zeta}^n\,d\mu_n(\zeta)}{1-\bar{\zeta}z}.$$

This, with (3.2), yields

$$\|\Phi\| = \inf\Bigl\{\sum_{n\geqslant 0}\|\nu_n\|_{M(\mathbb{T})}\colon (\nu)_{n\in\mathbb{N}} \in \mathfrak{M}_{\mathcal{K}_+\Phi}\Bigr\}. \quad \bullet$$

DEFINITION 3.2. Let $\Lambda \subset \mathbb{Z}$ and $d = (d_n)_{n\in\Lambda}$, $d \in S_0(\Lambda)$. We denote by P_d the linear operator defined by the equality

$$P_d f = \bigl(\hat{f}(n)d_n\bigr)_{n\in\Lambda}, \qquad f\in L^1(\mathbb{T}).$$

LEMMA 3.3. *Let $\mathfrak{E}$ be a Banach space of analytic functions satisfying* (1.2) *and* (1.3). *In addition, suppose that*:

a) $\mathfrak{E} \subset H^2$, *where the embedding is continuous*; *and*

b) *there are constants* p *and* κ $(0 < p < 1, \kappa > 0)$ *such that*

$$\kappa \| \mathcal{K}_+ \Phi \|_{H^p} \leqslant \| \Phi \|_{\mathfrak{E}^*}, \qquad \Phi \in \mathfrak{E}^*.$$

Then there are constants β_0 *and* β_1 *for which*

$$\beta_0 \| d \|_{l^2(\mathbb{N})} \leqslant \| P_d | \mathfrak{E} \to l^1(\mathbb{N}) \| \leqslant \beta_1 \| d \|_{l^2(\mathbb{N})}, \qquad d \in s_0(\mathbb{N}). \tag{3.3}$$

PROOF. Consider the operator P_d^* dual to P_d:

$$P_d^* \colon l^\infty(\mathbb{N}) \to \mathfrak{E}^*, \quad (P_d^* x)(f) = \langle P_d f, x \rangle = \sum_{n \geqslant 0} x_n \hat{f}(n) d_n, \qquad f \in \mathfrak{E},\ x \in l^\infty(\mathbb{N}).$$

Then $(\mathcal{K}_+ \Phi)(\eta) = \sum_{n \geqslant 0} x_n d_n \eta^n$, $\eta \in \mathbb{D}$, and $\Phi \stackrel{\text{def}}{=} P_d^* x$. By b), we have

$$\kappa \left\| \sum_{n \geqslant 0} x_n d_n z^n \right\|_{H^p} \leqslant \| \Phi \|_{\mathfrak{E}^*} \leqslant \| P_d^* \| \, \| x \|_{l^\infty(\mathbb{N})} = \| P_d \| \, \| x \|_{l^\infty(\mathbb{N})}.$$

Taking x to be the sequence $(r_n(t))_{n \in \mathbb{N}}$, $t \in [0, 1]$, where $(r_n)_{n \in \mathbb{N}}$ is the sequence of Rademacher functions, we obtain

$$\kappa^p \int_{\mathbb{T}} \left| \sum_{n \geqslant 0} r_n(t) d_n \zeta^n \right|^p dm \leqslant \| P_d \|^p, \qquad t \in [0, 1]. \tag{3.4}$$

It follows from (3.4) that

$$\kappa^p \int_0^1 \left(\int_{\mathbb{T}} \left| \sum_{n \geqslant 0} r_n(t) d_n \zeta^n \right|^p dm \right) dt \leqslant \| P_d \|^p.$$

If we apply Fubini's theorem and Khintchine's inequality (see [10]) to the left-hand side of this inequality, we obtain

$$\gamma_p^p \kappa^p \int_{\mathbb{T}} \left(\sum_{n \geqslant 0} | d_n |^2 | \zeta^n |^2 \right)^{p/2} dm \leqslant \| P_d \|^p,$$

where γ_p is the constant in Khintchine's inequality. Since $| \eta^n | = 1$ for $\eta \in \mathbb{T}$, we have

$$\gamma_p \kappa \| d \|_{l^2(\mathbb{N})} \leqslant \| P_d \|.$$

Moreover, if $f \in \mathfrak{E}$, then

$$\| P_d f \|_{l^1(\mathbb{N})} = \sum_{n \geqslant 0} | d_n | | \hat{f}(n) | \leqslant \| d \|_{l^2(\mathbb{N})} \| \hat{f} \|_{l^2(\mathbb{N})} \leqslant \beta_1 \| d \|_{l^2(\mathbb{N})} \| f \|_{\mathfrak{E}},$$

where β_1 is the norm of the operator embedding $\mathfrak{E}$ in H^2. ●

LEMMA 3.4. *Let $\mathfrak{E}$ satisfy the hypotheses of the preceding lemma. Suppose that*

$$\sigma \stackrel{\text{def}}{=} \sup\{ \| z^n f \|_{\mathfrak{E}} \colon \| f \|_{\mathfrak{E}} \leqslant 1, n \in \mathbb{N} \} < +\infty. \tag{3.5}$$

Then, for every $n \in \mathbb{N}$ *and every set of numbers* $d = (d_k)_0^n$,

$$2^{1-1/p} \frac{\beta_0}{\sigma + 1} \left(\sum_{k=0}^n | d_k |^2 \right)^{1/2} \leqslant \| P_d | \mathcal{P}_{\mathfrak{E}}^n \to l^1(\mathbb{N}_n) \| \leqslant \beta_1 \left(\sum_{k=0}^n | d_k |^2 \right)^{1/2},$$

where $\mathcal{P}_{\mathfrak{E}}^n$ *is the subspace of* $\mathfrak{E}$ *consisting of the polynomials of degree at most* n; p, β_0 *and* β_1 *are the constants in Lemma* 3.3; *and* $\mathbb{N}_n \stackrel{\text{def}}{=} \{0, 1, \ldots, n\}$, $n \in \mathbb{N}$.

PROOF. We first note that if $\Phi \in (\mathcal{P}_{\mathcal{E}}^n)^*$, then

$$\|\Phi\|_{(\mathcal{P}_{\mathcal{E}}^n)^*} = \inf\{\|\Psi\|_{\mathcal{E}^*}: \Psi \in \mathcal{E}^*;\ \hat{\Psi}(k) = \hat{\Phi}(k), k \in \mathbf{N}_n\}, \tag{3.6}$$

where $\hat{\Phi}(k) \stackrel{\text{def}}{=} \Phi(z^k)$, $k \in \mathbf{N}_n$. Moreover, if $\Psi \in \mathcal{E}^*$ and $\Psi_k(f) \stackrel{\text{def}}{=} \Psi(z^k)$ ($f \in \mathcal{E}$, $k \in \mathbf{N}$), then $\|\Psi_k\|_{\mathcal{E}^*} \leqslant \sigma\|\Psi\|_{\mathcal{E}^*}$, $k \in \mathbf{N}$. In addition,

$$\begin{aligned}\kappa\left\|\sum_{k=0}^{n} \hat{\Psi}(k)z^k\right\|_{H^p} &= \kappa\|\mathcal{K}_+\Psi - \mathcal{K}_+\Psi_{n+1}\|_{H^p} \\ &\leqslant \kappa 2^{1/p-1}(\|\mathcal{K}_+\Psi\|_{H^p} + \|\mathcal{K}_+\Psi_{n+1}\|_{H^p}) \\ &\leqslant 2^{1/p-1}(1+\sigma)\|\Psi\|_{\mathcal{E}^*}.\end{aligned} \tag{3.7}$$

Consequently, if $\Phi \in (\mathcal{P}_{\mathcal{E}}^n)^*$ we have

$$\kappa\left\|\sum_{k=0}^{n} \hat{\Phi}(k)z^k\right\|_{H^p} \leqslant 2^{1/p-1}(1+\sigma)\|\Phi\|_{\mathcal{E}^*} \tag{3.8}$$

(using (3.6) and (3.7)). Now we need only consider the operator P_d and repeat the reasoning of Lemma 3.3, replacing b) by (3.8). ●

DEFINITION 3.5. 1) We call a set $\Lambda \subset \mathbf{Z}$ *lacunary* if

$$\delta \stackrel{\text{def}}{=} \sup\left\{\frac{k}{n}: k, n \in \Lambda, 0 \leqslant k < n\right\} \cup \left\{\left|\frac{k}{n}\right|: k, n \in \Lambda, n < k < 0\right\} < 1.$$

2) We call a set $\Lambda \subset \mathbf{Z}$ a *Banach set* if, for every system $x = (x_n)_{n\in\Lambda}$, $x \in l^2(\Lambda)$, there is a function $f \in C(\mathbf{T})$ such that $\hat{f}(n) = x_n$, $n \in \Lambda$.

It is well known [10], [17] that every set $\Lambda \subset \mathbf{Z}$ that can be represented as a finite union of lacunary sets is a Banach set. More detailed information can be found, for example, in [18]. In particular, we shall need the following proposition.

LEMMA 3.6 [18]. *Let $p \in (0, 2)$. The following statements are equivalent.*

1) $\Lambda \subset \mathbf{Z}$ is a Banach set.

2) There is a constant $\alpha_p > 0$ such that

$$\alpha_p\|x\|_{l^2(\Lambda)} \leqslant \left\|\sum_{n\in\Lambda} x_n\zeta^n\right\|_p, \qquad x \in s_0(\Lambda).$$

This was proved by Banach [19] for $p = 1$.

LEMMA 3.7. *Let $\mathcal{E}$ be a Banach space of analytic functions satisfying the hypotheses of Lemma 3.3; let $\mathcal{E}$ be continuously embedded in C_A; let $\Lambda \subset \mathbf{N}$ be a Banach set; and let E be a closed subset of $\mathbf{T}$. Suppose that there is a constant $\omega(E) > 0$ such that*

$$\omega(E)\|\mu\|_{M(E_*)} \leqslant \|\Phi\|_{\mathcal{E}^*}, \qquad \mu \in M(E_*), \Phi \in \mathcal{E}^*, \tag{3.9}$$

if

$$E_* \stackrel{\text{def}}{=} \{\eta \in \mathbf{T}: \bar{\eta} \in E\} \quad \text{and} \quad \left(\mathcal{K}_+\Phi - \int_{E_*}\frac{d\mu(\zeta)}{1-\bar{\zeta}z}\right) \in H^1.$$

Then for every function $g \in C(E)$ and every system $x \in l^2(\Lambda)$, there is a function $f \in \mathcal{E}$, such that $f|E = g$ and $\hat{f}|\Lambda = x$.

PROOF. Consider the linear operator Q, $Qf \overset{\text{def}}{=} (f|E, \hat{f}|\Lambda)$, $f \in \mathcal{E}$. It is clear that Q operates continuously into the space $C(E) \times l^2(\Lambda)$. Let us show that the operator Q^* dual to Q has a bounded left inverse (once this is established, we obtain $Q(\mathcal{E}) = C(E) \times l^2(\Lambda)$ by Banach's theorem on the solvability of linear equations [20]). Note that

$$\langle f, Q^*(\mu, x)\rangle = \langle Qf, (\mu, x)\rangle = \int_{E_*} f(\bar{\zeta})\, d\mu + \sum_{n\in\Lambda} x_n \hat{f}(n),$$
$$f \in \mathcal{E},\ \mu \in M(E_*),\ x \in S_0(\Lambda).$$

Therefore

$$(\mathcal{K}_+ \Phi)(\eta) = \int_{E_*} \frac{d\mu}{1 - \bar{\zeta}\eta} + \sum_{n\in\Lambda} x_n \eta^n, \qquad \eta \in \mathbb{D},\ \Phi \overset{\text{def}}{=} Q^*(\mu, x).$$

In addition,

$$\left\| \sum_{n\in\Lambda} x_n z^n \right\|_{H^p} = \left\| \mathcal{K}_+ \Phi - \int_{E_*} \frac{d\mu(\bar{\zeta})}{1 - \bar{\zeta} z} \right\|_{H^p} \leqslant \left(\|\mathcal{K}_+ \Phi\|^p_{H^p} + \left\| \int_{E_*} \frac{d\mu(\zeta)}{1 - \bar{\zeta} z} \right\|^p_{H^p} \right)^{1/p}$$
$$\leqslant \left(\frac{1}{\kappa^p} \|\Phi\|^p_{\mathcal{E}^*} + \frac{4}{1-p} \|\mu\|^p_{M(E^*)} \right)^{1/p}$$
$$\leqslant \left(\frac{1}{\kappa^p} + \frac{4}{(1-p)(\omega(E))^p} \right)^{1/p} \|\Phi\|_{\mathcal{E}^*} \overset{\text{def}}{=} \beta_2 \|\Phi\|_{\mathcal{E}^*}$$

(using b), (3.9), and the Kolmogorov-Smirnov inequality (1.4)). If we apply this inequality together with (3.9) and Lemma 3.6, we obtain

$$\|\mu\|_{M(E_*)} \left\| \sum_{n\in\Lambda} x_n z^n \right\|_{H^p} \leqslant \left(\frac{1}{\omega(E)} + \beta_2 \right) \|\Phi\|_{\mathcal{E}^*},$$

$$\|\mu\|_{M(E)} + \alpha_p \|x\|_{l^2(\Lambda)} \leqslant (1/\omega(E) + \beta_2) \|\Phi\|_{\mathcal{E}^*}, \qquad \mu \in M(E_*),\ x \in S_0(\Lambda).$$

Consequently Q^* has a bounded left inverse. ●

The results given below (Theorems 3.8–3.11) are corollaries of (1.4), (2.9) and Lemmas 3.3, 3.4, and 3.7. We also note that proofs of 3.8 and 3.10 were published by the author [4] (but is a less transparent form).

THEOREM 3.8. *There is an absolute constant $c_4 > 0$ such that for every $n \in \mathbb{N}$ and every sequence of numbers $(d_k)_0^n$*

$$c_4 \left(\sum_{k=0}^{n} |d_k|^2 \right)^{1/2} \leqslant \sup\left\{ \sum_{k=0}^{n} |d_k| |\hat{f}(k)| : f \in \mathcal{P}^n,\ \|f\|_{C_A} \leqslant 1 \right\}$$
$$\leqslant \left(\sum_{k=0}^{n} |d_k|^2 \right)^{1/2}, \tag{3.10}$$

where $\mathcal{P}^n$ is the set of polynomials of degree at most n.

This is an old result of Paley and Stechkin [11], [6]. Paley proved (3.3) for $\mathcal{E} = C_A$ and Stechkin deduced (3.10) from Paley's result.

PROOF. It is evident that the space $\mathcal{E} = C_A$ satisfies (1.2) and (1.3). In addition, it is easy to prove that $\mathcal{K}_+(C_A^*)$ consists of the functions of the form

$$\int_{\mathbb{T}} \frac{d\mu(\zeta)}{1-\bar{\zeta}z}, \qquad \mu \in M(\mathbb{T}),$$

and

$$\|\Phi\|_{C_A^*} = \inf\left\{\|\mu\|_{M(\mathbb{T})}: \mu \in M(\mathbb{T}), \mathcal{K}_+\Phi = \int_{\mathbb{T}} \frac{d\mu(\zeta)}{1-\bar{\zeta}z}\right\}.$$

Therefore if we use (1.4) for $p = 1/2$ and Lemma 3.4 (here we still have to observe that $\|f\|_{C_A} \leqslant \|f\|_{H^2}$ and $\|z^n f\|_{C_A} = \|f\|_{C_A}$, $f \in C_A$, $n \in \mathbb{N}$), we obtain what was required. ●

The following result strengthens the Paley-Stechkin theorem.

THEOREM 3.9. *There is an absolute constant $c_5 > 0$ such that for every $n \in \mathbb{N}$ and every sequence of numbers $(d_k)_0^n$*

$$c_5\left(\sum_{k=0}^{n} |d_k|^2\right)^{1/2} \leqslant \sup\left\{\sum_{k=0}^{n} |d_k||\hat{f}(k)| : f \in \mathcal{P}^n, \|f\|_{U_A} \leqslant 1\right\}$$
$$\leqslant \left(\sum_{k=0}^{n} |d_k|^2\right)^{1/2}. \tag{3.11}$$

We emphasize that the upper bound in (3.11) is taken over all polynomials of degree at most n that are contained in the unit ball of U_A.

PROOF. It is clear that U_A satisfies (1.2) and (1.3), and

$$\|f\|_{U_A} \leqslant \|f\|_{H^2}, \quad \|z^n f\|_{U_A} = \|f\|_{U_A}, \qquad f \in U_A, n \in \mathbb{N}.$$

Consequently it remains only to apply Lemmas 3.4, 3.1, and Corollary 2.5 with $p = 1/2$. ●

In connection with Theorems 3.8 and 3.9 we call attention to the interesting paper by Jöricke [21], which completely solves the problem of estimating the norm of the operator P_d, $d \in s_0(\mathbb{N})$, acting from the space $\mathcal{P}^n$ with norm

$$\|f\| \stackrel{\text{def}}{=} \sup\{|f(\eta)| : \eta \in G\}, \qquad f \in \mathcal{P}^n,$$

where G is a region of the form $\mathbb{D} \cup \{\eta \in \mathbb{C}: |\eta| < r, \theta_0 < \arg\eta < \theta_1\}$, $r \in (1, +\infty)$, $0 < \theta_1 - \theta_0 < 2\pi$, to the space l^1. We note that Jöricke's estimate is essentially different from (3.10) and (3.11).

THEOREM 3.10. *Let E be a closed subset of $\mathbb{T}$, $mE = 0$, and Λ a Banach set, $\Lambda \subset \mathbb{N}$. Then for every function $g \in C(E)$ and every system $x \in l^2(\Lambda)$ there is a function $f \in C_A$ such that $f|E = g$ and $\hat{f}|\Lambda = x$.*

PROOF. We first note that C_A satisfies hypotheses (1.2), (1.3) and b) (see (1.4)). Hence in order to be able to apply Lemma 3.7 it is enough to verify that $\omega(E) > 0$. Let $\mu \in M(E_*)$, $\tilde{\mu}A = \mu(A \cap E_*)$, $A \in \mathfrak{B}(\mathbb{T})$,(2) and

(2) $\mathfrak{B}(\mathbb{T})$ is the σ-algebra of Borel subsets of $\mathbb{T}$.

$$\Phi \in C_A^* \quad \text{and} \quad h_0 \overset{\text{def}}{=} \left(\int_{\mathbf{T}} \frac{d\tilde{\mu}(\zeta)}{1 - \bar{\zeta} z} - \mathcal{K}_+ \Phi \right) \in H_+^1 .$$

Since C_A^* is linearly isometric to $M(\mathbf{T})/C_A^{\perp}$, $C_A^{\perp} \overset{\text{def}}{=} \{\nu \in M(\mathbf{T}) \colon (f * \nu)(1) = 0, \forall f \in C_A\}$, $C_A^{\perp} \cong H_-^1$ (by the M. and F. Riesz theorem [14]) and $\tilde{\mu}$ is singular with respect to Lebesgue measure m ($mE = 0$), we have

$$\|\Phi\|_{C_A^*} = \inf\left\{\|\tilde{\mu} - h_0\, dm + h\, dm\|_{M(\mathbf{T})} \colon h \in H_-^1\right\} \geqslant \|\tilde{\mu}\|_{M(\mathbf{T})} = \|\mu\|_{M(E_*)}.$$

Consequently we may apply Lemma 3.7. ●

Note that if we put $\Lambda = \varnothing$ in Theorem 3.10, it becomes the well-known Rudin-Carleson interpolation theorem (see, for example, [14]).

THEOREM 3.11. *Let E be a closed countable subset of $\mathbf{T}$, and let $\Lambda \subset \mathbf{N}$ be a Banach set. Then for every $g \in C(E)$ and every system $x \in l^2(\Lambda)$ there is a function $f \in U_A$ such that $f | E = g$ and $\hat{f} | \Lambda = x$.*

PROOF. It follows from Corollary 2.5 and Lemma 3.1 that the space U_A has property b); and the validity of (1.2) and (1.3) is evident. Let us verify that $\omega(E) > 0$. Let $\mu \in M(E_*)$, $\Phi \in U_A^*$ and

$$h_0 \overset{\text{def}}{=} \left(\int_{E_*} \frac{d\mu(\zeta)}{1 - \bar{\zeta} z} - \mathcal{K}_+ \Phi \right) \in H_+^1 ;$$

then

$$\int_{E_*} \frac{d\mu(\zeta)}{1 - \bar{\zeta} z} = \sum_{n \geqslant 0} z^n \int \frac{\bar{\zeta}^n\, d\nu_n(\zeta)}{1 - \bar{\zeta} z} + \int_{\mathbf{T}} \frac{h_0(\zeta)\, dm(\zeta)}{1 - \bar{\zeta} z},$$

where $(\nu_n)_{n \in \mathbf{N}} \in \mathfrak{M}_{\mathcal{K}_+ \Phi}$ (see Lemma 3.1).

If $\xi \in E$ and $r \in (0, 1)$, we have

$$\int_{E_*} \frac{(1 - r)\, d\mu}{(1 - r\xi\bar{\zeta})} = \sum_{n \geqslant 0} r^n \xi^n \int_{\mathbf{T}} \frac{(1 - r)\bar{\zeta}^n\, d\nu_n}{1 - r\xi\bar{\zeta}} + \int_{\mathbf{T}} \frac{h_0\, dm}{1 - r\xi\bar{\zeta}}.$$

Since

$$\left| \frac{(1 - r)\bar{\eta}^n}{1 - \bar{\eta} r \xi} \right| \leqslant 1, \quad \left| r^n \xi^n \int_{\mathbf{T}} \frac{(1 - r)\bar{\zeta}^n\, d\nu_n}{1 - r\xi\bar{\zeta}} \right| \leqslant \|\nu_n\|_{M(\mathbf{T})}$$

$$(\xi, \eta \in \mathbf{T}, 0 < r < 1, n \in \mathbf{N}),$$

$$\sum_{n \geqslant 0} \|\nu_n\|_{M(\mathbf{T})} < +\infty$$

and

$$\lim_{r \to 1-0} \frac{1 - r}{1 - \bar{\eta} r \xi} = \begin{cases} 1, & \eta = \xi, \\ 0, & \eta \neq \xi, \end{cases} \qquad (\xi, \eta \in \mathbf{T}),$$

we obtain, by Lebesgue's theorem on bounded convergence,

$$\mu(\{\xi\}) = \sum_{n \geqslant 0} \nu_n(\{\xi\}), \qquad \xi \in E_*.$$

Since E is countable, we have

$$\|\mu\|_{M(E_*)} = \sum_{\xi \in E_*} |\mu(\{\xi\})| \leqslant \sum_{\xi \in E_*} \Big(\sum_{n \geqslant 0} |\nu_n(\{\xi\})| \Big)$$
$$= \sum_{n \geqslant 0} \Big(\sum_{\xi \in E_*} |\nu_n(\{\xi\})| \Big) \leqslant \sum_{n \geqslant 0} \|\nu\|_{M(\mathbb{T})}.$$

Consequently

$$\|\mu\|_{M(E_*)} \leqslant \inf \Big\{ \sum_{n \geqslant 0} \|\nu_n\|_{M(\mathbb{T})} \colon (\nu_n)_{n \in \mathbb{N}} \in \mathfrak{M}_{\mathcal{K}_+ \Phi} \Big\} = \|\Phi\|_{U_A^*}.$$

It remains only to apply Lemma 3.7. ●

REMARK 3.12. 1) We note that when $\Lambda = \varnothing$ the conclusion of Theorem 3.11 is not new: in that case, Hruščev and the author [22] proved a stronger result.

2) When the present paper was being prepared for publication, we learned that Oberlin [29], using (0.1), which had been announced in [5], and the relation

$$\varlimsup_{t \to +\infty} t \cdot m \Big\{ \eta \in \mathbb{T} \colon \lim_{r \to 1-0} \Big| \int_{\mathbb{T}} \frac{d\mu}{1 - r\eta\bar{\zeta}} \Big| > t \Big\} = \frac{1}{\pi} \|\mu\|_{M(\mathbb{T})}, \tag{3.12}$$

where μ is a nonnegative measure on $\mathbb{T}$, singular with respect to m, had obtained the following analog of the Rudin-Carleson theorem for U_A:

If $E \subset \mathbb{T}$, E is closed, and $mE = 0$, then $R^E(U_A) = C(E)$, where $R^E f \overset{\text{def}}{=} f \,|\, E$.

In fact, (3.12) makes it quite easy to prove a proposition which includes Oberlin's theorem as a special case:

If $\mathcal{E}$ is a Banach space of analytic functions satisfying (1.2) *and* (1.3), $\mathcal{E} \subset C_A$ *and*

$$\kappa_0 \overset{\text{def}}{=} \varlimsup_{t \to +\infty} t \cdot m \Big\{ \eta \in \mathbb{T} \colon \varlimsup_{r \to 1-0} |\mathcal{K}_+ \Phi(r\eta)| > t \Big\} < +\infty,$$
$$\forall \Phi \in \mathcal{E}^*, \ \|\Phi\|_{\mathcal{E}^*} \leqslant 1,$$

where E is a closed subset of $\mathbb{T}$ and $mE = 0$, then $R^E \mathcal{E} = C(E)$.

To prove this, it is enough to note that:

a)

$$\mathcal{K}_+ ((R^E)^*(\lambda)) = \int_{E^*} \frac{d\lambda(\zeta)}{1 - \bar{\zeta} z},$$

where $\lambda \in M(E_*)$ and $(R^E)^*$ is the operator dual to R^E,

b) (3.12) holds even without the nonnegativity of μ, and

c)

$$\frac{1}{\pi} \|\lambda\|_{M(E_*)} \leqslant \varlimsup_{t \to +\infty} t \cdot m \Big\{ \eta \in \mathbb{T} \colon \lim_{r \to 1-0} \Big| \int_{E_*} \frac{d\lambda(\zeta)}{1 - r\eta\bar{\zeta}} \Big| > t \Big\}$$
$$\leqslant \kappa_0 \|(R^E)^*(\lambda)\|_{\mathcal{E}^*},$$

$\lambda \in M(E_*)$.

After this it remains only to apply Banach's theorem on the solvability of linear equations in Banach spaces [20].

3) *The conclusions of Theorems* 3.11, 4.10 *and* 4.14 *remain valid if instead of the countability of E we require that $mE = 0$.*

§4. Laurent coefficients and simultaneous interpolation of analytic functions with a small set of singularities

DEFINITION 4.1. 1) If E is a subset of $\mathbb{T}$, we denote by $\mathring{E}$ the interior of E with respect to $\mathbb{T}$, and by E_*, as in §3, the set $\{\eta \in \mathbb{T}\colon \bar{\eta} \in E\}$.

2) If μ is a complex Borel measure on $\mathbb{T}$, we denote by $\operatorname{supp}\mu$ the closed support of μ; that is, the smallest closed subset E of $\mathbb{T}$ for which $\mu(\mathbb{T}\setminus E) = 0$.

If $f \in L^1(\mathbb{T})$, then $\operatorname{supp} f$ denotes the closed support of the measure $f dm$.

3) Let Φ be a continuous functional on C^∞. We denote by $\operatorname{supp}\Phi$ the support of Φ; that is, the smallest closed subset E of $\mathbb{T}$ such that $\Phi(f) = 0$ for every function f ($f \in C^\infty$, $\operatorname{supp} f \subset \mathbb{T}\setminus E$).

Note that if $\mu \in M(\mathbb{T})$ and $\Phi(f) \overset{\text{def}}{=} (\mu * f)(1)$, $f \in C(\mathbb{T})$, then $\operatorname{supp}\Phi = (\operatorname{supp}\mu)_*$.

LEMMA 4.2 [20]. *Let Y be a subspace of the Banach space X, and let $\Phi \in X^*$. Then*

$$\sup\{|\Phi(x)| : x \in Y,\ \|x\|_X \leqslant 1\} = \inf\{\|\Phi + \Psi\|_{X^*}\colon \psi \in Y^\perp\},$$

where $Y^\perp = \{\Psi \in X^*\colon \Psi(y) = 0,\ \forall y \in Y\}$.

LEMMA 4.3 [23]. *Let $\Phi \in (C^\infty)^*$, $\hat{\Phi} \in l^\infty(\mathbb{Z})$, and let E be the smallest closed subset of $\mathbb{T}$ outside which the functions $\mathcal{K}_+\Phi$ and $-\mathcal{K}_-\Phi$ are analytic continuations of each other. Then* $\operatorname{supp}\Phi = E_*$.

LEMMA 4.4. *Let $\mathfrak{S}$ be a Banach space of analytic functions, and let E satisfy* (1.2), (1.3) *and* b) *of Lemma* 3.3. *If $\Gamma \subset \mathbb{T}$ is an arbitrary measurable set (measurable with respect to m), and if the functionals Φ and Ψ ($\Phi, \Psi \in \mathfrak{S}^*_\pm(\mathbb{T})$) satisfy*

$$\hat{\Phi} \in s_0(\mathbb{Z});\ \lim_{r\to 1-0} \Psi\left(\frac{1-r^2}{|1-r\eta\bar{\zeta}|^2}\right) = 0, \quad a.e.\ \eta \in \Gamma,$$

then

$$\kappa 2^{1-1/p}\left\|\sum_{n\in\mathbb{Z}} \hat{\Phi}(n)\zeta^n\right\|_{L^p(\Gamma)} \leqslant \kappa\left(\|\mathcal{K}_+(\Phi+\Psi)\|_{H^p_+} + \|\mathcal{K}_-(\Phi+\Psi)\|_{H^p_-}\right)$$
$$\leqslant \|\Phi + \Psi\|_{\mathfrak{S}^*_\pm}, \tag{4.1}$$

where p and κ ($p \in (0,1)$, $\kappa > 0$) are the constants of hypothesis b) *of Lemma* 3.3.

PROOF. Let Φ and Ψ be functionals satisfying the hypotheses of the lemma. Then

$$\begin{aligned}\|\Phi + \Psi\|_{\mathfrak{S}^*_\pm} &= \|\Phi_+ + \Psi_+\|_{\mathfrak{S}^*_+} + \|\Phi_- + \Psi_-\|_{\mathfrak{S}^*_-}\\ &\geqslant \kappa\left(\|\mathcal{K}_+(\Phi+\Psi)\|_{H^p_+} + \|\mathcal{K}_-(\Phi+\Psi)\|_{H^p_-}\right)\\ &= \kappa\left(\left\|\sum_{n\geqslant 0}\hat{\Phi}(n)\zeta^n + h_+\right\|_{L^p(\mathbb{T})} + \left\|\sum_{n<0}\hat{\Phi}(n)\zeta^n + h_-\right\|_{L^p(\mathbb{T})}\right)\\ &\geqslant \kappa 2^{1-1/p}\left\|\sum_{n\in\mathbb{Z}}\hat{\Phi}(n)\zeta^n + (h_+ + h_-)\right\|_{L^p(\Gamma)},\end{aligned}$$

where h_+ denotes the radial boundary values of $\mathcal{K}_+\Psi$ and h_- denotes the radial boundary values of $\mathcal{K}_-\Psi$. Since

$$\lim_{r\to 1-0}\Psi\left(\frac{1-r^2}{|1-r\eta\bar{\zeta}|^2}\right) = 0, \quad \text{a.e. } \eta \in \mathbb{T},$$

we have $h_+(\eta) + h_-(\eta) = 0$, a.e. $\eta \in \mathbb{T}$. Consequently

$$\kappa 2^{1-1/p}\Big\| \sum_{n\in\mathbf{Z}} \hat\Phi(n)\zeta^n \Big\|_{L^p(\Gamma)} \leqslant \kappa\big(\|\mathcal{K}_+(\Phi+\Psi)\|_{H^p_+} + \|\mathcal{K}_-(\Phi+\Psi)\|_{H^p_-}\big)$$
$$\leqslant \|\Phi+\Psi\|_{\mathcal{E}^*_\pm}. \quad \bullet$$

LEMMA 4.5. *Let Γ be a finite union of closed arcs, $\Gamma \subset \mathbb{T}$, $\mathcal{E}$ a Banach space of analytic functions satisfying the hypotheses of Lemma* 3.3, *and*

$$\mathcal{E}_\pm(\Gamma) \stackrel{\text{def}}{=} \{f \in \mathcal{E}_\pm(\mathbb{T}) : \operatorname{supp} f \subset \Gamma\}.$$

Then

$$\beta_0(m\Gamma)^{1/p}\|d\|_{l^2(\mathbf{Z})} \leqslant \|P_d \mid \mathcal{E}_\pm(\Gamma) \to l^1(\mathbf{Z})\| \leqslant \beta_1\|d\|_{l^2(\mathbf{Z})}, \qquad d \in s_0(\mathbf{Z}),$$

where P_d is the operator defined in 3.2, *and β_0, β_1 and p are the constants from Lemma* 3.3.

PROOF. Let $x \in S_0(\mathbf{Z})$ and $\|x\|_{l^\infty(\mathbf{Z})} \leqslant 1$. Applying Lemma 4.2, we obtain

$$\|P_d\| \geqslant \sup\{|\langle P_d f, x\rangle| : f \in \mathcal{E}_\pm(\Gamma),\ \|f\|_{\mathcal{E}_\pm} \leqslant 1\}$$
$$= \sup\Big\{\Big|\sum_{n\in\mathbf{Z}} x_n d_n \hat f(n)\Big| : f \in \mathcal{E}_\pm(\Gamma),\ \|f\|_{\mathcal{E}_\pm} \leqslant 1\Big\}$$
$$= \inf\{\|\Phi+\Psi\|_{\mathcal{E}^*_\pm} : \Psi \in \mathcal{E}^\perp_\pm(\Gamma) \subset \mathcal{E}^*_\pm(\mathbb{T})\},$$

where $\hat\Phi(n) = x_n d_n$, $n \in \mathbf{Z}$, and $d = (d_n)_{n\in\mathbf{Z}} \in s_0(\mathbf{Z})$.

It follows from Lemmas 4.3, 4.4 that

$$\|P_d\| \geqslant \inf\{\|\Phi+\Psi\|_{\mathcal{E}^*_\pm} : \Psi \in \mathcal{E}^*_\pm(\mathbb{T}),\ \operatorname{supp}\Psi \subset \mathbb{T}\setminus\mathring{\Gamma}\}$$
$$\geqslant \kappa 2^{1-1/p}\Big\|\sum_{n\in\mathbf{Z}} x_n d_n \zeta^n\Big\|_{L^p(\Gamma_*)} \qquad (0<p<1),$$

where κ is the constant from Lemma 3.3.

Moreover, if we reason as in the proof of Lemma 3.3, we obtain

$$\|P_d\|^p \geqslant \kappa^p \int_0^1 \left(\int_{\Gamma_*}\Big|\sum_{n\in\mathbf{Z}} d_n r_{k_n}(t)\zeta^n\Big|^p dm(\zeta)\right) dt$$
$$\geqslant (\gamma_p\kappa)^p \int_{\Gamma_*}\Big(\sum_{n\in\mathbf{Z}} |d_n|^2|\zeta^n|^2\Big)^{p/2} dm = (\gamma_p\kappa)^p m\Gamma\Big(\sum_{n\in\mathbf{Z}}|d_n|^2\Big)^{p/2}$$

($(k_n)_{n\in\mathbf{Z}}$ is a rearrangement of the nonnegative integers). Consequently

$$\kappa\gamma_p(m\Gamma)^{1/p}\Big(\sum_{n\in\mathbf{Z}}|d_n|^2\Big) \leqslant \|P_d\|.$$

The inequality $\|P_d\| \leqslant \beta_1(\sum_{n\in\mathbf{Z}}|d_n|^2)^{1/2}$ is evident (see the proof of Lemma 3.3).$\bullet$

LEMMA 4.6. *Let $\mathcal{E}$, Γ and $\mathcal{E}_\pm(\Gamma)$ be the same as in Lemma* 4.5, *$m\Gamma > 0$, $\mathcal{E} \subset C_A$, and let E and F be closed subsets of $\mathring{\Gamma}$ and Λ a lacunary set of integers. Consider the operator Q defined by*

$$Qf = (f_+ \mid E, f_- \mid F, \hat f \mid \Lambda), \qquad f \in \mathcal{E}_\pm(\Gamma).$$

Suppose that $\mathcal{E}$ satisfies the hypotheses of Lemma 3.3 and, in addition, that there is a constant $\omega = \omega(E, F)$, $\omega > 0$, for which

$$\omega\left(\|\mu\|_{M(E_*)} + \|\nu\|_{M(F_*)}\right) \leqslant \|\Phi + \Psi\|_{\mathcal{E}^*_\pm}, \tag{4.2}$$

if $\Phi, \Psi \in \mathcal{E}^*_\pm(\mathbb{T})$, $\mu \in M(E_*)$, $\nu \in M(F_*)$, and $\operatorname{supp}\Psi \subset \mathbb{T}\setminus\mathring{\Gamma}$,

$$\left(\int_{E_*}\frac{d\mu(\zeta)}{1-\bar{\zeta}z} - \mathcal{K}_+\Phi\right) \in H^1_+, \quad \left(\int_{F_*}\frac{d\nu(\zeta)}{1-\bar{\zeta}z} - \mathcal{K}_-\Phi\right) \in H^1_-.$$

Then for all functions g, h ($g \in C(E)$, $h \in C(F)$) and every system $x = (x_n)_{n\in\Lambda}$, $x \in l^2(\Lambda)$, there is a function $f \in \mathcal{E}_\pm(\Gamma)$ satisfying

$$Qf = (g, h, x), \quad \|f\|_{\mathcal{E}_\pm(\Gamma)} \leqslant b^{-1/p}\beta_3\left(\|g\|_{C(E)} + \|h\|_{C(F)} + \|x\|_{l^2(\Lambda)}\right),$$

where the constant $b = b(\Gamma, \Lambda)$ depends only on Γ and Λ, and β_3 depends only on $\omega(E, F)$ and on the constants p and κ in hypothesis b) *of Lemma* 3.3.

Proof. It is clear that Q operates continuously into $C(E) \times C(F) \times l^2(\Lambda)$.
Note that

$$\begin{aligned}\langle f, Q^*(\mu, \nu, x)\rangle &= \langle Qf, (\mu, \nu, x)\rangle \\ &= \int_{E_*} f_+(\bar{\zeta})\,d\mu + \int_{F_*} f_-(\bar{\zeta})\,d\nu + \sum_{n\in\Lambda} x_n\hat{f}(n) = \langle f, \Phi\rangle,\end{aligned}$$

where

$$(\mathcal{K}_+\Phi)(\eta) = \sum_{n\in\Lambda,\, n\geqslant 0} x_n\eta^n + \int_{E_*}\frac{d\mu}{1-\bar{\zeta}\eta}, \qquad |\eta| < 1;$$

$$(\mathcal{K}_-\Phi)(\eta) = \sum_{n\in\Lambda,\, n<0} x_n\eta^n + \int_{F_*}\frac{d\nu}{1-\bar{\zeta}\eta}, \qquad |\eta| > 1;$$

$$f \in \mathcal{E}_\pm(\Gamma),\ \mu \in M(E_*),\ \nu \in M(F_*),\ x \in s_0(\mathbb{Z}).$$

Moreover, it follows from Lemma 4.2 that

$$\begin{aligned}\|Q^*(\mu, \nu, x)\|_{\mathcal{E}^*_\pm(\Gamma)} &= \sup\left\{|\langle f, \Phi\rangle| : f \in \mathcal{E}_\pm(\Gamma),\ \|f\|_{\mathcal{E}_\pm} \leqslant 1\right\} \\ &\geqslant \inf\left\{\|\Phi + \Psi\|_{\mathcal{E}^*_\pm} : \Psi \in \mathcal{E}^*_\pm(\mathbb{T}),\ \operatorname{supp}\Psi \subset \mathbb{T}\setminus\mathring{\Gamma}\right\}.\end{aligned}$$

Let $\Psi \in \mathcal{E}^*_\pm(\mathbb{T})$, $\operatorname{supp}\Psi \subset \mathbb{T}\setminus\mathring{\Gamma}$. Using Lemma 4.3 and inequalities (4.1) and (1.4), we obtain

$$\begin{aligned}\left\|\sum_{n\in\Lambda} x_n\zeta^n\right\|_{L^p(\Gamma)} &\leqslant 2^{1/p-1}\Bigg(\left\|\sum_{n\in\Lambda,\, n\geqslant 0} x_n z^n + \mathcal{K}_+\Psi\right\|_{H^p_+} \\ &\qquad + \left\|\sum_{n\in\Lambda,\, n<0} x_n z^n + \mathcal{K}_-\Psi\right\|_{H^p_-}\Bigg) \\ &= 2^{1/p-1}\left(\left\|\mathcal{K}_+(\Phi+\Psi) - \int_{E_*}\frac{d\mu(\zeta)}{1-\bar{\zeta}z}\right\|_{H^p_+} + \left\|\mathcal{K}_-(\Phi+\Psi) - \int_{F_*}\frac{d\nu(\zeta)}{1-\bar{\zeta}z}\right\|_{H^p_-}\right)\end{aligned}$$

$$\leqslant 4^{1/p-1}\left(\|\mathcal{K}_+(\Phi+\Psi)\|_{H^p_+}+\left(\frac{4}{1-p}\right)^{1/p}\|\mu\|_{M(E_*)}+\|\mathcal{K}_-(\Phi+\Psi)\|_{H^p_-}\right.$$
$$\left.+\left(\frac{4}{1-p}\right)^{1/p}\|\nu\|_{M(F_*)}\right)$$
$$\leqslant \frac{1}{\kappa}4^{1/p-1}\|\Phi+\Psi\|_{\mathcal{E}^*_\pm}+\frac{1}{4}\left(\frac{16}{1-p}\right)^{1/p}\left(\|\mu\|_{M(E_*)}+\|\nu\|_{M(F_*)}\right).$$

By Lemma 5.2 (see §5)

$$b^{1/p}\|x\|_{l^2(\Lambda)}\leqslant\Big\|\sum_{n\in\Lambda}x_n\zeta^n\Big\|_{L^p(\Gamma)},\qquad x\in s_0(\Lambda),$$

and therefore

$$b^{1/p}\|x\|_{l^2(\Lambda)}+\|\mu\|_{M(E_*)}+\|\nu\|_{M(F_*)}\leqslant\Big\|\sum_{n\in\Lambda}x_n\zeta^n\Big\|_{L^p(\Gamma)}+\|\mu\|_{M(E_*)}+\|\nu\|_{M(F_*)}$$
$$\leqslant\frac{4^{1/p}}{\kappa}\|\Phi+\Psi\|_{\mathcal{E}^*_\pm}+\left(\frac{16}{1-p}\right)^{1/p}\left(\|\mu\|_{M(E_*)}+\|\nu\|_{M(F_*)}\right)$$
$$\leqslant\left(\frac{4^{1/p}}{\kappa}+\left(\frac{16}{1-p}\right)^{1/p}\frac{1}{\omega}\right)\|\Phi+\Psi\|_{\mathcal{E}^*_\pm},$$

$\Psi\in\mathcal{E}^*_\pm(\mathbb{T})$ and $\operatorname{supp}\Psi\subset\mathbb{T}\backslash\mathring{\Gamma}$ (in the last inequality we used (4.2)).
Consequently

$$b^{1/p}\left(\|x\|_{l^2(\Lambda)}+\|\mu\|_{M(E_*)}+\|\nu\|_{M(F_*)}\right)\leqslant\tfrac{1}{2}\beta_3\|Q^*(\mu,\nu,x)\|_{\mathcal{E}^*_\pm(\Gamma)},$$
$$\beta_3\overset{\text{def}}{=}2\left(\frac{4^{1/p}}{\kappa}+\left(\frac{16}{1-p}\right)^{1/p}\frac{1}{\omega}\right).$$

It remains only to apply Banach's theorem [20] on the solvability of linear equations in Banach spaces. ●

Theorem 4.7. *Let Γ be a finite union of closed arcs, $\Gamma\subset\mathbb{T}$, $m\Gamma>0$. Then there is an absolute constant $c_6>0$ such that*

$$c_6(m\Gamma)^2\|d\|_{l^2(\mathbf{Z})}\leqslant\|P_d\,|\,U_\pm(\Gamma)\to l^1\|\leqslant\|d\|_{l^2(\mathbf{Z})},\qquad d\in s_0(\mathbf{Z}),$$

where $U_\pm(\Gamma)\overset{\text{def}}{=}(U_A)_\pm(\Gamma)$.

Proof. Since the space $\mathcal{E}=U_A$ satisfies the hypotheses of Lemma 3.3 (see Corollary 2.5 and Lemma 3.1), and $\|f\|_{U_A}\leqslant\|f\|_{H^2}$, $f\in U_A$, we obtain what was required by applying Lemma 4.5 with $p=1/2$. ●

Lemma 4.8. *Let Γ be a finite union of closed arcs, $\Gamma\subset\mathbb{T}$, E and F closed subsets of $\mathbb{T}$, $E\cup F\subset\mathring{\Gamma}$; $\mu,\nu\in M(\mathbb{T})$, $\operatorname{supp}\mu\subset E_*$ and $\operatorname{supp}\nu\subset F_*$. Consider a functional $\Phi\in(C^\infty)^*$ satisfying*

$$\hat{\Phi}(k)=\hat{\mu}(k)+\hat{h}(k)\quad(k\in\mathbb{N}),\qquad\hat{\Phi}(k)=\hat{\nu}(k)+\hat{h}(k)\quad(k\in\mathbf{Z}\backslash\mathbb{N}),\quad(4.3)$$

where $h \in L^1(\mathbb{T})$. *Then the following assertions are true*:

1) *If* $\Psi \in C^*_{\pm}(\mathbb{T})$, $C_{\pm}(\mathbb{T}) \overset{\text{def}}{=} (C_A)_{\pm}$, $mE = mF = 0$ *and* $\operatorname{supp}\Psi \subset \mathbb{T}\backslash\mathring{\Gamma}$, *then*

$$\|\mu\|_{M(\mathbb{T})} + \|\nu\|_{M(\mathbb{T})} \leqslant \|\Phi + \Psi\|_{C^*_{\pm}}.$$

2) *If* $\Psi \in U^*_{\pm}(\mathbb{T})$, $U_{\pm}(\mathbb{T}) \overset{\text{def}}{=} (U_A)_{\pm}$, $\operatorname{supp}\Psi \subset \mathbb{T}\backslash\mathring{\Gamma}$ *and* E *and* F *are countable sets, then*

$$\|\mu\|_{M(\mathbb{T})} + \|\nu\|_{M(\mathbb{T})} \leqslant \|\Phi + \Psi\|_{U^*_{\pm}}.$$

PROOF. 1) Let $\Psi \in C^*_{\pm}(\mathbb{T})$ and $\operatorname{supp}\Psi \subset \mathbb{T}\backslash\mathring{\Gamma}$; then there are complex measures λ and ρ ($\lambda, \rho \in M(\mathbb{T})$) such that

$$\hat{\Psi}(k) = \hat{\lambda}(k) \quad (k \in \mathbb{N}), \qquad \hat{\Psi}(k) = \hat{\rho}(k) \quad (k \in \mathbb{Z}\backslash\mathbb{N}).$$

We find a lower bound for the norm $\|\Phi + \Psi\|_{C^*_{\pm}}$:

$$\begin{aligned}\|\Phi + \Psi\|_{C^*_{\pm}} &= \|\Phi_+ + \Psi_+\|_{C^*_A} + \|\Phi_- + \Psi_-\|_{(\tau C_A)^*}\\ &= \inf\left\{\|\mu + \lambda + h\,dm + g\,dm\|_{M(\mathbb{T})} : g \in H^1_+\right\}\\ &\quad + \inf\left\{\|\nu + \rho + h\,dm + g\,dm\|_{M(\mathbb{T})} : g \in H^1_+\right\}. \end{aligned} \tag{4.4}$$

Since

$$\operatorname{supp}\Psi \subset \mathbb{T}\backslash\mathring{\Gamma}, \quad (\mathcal{K}_+\Psi)(\eta) = \int_{\mathbb{T}} \frac{d\lambda}{1 - \bar{\zeta}\eta} \qquad (\eta \in \mathbb{D}),$$

$$(\mathcal{K}_-\Psi)(\eta) = \int \frac{d\rho}{1 - \bar{\zeta}\eta} \qquad (\eta \in \overline{\mathbb{C}}\backslash\overline{\mathbb{D}})$$

and the function $\mathcal{K}_+\Psi$ has an analytic continuation across $\mathring{\Gamma}_*$ into $-\mathcal{K}_-\Psi$ (see Lemma 4.3), it follows from Lemma 5.3 (see §5) that the supports of the singular parts of the measures λ and ρ are contained in $\mathbb{T}\backslash\mathring{\Gamma}_*$. Therefore, if we take account of (4.4) and the fact that μ and ν are singular (indeed, $mE = mE_* = mF = mF_* = 0$ and $\operatorname{supp}\mu \subset E_*$, $\operatorname{supp}\nu \subset F_*$), we obtain what was required:

$$\|\Phi + \Psi\|_{C^*_{\pm}} \geqslant \|\mu\|_{M(\mathbb{T})} + \|\nu\|_{M(\mathbb{T})}.$$

2) Let $\Psi \in U^*_{\pm}(\mathbb{T})$ and $\operatorname{supp}\Psi \subset \mathbb{T}\backslash\mathring{\Gamma}$; then for every $\xi \in \mathring{\Gamma}_*$ there are finite limits

$$\lim_{r \to 1-0} (\mathcal{K}_+\Psi)(r\xi) \quad \text{and} \quad \lim_{r \to 1+0} (\mathcal{K}_-\Psi)(r\xi),$$

because $\mathcal{K}_+\Psi$ has an analytic continuation across $\mathring{\Gamma}_*$ into $-\mathcal{K}_-\Psi$ (see Lemma 4.3). Consequently,

$$\lim_{r \to 1-0} (1 - r)(\mathcal{K}_+\Psi)(r\xi) = \lim_{r \to 1+0} (r - 1)(\mathcal{K}_-\Psi)(r\xi) = 0, \qquad \xi \in \mathring{\Gamma}_*. \tag{4.5}$$

Since $(\Phi + \Psi)_+ \in U^*_A$, there exists (see Lemma 3.1) a sequence of complex measures $\lambda = (\lambda_n)_{n \in \mathbb{N}}$, $\lambda \in l^1(M(\mathbb{T}))$, such that

$$\mathcal{K}_+(\Phi + \Psi)(\eta) = \sum_{n \geqslant 0} \eta^n \int_{\mathbb{T}} \frac{\bar{\zeta}^n\,d\lambda_n}{1 - \bar{\zeta}\eta}, \qquad \eta \in \mathbb{D}.$$

Reasoning as in the proof of Theorem 3.11, we obtain

$$(1-r)\mathcal{K}_+(\Phi+\Psi)(r\xi)=\sum_{n\geqslant 0}(r\xi)^n\int_{\mathbb{T}}\frac{(1-r)\bar{\zeta}^n\,d\lambda_n}{1-\bar{\zeta}r\xi},\qquad \xi\in\mathbb{T},r\in(0,1);$$

$$\lim_{r\to 1-0}(1-r)\mathcal{K}_+(\Phi+\Psi)(r\xi)=\sum_{n\geqslant 0}\lambda_n(\{\xi\}),\qquad \xi\in\mathbb{T}.$$

On the other hand (see (4.3) and (4.5)),

$$(1-r)\mathcal{K}_+(\Phi+\Psi)(r\xi)=\int_{\mathbb{T}}\frac{(1-r)\,d\mu}{1-r\xi\bar{\zeta}}+\int_{\mathbb{T}}\frac{(1-r)h\,dm}{1-r\xi\bar{\zeta}}$$
$$+(1-r)(\mathcal{K}_+\Psi)(r\xi),\qquad \xi\in\mathbb{T},r\in(0,1);$$

$$\lim_{r\to 1-0}(1-r)\mathcal{K}_+(\Phi+\Psi)=\mu(\{\xi\}),\qquad \xi\in\mathring{\Gamma}_*.$$

Since E is countable and $\operatorname{supp}\mu\subset E_*$, we obtain

$$\|\mu\|_{M(\mathbb{T})}=\sum_{\xi\in E_*}|\mu(\{\xi\})|=\sum_{\xi\in E_*}\Big|\sum_{n\geqslant 0}\lambda_n(\{\xi\})\Big|\leqslant\sum_{\xi\in\mathbb{T}}\sum_{n\geqslant 0}|\lambda_n(\{\xi\})|$$
$$=\sum_{n\geqslant 0}\sum_{\xi\in\mathbb{T}}|\lambda_n(\{\xi\})|=\sum_{n\geqslant 0}\|\lambda_n\|_{M(\mathbb{T})}.$$

Consequently

$$\|\mu\|_{M(\mathbb{T})}\leqslant\inf\Big\{\sum_{n\geqslant 0}\|\lambda_n\|_{M(\mathbb{T})}:\lambda\in\mathfrak{M}_{\mathcal{K}_+(\Phi+\Psi)}\Big\}=\|(\Phi+\Psi)_+\|_{U_A^*}$$

(see Lemma 3.1). We can show similarly that $\|\nu\|_{M(\mathbb{T})}\leqslant\|(\Phi+\Psi)_-\|_{(\tau U_A)^*}$. Thus,

$$\|\mu\|_{M(\mathbb{T})}+\|\nu\|_{M(\mathbb{T})}\leqslant\|(\Phi+\Psi)_+\|_{U_A^*}+\|(\Phi+\Psi)_-\|_{(\tau U_A)^*}=\|\Phi+\Psi\|_{U_\pm^*}.\quad\bullet$$

THEOREM 4.9. *Let Γ be a finite union of closed arcs, $\Gamma\subset\mathbb{T}$, $m\Gamma>0$; let E and F be sets closed in $\overline{\mathbb{C}}$ and contained in $\mathring{\Gamma}$, with $mE=mF=0$; let Λ be a lacunary set of integers. Then for every g and h ($g\in C(E)$, $h\in C(F)$) and every system of numbers $x\in l^2(\Lambda)$ there is an $f\in C_\pm(\Gamma)$ such that*

$$f_+|E=g,\quad f_-|F=h,\quad \hat{f}|\Lambda=x,$$

where $C_\pm(\Gamma)\overset{\text{def}}{=}(C_A)_\pm(\Gamma)$.

PROOF. It is evident that $\mathcal{E}=C_A$ satisfies the hypotheses of Lemma 3.3 (see (1.4)). Therefore it remains only to apply Lemma 4.6 (with $p=1/2$) and Lemma 4.8 (conclusion 1)). $\bullet$

THEOREM 4.10. *Let Γ and Λ be the same as in Theorem 4.9, E and F closed countable sets, $E\subset\mathring{\Gamma}$ and $F\subset\mathring{\Gamma}$. Then for every g and h ($g\in C(E)$, $h\in C(F)$) and every system of numbers $x\in l^2(\Lambda)$ there is an $f\in U_\pm(\Gamma)$ such that*

$$f_+|E=g,\quad f_-|F=h,\quad \hat{f}|\Lambda=x,$$
$$\|f\|_{U_\pm(\Gamma)}\leqslant c_7b^{-2}\max\{\|g\|_{C(E)},\|h\|_{C(F)},\|x\|_{l^2\Lambda}\}$$

(c_7 is an absolute constant, $c_7\in(0,+\infty)$, and $b=b(\Gamma,\Lambda)$ is the constant from Lemma 4.6).

The space $\mathfrak{S} = U_A$ satisfies the hypotheses of Lemma 3.3 (see Corollary 2.5 and Lemma 3.1). Therefore we obtain what is required by applying Lemma 4.6 (with $p = 1/2$) and Lemma 4.8 (assertion 2)). ●

DEFINITION 4.11. Let E be a closed subset of $\mathbb{T}$. By U_A^B we denote the space of the functions f that are analytic in $\mathbb{D}$ and such that the norm

$$\|f\|_{U_A^B} = \sup\left\{\left\|\sum_{k\geqslant n} \hat{f}(k)z^k\right\|_{H^\infty} : n \in \mathbb{N}\right\}$$

is finite; $U_\pm^B(E) \stackrel{\text{def}}{=} \{f \in (U_A^B)_\pm : \operatorname{supp} f \subset E\}$.

THEOREM 4.12. *Let E be a closed subset of $\mathbb{T}$, $0 < mE < 1$. Then the following assertions hold*:

1) *There is an absolute constant $c_6 > 0$ such that*

$$c_6(mE)^2\|d\|_{l^2(\mathbb{Z})} \leqslant \|P_d \mid U_\pm^B(E) \to l^1\| \leqslant \|d\|_{l^2(\mathbb{Z})}, \tag{4.6}$$

for every system $d = (d_n)_{n\in\mathbb{Z}}$, $d \in s_0(\mathbb{Z})$.

2) *If Λ is a lacunary set of integers, then for every system $x \in l^2(\Lambda)$ there is a function $f \in U_\pm^B(E)$ such that $\hat{f}(n) = x_n$, $n \in \Lambda$.*

PROOF. 1) We first note that

$$\|P_d \mid U_\pm^B(E) \to l^1\| \leqslant \|P_d \mid L^2(\mathbb{T}) \to l^1\| = \|d\|_{l^2(\mathbb{Z})}, \qquad d \in s_0(\mathbb{Z}).$$

Let us establish the left-hand part of (4.6). Let $d = (d_n)_{n\in\mathbb{Z}}$ be a system of numbers, $d \in s_0(\mathbb{Z})$, and let $(\Gamma_j)_{j\geqslant 0}$ be a sequence of closed sets satisfying the following conditions:

a) $\Gamma_j \subset \mathbb{T}, j \in \mathbb{N}$; $E = \bigcap_{j\geqslant 0}\Gamma_j$.

b) Γ_j is a finite union of closed arcs of the unit circle, $j \in \mathbb{N}$.

By Theorem 4.7 we can assert that, for each $j \in \mathbb{N}$, there is a function φ_j such that

$$\varphi_j \in U_\pm(\Gamma_j), \qquad \|\varphi_j\|_{U_\pm(\Gamma_j)} \leqslant 1,$$

$$c_6\left(1 - \frac{1}{j+1}\right)(m\Gamma_j)^2\|d\|_{l^2(\mathbb{Z})} \leqslant \sum_{n\in\mathbb{Z}} |\varphi_j(n)|\,|d_n| \tag{4.7}$$

(c_6 is the absolute constant from Theorem 4.7). Since

$$\sup\left\{\left|\int_{\Gamma_j}\frac{\varphi_j\,dm}{1-\eta\bar{\xi}}\right| : \eta \in \mathbb{C}\setminus\Gamma_j, j \in \mathbb{N}\right\} < +\infty,$$

there are an increasing sequence of indices $(j_n)_{n\geqslant 0}$ and a function F analytic in $\overline{\mathbb{C}}\setminus E$, such that

$$\lim_{n\to\infty}\tilde{\varphi}_{j_n}(\eta) = F(\eta), \qquad \eta \in \mathbb{C}\setminus E;$$

$$\tilde{\varphi}_k(\eta) \stackrel{\text{def}}{=} \begin{cases} \displaystyle\int_{\Gamma_k}\frac{\varphi_k\,dm}{1-\eta\bar{\xi}}, & \eta \in \mathbb{C}\setminus\Gamma_k, \\ 0, & \eta \in \Gamma_k \end{cases}$$

(we have applied the theorems of Montel and Vitali on bounded families of analytic functions (see [24])).

Since

$$\Big|\sum_{k\geqslant n}\hat{\varphi}_j(k)\eta^k\Big|\leqslant\|\varphi_j\|_{U_\pm(\Gamma_j)}\leqslant 1,\qquad n,j\in\mathbb{N},\eta\in\mathbb{D};$$

$$\Big|\sum_{k\leqslant -(n+1)}\hat{\varphi}_j(k)\eta^k\Big|\leqslant\|\varphi_j\|_{U_\pm(\Gamma_j)}\leqslant 1,\qquad n,j\in\mathbb{N},\eta\in\mathbb{C}\setminus\overline{\mathbb{D}},$$

we obtain

$$\|f\|_{U^B_\pm(E)}\leqslant 1,\qquad f\stackrel{\text{def}}{=}f_+ + f_-,$$

$$f_+(\xi)\stackrel{\text{def}}{=}\lim_{r\to 1-0}F(r\xi),\quad f_-(\xi)\stackrel{\text{def}}{=}-\lim_{r\to 1+0}F(r\xi),\qquad \xi\in\mathbb{T}.$$

In addition, it is clear that taking limits in (4.7) yields

$$c_6(mE)^2\|d\|_{l^2(\mathbb{Z})}\leqslant\sum_{n\in\mathbb{Z}}|\hat{f}(n)|\,|d_n|\leqslant\|P_d\,|\,U^B_\pm(E)\to l^1\|.$$

2) Let $(\Gamma_j)_{j\geqslant 0}$ be a sequence of arcs satisfying conditions a) and b), and let $x=(x_n)_{n\in\Lambda}$ be a system of numbers from $l^2(\Lambda)$. It follows from Theorem 4.10 that there is a sequence of functions $(\varphi_j)_{j\geqslant 0}$ such that

$$\varphi_j\in U_\cdot(\Gamma_j),\quad j\in\mathbb{N};\qquad b^2(\Gamma_j,\Lambda)\cdot\|\varphi_j\|_{U_\cdot}\leqslant 2c_7\|x\|_{l^2(\Lambda)};\tag{4.8}$$

$$\hat{\varphi}_j(k)=x_k,\qquad k\in\Lambda,j\in\mathbb{N}\tag{4.9}$$

(c_7 and $b(\Gamma,\Lambda)$ are the constants from Theorem 4.10).

In addition, we need, as in the proof of part 1), to select from $(\varphi_j)_{j\geqslant 0}$ a corresponding subsequence $(\varphi_{j_n})_{n\geqslant 0}$ and then take limits over $(\varphi_{j_n})_{n\geqslant 0}$ in (4.8) and (4.9), making use of the conditions $0<b(E,\Lambda)\leqslant b(\Gamma_j,\Lambda)$, $j\in\mathbb{N}$ (see Lemma 5.2 in §5). ●

REMARK 4.13. 1) If E is a closed nowhere dense set (with respect to $\mathbb{T}$), $E\subset\mathbb{T}$, and $mE>0$, the conclusions of Theorem 4.12 fail for the spaces $C_\pm(E)$ and $U_\cdot(E)$, since in those cases $C_\pm(E)=U_\pm(E)=\mathbb{C}$.

2) In Theorem 4.12 the condition $mE>0$ cannot be removed, since if $mE=0$ the space $H^\infty(\overline{\mathbb{C}}\setminus E)$ consists only of constants.

THEOREM 4.14. *Let Λ be a lacunary set of nonnegative integers, E a nonempty closed and at most countable subset of the unit circle. Then for every system $x\in l^2(\Lambda)$ and every function $g\in C(E)$ there is a function f that is analytic in $\overline{\mathbb{C}}\setminus E$ and such that*

$$f\,|\,\mathbb{D}\in U_A,\quad f\,|\,E=g,\quad \hat{f}\,|\,\Lambda=x.$$

The proof is based on the following proposition [4].

THEOREM 4.15. *Let $(X_n)_{n\in\mathbb{N}}$ be a sequence of Banach spaces with the following properties:*

1) *$X_{n+1}\subset X_n$, $n\in\mathbb{N}$, where these embeddings are continuous.*

2) *X_{n+1} is dense in X_n, $n\in\mathbb{N}$.*

In addition, let R be a bounded linear operator from X_0 to the Banach space Y. Suppose that $R(X_n)=Y$, $n\in\mathbb{N}$. Then

$$R\Big(\bigcap_{n\geqslant 0}X_n\Big)=Y.$$

The proof of this (and even of a more general proposition) is given in §5 (see Theorem 5.8).

PROOF OF THEOREM 4.14. Let $(\Gamma_n)_{n\geqslant 0}$ be a sequence of closed subsets of the complex plane, subject to the following conditions: Γ_n is a finite union of closed arcs, $\Gamma_n \subset \mathbb{T}$, $E \subset \mathring{\Gamma}_n$, $\Gamma_{n+1} \subset \mathring{\Gamma}_n$ $(n \in \mathbb{N})$ and $E = \bigcap_{n\geqslant 0} \Gamma_n$. It is clear that such a sequence exists. In addition, put

$$G_n \stackrel{\text{def}}{=} \overline{\mathbb{C}} \setminus \left\{ r\eta \colon \eta \in \Gamma_n,\, r \in \left[1, \frac{n+2}{n+1}\right],\, n \in \mathbb{N} \right\}.$$

We denote by $U(G_n)$ the closure of the set $A(\overline{G}_n)$ of functions that are analytic on $\overline{G}_n$(³) with respect to the norm

$$\| f \|_{U(G_n)} \stackrel{\text{def}}{=} \sup\{ |f(\eta)| : \eta \in G_n \} + \| f | \mathbb{D} \|_{U_A}.$$

It is clear that $f|\mathbb{D} \in U_A$ and $f \in A(\overline{\mathbb{C}} \setminus E)$ if $f \in \bigcap_{n\geqslant 0} U(G_n)$. Since $U(G_{n+1})$ is continuously embedded in $U(G_n)$, $U(G_{n+1})$ is dense in $U(G_n)$ (see §5, Lemma 5.4), $U_{\pm}(\Gamma_n)$ is naturally embedded in $U(G_n)$, $n \in \mathbb{N}$ (see §5, Lemma 5.5), and $Q(U(G_n)) = C(E) \times l^2(\Lambda)$ (see Theorem 4.10), where $Qf = (f|E, \hat{f}|\Lambda)$, $f \in C_A$, it follows from Theorem 4.15 that

$$Q\Big(\bigcap_{n\geqslant 0} U(G_n) \Big) = C(E) \times l^2(\Lambda). \quad \bullet$$

REMARK 4.16. 1) If we take E in Theorem 4.14 to be a singleton, we obtain the following proposition: for every lacunary set Λ and every system $x \in l^2(\Lambda)$ there is a function $f \in A(\overline{\mathbb{C}} \setminus \{\xi\})$, where $\xi \in \mathbb{T}$, such that $f|\mathbb{D} \in U_A$ and $\hat{f}|\Lambda = x$.

2) In Theorem 4.14, the requirement that Λ is lacunary can be considerably weakened (see [4]).

THEOREM 4.17. 1) *For every sequence of nonnegative numbers* $d = (d_n)_{n\in\mathbb{Z}}$ *satisfying* $\sum_{n\in\mathbb{Z}} d_n^2 = +\infty$ *and for every closed arc* $I \subset \mathbb{T}$, $mI > 0$, *there is a function* F, *analytic in* $\overline{\mathbb{C}} \setminus I$ *and such that*

$$F|\mathbb{D} \in U_A, \quad F|\overline{\mathbb{C}} \setminus \overline{\mathbb{D}} \in \tau U_A, \quad \sum_{n\in\mathbb{Z}} |\hat{F}(n)|\, d_n = +\infty,$$

where

$$\hat{F}(n) = \int_{\mathbb{T}} F(r\zeta) \frac{dm}{(r\zeta)^n}, \qquad n \in \mathbb{N}, 0 < r < 1;$$

$$\hat{F}(n) = \int_{\mathbb{T}} F(r\zeta) \frac{dm}{(r\zeta)^n}, \qquad n \in \mathbb{Z} \setminus \mathbb{N}, r > 1.$$

2) *For every sequence* $d = (d_n)_{n\in\mathbb{N}}$ *of nonnegative numbers satisfying* $\sum_{n\geqslant 0} d_n^2 = +\infty$, *there is a function* F *that is analytic in* $\overline{\mathbb{C}} \setminus \{1\}$ *and such that*

$$F|\mathbb{D} \in U_A, \qquad \sum_{n\geqslant 0} |\hat{F}(n)|\, d_n = +\infty.$$

(³)$A(W)$ is the class of functions that are analytic on $W \subset \overline{\mathbb{C}}$ (f is analytic on W if it is analytic in some neighborhood of W).

PROOF. 1) If $\Sigma_{n\in\mathbb{Z}} d_n^2 = +\infty$, it follows from Theorem 4.7 that

$$\lim_{n\to\infty} \| P_{d^{(n)}} \mid U_\pm(I) \to l^1 \| = +\infty,$$

where
$$d_k^{(n)} = \begin{cases} d_k, & k \in [-n, n] \cap \mathbb{Z}, \\ 0, & k \in \mathbb{Z}\backslash[-n, n], \end{cases} \qquad n \in \mathbb{N}.$$

Hence, by the Banach-Steinhaus theorem [20], there is a function $f \in U_\pm(I)$ such that $\Sigma_{k\in\mathbb{Z}} |\hat{f}(k)| d_k = +\infty$. It is evident that the function F defined by

$$F(\eta) \overset{\text{def}}{=} \int_I \frac{f\,dm}{1 - \bar{\zeta}\eta} \qquad (\eta \in \overline{\mathbb{C}}\backslash I)$$

has the required properties.

Conclusion 2) can be proved in either of two ways: either by repeating the reasoning of [4] concerning a similar fact for the space C_A; or, on the basis of conclusion 1), by using a theorem of Khrushchev [25], [26] on analytic singularities of linear operators. ●

REMARK 4.18. Conclusion 1) of Theorem 4.17 remains valid if the interval I considered in that theorem is replaced by an arbitrary closed set $E \subset \mathbb{T}$, $mE > 0$, and U_A is replaced by U_A^B.

§5. Appendix

LEMMA 5.1. *Let $s > 0, f \in L^\infty(\mathbb{T})$ and $\| f \|_\infty \leqslant 2s$. Then*

$$\| f \|_{p1} \leqslant 2\left(\frac{p+1}{p-1}\right) s^{1-1/p} \| f \|_1^{1/p}, \qquad p \in (1, +\infty).$$

PROOF. Let f^* be the nonincreasing rearrangement of $f \in L^\infty(\mathbb{T})$:

$$f^*(t) \overset{\text{def}}{=} \inf\{s\colon \lambda_f(s) \leqslant t\}, \quad t > 0; \qquad \lambda_f(s) \overset{\text{def}}{=} m\{\eta \in \mathbb{T}\colon |f(\eta)| > s\}, \quad s > 0.$$

Since (see [12], Chapter V)

$$\| f \|_{p1} \leqslant \frac{p}{p-1} \| f \|_{p1}^*, \qquad \| f \|_{p1}^* \overset{\text{def}}{=} \frac{1}{p} \int_0^\infty t^{1/p} f^*(t) \frac{dt}{t}, \qquad p \in (1, +\infty),$$

and $\| f \|_\infty = \| f^* \|_\infty$, we have

$$\begin{aligned} \| f \|_{p1} &\leqslant \frac{1}{p-1} \left(\int_0^a f^*(t) \frac{dt}{t^{1-1/p}} + \int_a^\infty f^*(t) \frac{dt}{t^{1-1/p}} \right) \\ &\leqslant \frac{1}{p-1} \left(2s \int_0^a \frac{dt}{t^{1-1/p}} + \frac{1}{a^{1-1/p}} \int_a^\infty f^*(t)\,dt \right) \\ &\leqslant \frac{1}{p-1} \left(2psa^{1/p} + \frac{\| f \|_1}{a^{1-1/p}} \right), \qquad a > 0. \end{aligned}$$

Putting $a = \| f \|_1 / s$, we obtain

$$\| f \|_{p1} \leqslant \frac{1}{p-1} (2ps^{1-1/p} + s^{1-1/p}) \| f \|_1^{1/p} \leqslant 2\left(\frac{p+1}{p-1}\right) s^{1-1/p} \| f \|_1^{1/p}. \quad ●$$

LEMMA 5.2. *Let Λ be a lacunary set of integers, and let $E \subset \mathbb{T}$, $mE > 0$. Then there is a constant $b = b(E, \Lambda)$, $b \in (0, 1]$, depending only on Λ and E, such that*

$$b^{1/p} \| x \|_{l^2(\Lambda)} \leqslant \left\| \sum_{n\in\Lambda} x_n \zeta^n \right\|_{L^p(E)}, \qquad x \in s_0(\Lambda), p \in (0, 2]. \tag{5.1}$$

PROOF. Since Λ is lacunary, there exists (see [10]) a positive integer $n_0 = n_0(E, \Lambda)$ satisfying

$$\frac{1}{2}(mE)^{1/2}\|x\|_{l^2(\Lambda)} \leqslant \Big\| \sum_{n\in\Lambda, |n|\geqslant n_0} x_n\zeta^n \Big\|_{L^2(E)}, \qquad x \in l^2(\Lambda). \tag{5.2}$$

Suppose that the conclusion of the lemma is false for $p = 2$. Then there is a sequence $(f_n)_{n\in\mathbb{N}}$ such that

$$f_n \in L^2(\mathbb{T}); \quad \|f_n\|_{L^2(\mathbb{T})} = 1; \quad \hat{f}_n(k) = 0, \qquad k \in \mathbb{Z}\setminus\Lambda, n \in \mathbb{N};$$
$$\lim_{n\to\infty} \|f_n\|_{L^2(E)} = 0; \qquad \lim_{n,n\to\infty}\big(\hat{f}_n(k) - \hat{f}_j(k)\big) = 0, \qquad k \in \Lambda \cap [-n_0, n_0].$$

Using (5.2), we obtain

$$\begin{aligned}\|f_n - f_j\|_{L^2(\mathbb{T})} &\leqslant \Big\| \sum_{|k|\geqslant n_0} (\hat{f}_n(k) - \hat{f}_j(k))\zeta^k \Big\|_{L^2(\mathbb{T})} + \sum_{|k|\leqslant n_0} |\hat{f}_n(k) - \hat{f}_j(k)| \\ &\leqslant \frac{2}{\sqrt{mE}} \Big\| \sum_{|k|\geqslant n_0} (\hat{f}_n(k) - \hat{f}_j(k))\zeta^k \Big\|_{L^2(E)} + \sum_{|k|\leqslant n_0} |\hat{f}_n(k) - \hat{f}_j(k)| \\ &\leqslant \frac{2}{\sqrt{mE}}\|f_n - f_j\|_{L^2(E)} + \left(\frac{2}{\sqrt{mE}} + 1\right) \sum_{|k|\leqslant n_0} |\hat{f}_n(k) - \hat{f}_j(k)|.\end{aligned}$$

Therefore the sequence $(f_n)_{n\in\mathbb{N}}$ converges in $L^2(\mathbb{T})$ to a function $\varphi \in L^2(\mathbb{T})$. It is clear that $\|\varphi\|_{L^2(\mathbb{T})} = 1$, $\hat{\varphi}(k) = 0$, $k \in \mathbb{Z}\setminus\Lambda$, and $\varphi(\eta) = 0$ a.e. on E, since

$$\|\varphi\|_{L^2(E)} \leqslant \|\varphi - f_n\|_{L^2(E)} + \|f_n\|_{L^2(E)}, \quad \lim_{n\to\infty}\|\varphi - f_n\|_{L^2(E)} = \lim_{n\to\infty}\|f_n\|_{L^2(E)} = 0.$$

It follows from the uniqueness theorem for lacunary trigonometric series (see [17]) that $\varphi(\eta) = 0$ almost everywhere on $\mathbb{T}$. The resulting contradiction (on the one hand, $\|\varphi\|_{L^2(\mathbb{T})} = 1$, but on the other, $\varphi(\eta) = 0$ a.e. on $\mathbb{T}$) shows that there is a constant $b_0 = b_0(E, \Lambda)$ that satisfies

$$b_0\|x\|_{l^2(\Lambda)} \leqslant \Big\| \sum_{n\in\Lambda} x_n\zeta^n \Big\|_{L^2(E)}, \qquad x \in l^2(\Lambda).$$

Let us now verify (5.1) for the other values of p, $0 < p < 2$. Let $x \in l^2(\Lambda)$ and $p \in (0, 2)$; then

$$\begin{aligned} b_0^2\|x\|^2_{l^2(\Lambda)} &\leqslant \Big\| \sum_{n\in\Lambda} x_n\zeta^n \Big\|^2_{L^2(E)} = \int_E \Big| \sum_{n\in\Lambda} x_n\zeta^n \Big|^{p/2} \Big| \sum_{n\in\Lambda} x_n\zeta^n \Big|^{2-p/2} dm \\ &\leqslant \left(\int_E \Big| \sum_{n\in\Lambda} x_n\zeta^n \Big|^p dm\right)^{1/2} \left(\int_E \Big| \sum_{n\in\Lambda} x_n\zeta^n \Big|^{4-p} dm\right)^{1/2} \\ &\leqslant \Big\| \sum_{n\in\Lambda} x_n\zeta^n \Big\|^{p/2}_{L^p(E)} \Big\| \sum_{n\in\Lambda} x_n\zeta^n \Big\|^{2-p/2}_{L^{4-p}(\mathbb{T})}. \end{aligned}$$

Since Λ is a lacunary set and $4 - p \in (2, 4)$, there is a constant $b_1 \in [1, +\infty)$, depending only on Λ (see [10]), such that

$$\Big\| \sum_{n\in\Lambda} x_n\zeta^n \Big\|_{L^{4-p}(\mathbb{T})} \leqslant b_1\|x\|_{l^2(\Lambda)}, \qquad x \in l^2(\Lambda).$$

Therefore

$$b_0^2\|x\|_{l^2(\Lambda)}^2 \leqslant \Big\|\sum_{n\in\Lambda} x_n\zeta^n\Big\|_{L^p(E)}^{p/2} b_1^{2-p/2}\|x\|_{l^2(\Lambda)}^{2-p/2} \leqslant b_1^2\Big\|\sum_{n\in\Lambda} x_n\zeta^n\Big\|_{L^p(E)}^{p/2}\|x\|_{l^2(\Lambda)}^{2-p/2},$$

$$b^{1/p}\|x\|_{l^2(\Lambda)} \leqslant \Big\|\sum_{n\in\Lambda} x_n\zeta^n\Big\|_{L^p(E)}, \qquad b \stackrel{\text{def}}{=} \left(\frac{b_0}{b_1}\right)^4, p\in(0,2]. \quad \bullet$$

LEMMA 5.3. *Let μ be a complex Borel measure on $\mathbb{T}$,*

$$F(\eta) \stackrel{\text{def}}{=} \int_{\mathbb{T}}\frac{d\mu}{1-\bar{\zeta}\eta}, \qquad \eta\in\mathbb{C}\setminus\mathbb{T};$$

$$F_+(\xi) \stackrel{\text{def}}{=} \lim_{r\to 1-0} F(r\xi), \quad F_-(\xi) \stackrel{\text{def}}{=} \lim_{r\to 1+0} F(r\xi), \qquad \xi\in\mathbb{T};$$

and let I be an arc of $\mathbb{T}$. If $F_+|I\in L^\infty(I)$ or $F_-|I\in L^\infty(I)$, then $\operatorname{supp}\mu_s\subset\mathbb{T}\setminus\mathring{I}$, *where μ_s is the singular component of μ (with respect to m).*

PROOF. Suppose that $F_+|I\in L^\infty(I)$. Let I_1 be a closed arc lying in $\mathring{I}$, and $G \stackrel{\text{def}}{=} \{r\xi\colon 0<r<1,\ \xi\in\mathring{I}_1\}\cup\{\eta\in\mathbb{C}\colon |\eta|<1/2\}$. Since $F|\mathbb{D}\in H^p$, $p\in(0,1)$, and $F_+|I\in K^\infty(I)$, it is easily verified that F is bounded in G. It is evident that

$$\int_{\mathbb{T}}\frac{\xi\,d\mu(\xi)}{\xi-\eta} - \frac{1}{2\pi i}\int_{\partial G}\frac{F(\xi)\,d\xi}{\xi-\eta} = 0, \qquad \eta\in G.$$

Consequently

$$\int_{\partial G} f(\xi)F(\xi)\frac{d\xi}{\xi} - \int_{\mathbb{T}} f\,d\mu = 0, \qquad f\in C_A(\overline{\mathbb{C}}\setminus\overline{G}), \tag{5.3}$$

where $C_A(\overline{\mathbb{C}}\setminus\overline{G})$ is the space of functions that are analytic in $\overline{\mathbb{C}}\setminus\overline{G}$ and continuous in $\overline{\mathbb{C}}\setminus G$.

Let E be a closed subset of $\mathbb{T}$, with $mE=0$ and $E\subset I_1$. Then there is a function g (see [14]) satisfying

$$g\in C_A(\overline{\mathbb{C}}\setminus\overline{G}); \qquad g(\xi)=1,\ \ \xi\in E; \qquad |g(\xi)|<1,\ \ \xi\in\partial G\setminus E.$$

Substituting $f=g^n$ ($n\in\mathbb{N}$) into (5.3) and taking the limit as $n\to\infty$, we find that $\mu E=0$. The case $F_-|I\in L^\infty(I)$ is disposed of in a similar way. $\bullet$

LEMMA 5.4. *Let Γ be a finite union of closed nondegenerate arcs of $\mathbb{T}$,*

$$G \stackrel{\text{def}}{=} \overline{\mathbb{C}}\setminus\{\eta\in\overline{\mathbb{C}}\colon 1\leqslant|\eta|\leqslant\rho,\ \eta/|\eta|\in\Gamma\}, \qquad \rho\in(1,2),$$

and let Ω be open in $\overline{\mathbb{C}}$ and contain the closure of G. Then the set $A(\Omega)$ is dense in $U(G)$ (for the definition of $U(G)$ see the proof of Theorem 4.14).

PROOF. Since $A(\overline{G})$ is dense in $U(G)$ (see the proof of Theorem 4.14), it is enough to show that the closure (in $U(G)$) of $A(\Omega)$ contains $A(\overline{G})$. Let $f\in A(\overline{G})$; then there is an open set $G_0\subset\overline{\mathbb{C}}$ such that $\overline{G}_0\subset\Omega$, $\overline{G}\subset G_0$ and f is analytic in some open neighborhood of $\overline{G}_0$. We note that

$$\|h\|_{U(G)} \leqslant l\sup\{|h(\eta)|:\eta\in\overline{G}_0\}, \qquad h\in A(\overline{G}_0),$$

where l, $0<l<+\infty$, is independent of $h\in A(\overline{G}_0)$.

Moreover, if $s > 0$ there is a function $g \in A(\Omega)$ such that

$$\sup\{|g(\eta) - f(\eta)| : \eta \in G_0\} \leqslant s/l.$$

Consequently $\|g - f\|_{U(G)} \leqslant \varepsilon$. ●

LEMMA 5.5. *Let Γ and G be as in the preceding lemma, and let Δ be a finite union of nondegenerate closed arcs of $\mathbb{T}$, $\Delta \subset \mathring{\Gamma}$. Then there is a natural embedding of $U_\pm(\Delta)$ in $U(G)$, i.e. if $f \in U_\pm(\Delta)$ then the function F defined by*

$$F(\eta) \overset{\text{def}}{=} \int_\Delta \frac{f\,dm}{1 - \bar{\xi}\eta}, \qquad \eta \in \overline{\mathbb{C}} \setminus \Delta,$$

belongs to $U(G)$.

PROOF. Let $f \in U_\pm(\Delta)$ and

$$F(\eta) \overset{\text{def}}{=} \int_\Delta \frac{f\,dm}{1 - \bar{\xi}\eta}, \qquad \eta \in \overline{\mathbb{C}} \setminus \Delta,$$

$$F_r(\eta) \overset{\text{def}}{=} F(r\eta), \qquad \eta \in G,\ 1/\rho < r < 1.$$

It is easy to see that $F_r \in A(G)$, $r \in (1/\rho, 1)$, and

$$\lim_{r \to 1-0} \|(F - F_r) \,|\, D\|_{U_A} = 0, \qquad \limsup_{r \to 1-0} \{|F(\eta) - F_r(\eta)| : \eta \in G\} = 0.$$

Consequently $F \in U(G)$. ●

LEMMA 5.6. *Let the F-spaces*(4) *X_0, X_1 and Y and the operators R_0 and R_1 satisfy the following conditions*:

1) *$X_1 \subset X_0$, and $\|x\|_0 \leqslant \|x\|_1$ if $x \in X_1$ ($\|\cdot\|_j$ is the quasinorm in X_j, $j \in \{0, 1\}$).*
2) *X_1 is dense in X_0.*
3) *R_0 and R_1 are continuous linear operators from X_0 and X_1, respectively, onto Y.*
4) *$R_1 = R_0 | X_1$.*

Then the closure of the set $\mathfrak{N}_1 \overset{\text{def}}{=} \{x \in X_1\colon R_1 x = \mathbf{0}\}$ in the space X_0 coincides with the set $\mathfrak{N}_0 \overset{\text{def}}{=} \{x \in X_0\colon R_0 x = \mathbf{0}\}$.

PROOF. We first prove the lemma under the hypothesis that R_1 is a linear homeomorphism. To do this, we show that X_1 is, in this case, closed in X_0. In fact, if $\lim_{n\to\infty} \|x_n - a\|_0 = 0$ and $x_n \in X_1$ ($n \in \mathbb{N}$), $a \in X_0$, then

$$X_1 \lim_{n\to\infty} x_n = X_1 \lim_{n\to\infty} R_1^{-1}(R_0 x_n) = R_1^{-1}(R_0 a).$$

Consequently $a = R_1^{-1}(R_0 a)$ and $a \in X_1$. Thus, if R_1 is a linear homeomorphism, then X_1 coincides with X_0 (since X_1 is dense and closed in X_0), and R_0 is a linear homeomorphism of X_0 on Y.

We now turn to the proof of the general case. Let $\overline{\mathfrak{N}}_1$ be the closure of $\mathfrak{N}_1$ in X_0. Consider the quotient space $\dot{X}_0 \overset{\text{def}}{=} X_0/\overline{\mathfrak{N}}_1$ with the usual norm,

$$\|\dot{x}\|_2 \overset{\text{def}}{=} \inf\{\|x\|_0\colon x \in \dot{x}\}, \qquad \dot{x} \in \dot{X}_0.$$

---

(4)An F-space is a complete linear space with an invariant metric ρ, $\rho(x, y) = \|x - y\|$, where $\|x\| \overset{\text{def}}{=} \rho(x, 0)$ is the (quasi) norm in the space X.

It is clear that $\dot{X}_0$ is an F-space. We put

$$\dot{X}_1 \stackrel{\text{def}}{=} \{\dot{x} \in \dot{X}_0 : \dot{x} \cap X_1 \neq \varnothing\}, \quad \|\dot{x}\|_2 \stackrel{\text{def}}{=} \inf\{\|x\|_1 : x \in \dot{x} \cap X_1\}, \qquad \dot{x} \in \dot{X}_1.$$

It is easily verified that $\|\cdot\|_3$ is a norm on $\dot{X}_1$ and that $\dot{X}_1$ is an F-space with respect to $\|\cdot\|_3$ (it is enough to verify that the mapping φ: $X_1/\mathcal{N}_1 \to \dot{X}_1$, $\varphi(x + \mathcal{N}_1) \stackrel{\text{def}}{=} x + \overline{\mathcal{N}}_1$, $x \in X_1$, is a linear homeomorphism).

We define the operators $\dot{R}_j$ ($j \in \{0, 1\}$) by

$$\dot{R}_0 \dot{x} = R_0 x, \quad \dot{x} \in \dot{X}_0, x \in \dot{x}; \qquad \dot{R}_1 = \dot{R}_0 \,|\, \dot{X}_1.$$

It is clear that the definition of $\dot{R}_0$ is consistent ($R_0 x = R_0 y$ if $x - y \in \mathcal{N}_0$, $\overline{\mathcal{N}}_1 \subset \mathcal{N}_0$). In addition, it is readily verified that:

a) $\dot{R}_0$ is a continuous linear operator from $\dot{X}_0$ onto Y,

b) $\dot{R}_1$ is a linear topological isomorphism of $\dot{X}_1$ on Y (this follows from Banach's open mapping theorem [27]),

c) $\|\dot{x}\|_2 \leqslant \|\dot{x}\|_3$ for all $\dot{x} \in \dot{X}_1$, and

d) $\dot{X}_1$ is dense in $\dot{X}_0$ (we need to use the fact that X_1 is dense in X_0).

It follows from what was proved above that $\dot{X}_0$ coincides with $\dot{X}_1$ and that $\dot{R}_0$ is a topological linear isomorphism of $\dot{X}_0$ on Y. Since

$$\dot{R}^{-1}(\mathbf{0}) = \{\dot{x} \in \dot{X}_0 : \dot{R}_0 \dot{x} = \mathbf{0}\} = \{\dot{x} \in \dot{X}_0 : \dot{x} \cap \mathcal{N}_0 \neq \varnothing\}$$

and $\dot{R}^{-1}(\mathbf{0})$ consists of the single element $\overline{\mathcal{N}}_1$ (the zero element of $\dot{X}_0$), then $\overline{\mathcal{N}}_1 = \mathcal{N}_0$. ●

LEMMA 5.7 [28]. *Let the sequence of complete metric spaces* $(X_n, \rho_n)_{n \in \mathbb{N}}$ *satisfy* 1) $X_{n+1} \subset X_n$, $n \in \mathbb{N}$; 2) $\rho_n(x, y) \leqslant \rho_{n+1}(x, y)$ $(x, y \in X_{n+1})$, $n \in \mathbb{N}$; *and* 3) X_{n+1} *is dense in* X_n, $n \in \mathbb{N}$. *Then* $\bigcap_{n \geqslant 0} X_n$ *is dense in* X_n, $n \in \mathbb{N}$.

PROOF. Clearly it is enough to establish the conclusion of the lemma for $n = 0$. Let $\varepsilon > 0$ and $x_0 \in X_0$. Since X_1 is dense in X_0, there is an element $x_1 \in X_1$ such that $\rho_0(x_0, x_1) < \varepsilon/2$; since X_2 is dense in X_1, there is an element $x_2 \in X_2$ such that $\rho_1(x_1, x_2) < \varepsilon/2^2$. It is easy to give an inductive construction of a sequence $(x_n)_{n \geqslant 1}$ with the following properties:

a) $x_n \in X_k$ if $k \leqslant n$, and

b) $\rho_{n-1}(x_{n-1}, x_n) < \varepsilon/2^n$, $n \in \mathbb{N} \backslash \{0\}$.

Let us show that $(x_n)_{n \geqslant k+1}$ is a Chauchy sequence in the space X_k: if $n \geqslant k + 1$, we have, for $n \in \mathbb{N} \cap [k + 1, +\infty)$ and $p \in \mathbb{N} \backslash \{0\}$,

$$\rho_k(x_n, x_{n+p}) \leqslant \sum_{j=0}^{p-1} \rho_k(x_{n+j}, x_{n+j+1}) \leqslant \sum_{j=0}^{p-1} \rho_{n+j}(x_{n+j}, x_{n+j+1})$$
$$< \sum_{j \geqslant 0} \frac{\varepsilon}{2^{n+j+1}} = \frac{\varepsilon}{2^n}.$$

Therefore the sequence $(x_n)_{n\geqslant k+1}$ has a limit $\bar{x}_k \in X_k$, $k \in \mathbb{N}$. Evidently $\bar{x}_k = \bar{x}_{k+1}$, $k \in \mathbb{N}$. Denote this common limit by $\bar{x}$. Since

$$\rho_0(x_0, x_n) \leqslant \sum_{j=0}^{n-1} \rho_0(x_j, x_{j+1}) \leqslant \sum_{j=0}^{n-1} \rho_j(x_j, x_{j+1})$$
$$< \sum_{j\geqslant 0} \frac{\varepsilon}{2^{j+1}} = \varepsilon, \qquad n \in \mathbb{N}\setminus\{0\},$$

we have $\rho_0(x_0, \bar{x}) \leqslant \varepsilon$. ●

THEOREM 5.8. *Let the sequence of F-spaces* $(X_n)_{n\in\mathbb{N}}$ *satisfy* 1) $X_{n+1} \subset X_n$, $n \in \mathbb{N}$; 2) $\rho_n(x, y) \leqslant \rho_{n+1}(x, y)$ $(x, y \in X_{n+1}, n \in \mathbb{N})$, *where* ρ_n *is the metric of* X_n; *and* 3) X_{n+1} *is dense in* X_n, $n \in \mathbb{N}$.

Then, if R *is a continuous linear operator from* X_0 *onto the F-space* Y, *and* $R(X_n) = Y$ *for all* $n \in \mathbb{N}$, *it follows that*

$$R\left(\bigcap_{n\geqslant 0} X_n\right) = Y.$$

PROOF. Let R_n denote the restriction of R to X_n. Let $y \in Y$; then it follows from Lemma 5.6 that the closure of $R_{n+1}^{-1}(y)$ in X_n coincides with $R_n^{-1}(y)$, where $R_n^{-1}(y) \overset{\text{def}}{=} \{x \in X_n \colon R_n x = y\}$, $n \in \mathbb{N}$. Moreover, $R_{n+1}^{-1}(y) \subset R_n^{-1}(y)$ and $R_n^{-1}(y)$ is closed in X_n, $n \in \mathbb{N}$. Applying Lemma 5.7 to the sequence $(R_n^{-1}(y))_{n\in\mathbb{N}}$, we find that $\bigcap_{n\geqslant 0} R_n^{-1}(y) \neq \varnothing$. Let $x \in \bigcap_{n\geqslant 0} R_n^{-1}(y)$; then $x \in \bigcap_{n\geqslant 0} X_n$ and $R_n x = y$, $n \in \mathbb{N}$, i.e. $x \in \bigcap_{n\geqslant 0} X_n$ and $Rx = y$. ●

BIBLIOGRAPHY

1. V. Smirnoff [V. I. Smirnov], *Sur les valeurs limites des fonctions régulières à l'intérieur d'un cercle*, Zh. Leningrad. Fiz.-Mat. Obshch. **2** (1928), no. 2, 22–37.

2. A. Kolmogoroff [A. N. Kolmogorov], *Sur les fonctions harmoniques conjuguées at les séries de Fourier*, Fund. Math **7** (1925), 23–28.

3. S. A. Vinogradov, *Paley singularities and Rudin-Carleson interpolation theorems for some classes of analytic functions*, Dokl. Akad. Nauk SSSR **178** (1968), 511–514; English transl. in Math. USSR Dokl. **9** (1968).

4. ______, *Interpolation theorems of Banach-Rudin-Carleson and norms of imbedding operators for certain classes of analytic functions*, Zap. Nauchn. Sem. Leningrad. Otdel. Mat. Inst. Steklov. (LOMI) **19** (1970), 6–54; English transl. in Sem. Math. V. A. Steklov Math. Inst. Leningrad **19** (1970).

5. ______, *Convergence almost everywhere of Fourier series of functions in* L^2 *and the behavior of the coefficients of uniformly convergent Fourier series*, Dokl. Akad, Nauk SSSR **230** (1976), 508–511; English transl. in Soviet Math. Dokl. **17** (1976).

6. S. B. Stechkin, *An extremal problem for polynomials*, Izv. Akad. Nauk SSSR Ser. Mat. **20** (1956), 765–774; English transl. in Amer. Math. Soc. Transl. (2) **14** (1960).

7. Lars Hörmander, *Estimates for translation invariant operators in* L^p*-spaces*, Acta Math. **104** (1960), 93–140.

8. Richard A. Hunt, *On the convergence of Fourier series*, Orthogonal Expansions and Their Continuous Analogues (Proc. Conf., Edwardsville, Ill., 1967), Southern Illinois Univ. Press. Carbondale, Ill., 1968, pp. 235–255.

9. Lennart Carleson, *On convergence and growth of partial sums of Fourier series*, Acta Math. **116** (1966), 135–157.

10. A. Zygmund, *Trigonometric series*, 2nd rev. ed., Vols. I, II, Cambridge Univ. Press, 1959.

11. R. E. A. C. Paley, *A note on power series*, J. London Math. Soc. **7** (1932), 122–130.

12. Elias M. Stein and Guido Weiss, *Introduction to Fourier analysis on Euclidean spaces*, Princeton Univ. Press, Princeton, N.J., 1971.

13. S. G. Kreĭn and Yu. I. Petunin, *Scales of Banach spaces*, Uspekhi Mat. Nauk **21** (1966), no. 2(128), 89–168; English transl. in Russian Math. Surveys **21** (1966).

14. Kenneth Hoffman, *Banach spaces of analytic functions*, Prentice-Hall, Englewood Cliffs, N.J., 1962.

15. I. I. Privalov, *Boundary properties of analytic functions*, 2nd ed., GITTL, Moscow, 1950; German transl., VEB Deutscher Verlag Wiss., Berlin, 1956.

16. N. K. Nikol'skiĭ, *Selected problems of weighted approximation and spectral analysis*, Trudy Mat. Inst. Steklov. **120** (1974); English transl., Proc. Steklov Inst. Math. **120** (1974).

17. N. K. Bari, *Trigonometric series*, Fizmatgiz, Moscow, 1961; English transl., Vols. I, II, Macmillan, New York, and Pergamon Press, Oxford, 1964.

18. Walter Rudin, *Trigonometric series with gaps*, J. Math. Mech. **9** (1960), 203–227.

19. S. Banach, *Über einige Eigenschaften der lacunären trigonometrischen Reihen*, Studia Math. **2** (1930), 207–220.

20. L. V. Kantorovich and G. P. Akilov, *Functional analysis in normed spaces*, Fizmatgiz, Moscow, 1959; English transl., Macmillan, 1964.

21. B. Ërikke [Jöricke], *A comparison of the maximum of the modulus of a polynomial with the weighted sum of the moduli of its coefficients*, Vestnik Leningrad. Univ. **1978**, no. 1 (Ser. Mat. Meh. Astr. vyp. 1), 142–144. (Russian)

22. S. A. Vinogradov and V. P. Havin, *Free interpolation in H^∞ and in certain other classes of functions*. II, Zap. Nauchn. Sem. Leningrad. Otdel. Mat. Inst. Steklov. (LOMI) **56** (1976), 12–58; English transl. in J. Soviet Math. **14** (1980), no. 2.

23. Jean-Pierre Kahane and Raphael Salem, *Ensembles parfaits et séries trigonométriques*, Actualités Sci. Indust., No. 1301, Hermann, Paris, 1963.

24. A. I. Markushevich, *Theory of analytic functions*, GITTL, Moscow, 1950; English transl., *Theory of functions of a complex variable*, Vols. I, II, III, Prentice-Hall, Englewood Cliffs, N.J., 1965, 1967.

25. S. V. Khrushchev, *Singularities of linear operators in spaces of analytic functions*, Math. Programming and Related Questions (Proc. Seventh Winter School, Drogobych, 1974): Theory of Functions and Functional Anal., Tsentral. Èkonom.-Mat. Inst. Akad. Nauk SSSR, Moscow, 1976, pp. 87–114; English transl. in Amer. Math. Soc. Transl. (2) **115** (1981).

26. ______, *Paley singularities of power series*, Zap. Nauchn. Sem. Leningrad. Otdel. Mt. Inst. Steklov. (LOMI) **30** (1972), 176–179; English transl. in J. Soviet Math. **4** (1975), no. 4.

27. Nelson Dunford and Jacob T. Schwartz, *Linear operators*. Vol. I, Interscience, 1958.

28. G. P. Akilov and A. M. Veršik, *On the reciprocal continuous extension of linear operations*, Vestnik Leningrad. Univ. **1958**, no. 7 (Ser. Mat. Meh. Astr. vyp. 2), 27–33. (Russian)

29. Daniel M. Oberlin, *Uniformly convergent Fourier series and sets of measure zero*, preprint, 1978.

Translated by R. P. BOAS

Труды Матем. Инст. Стеклов.
155(1981)

Proceedings of the Steklov
Institute of Mathematics
1983, Issue 1

A CONSTRUCTIVE CHARACTERIZATION OF THE SOBOLEV AND BESOV CLASSES

UDC 517.54

E. M. DYN'KIN

ABSTRACT. The article is devoted to the development of the theory of pseudoanalytic continuation of functions from the spaces $B_p^s[-1,1]$, $s>0$, $1 \leqslant p \leqslant \infty$, and $W_p^l[-1,1]$, $l=1,2,\ldots$, $1<p<\infty$. Three methods of constructing such a continuation are described, namely by means of quasiconformal symmetry, local polynomial approximations, and global polynomial approximations. The results are applied to a constructive characterization of classes B_p^s and W_p^l by means of approximations by algebraic polynomials.

THEOREM 1. *$f \in B_p^s[-1,1] \Leftrightarrow \sum_{n=1}^{\infty} n^{-1} E_n(f)^p < \infty$, where*

$$E_n(f) = \inf\left\{ \left(\int_{-1}^{1} |f(x) - P_n(x)|^p \left(\frac{1}{n}\sqrt{1-x^2} + \frac{1}{n^2} \right)^{-ps} dx \right)^{1/p}, \quad P_n \text{ a polynomial of degree } n \right\}.$$

THEOREM 2.

$$f \in W_p^l[-1,1] \Leftrightarrow \int_{-1}^{1} \left\{ \sum_{n=0}^{\infty} |f(x) - P_{2^n}(x)|^2 \left(2^{-n}\sqrt{1-x^2} + 4^{-n} \right)^{-2l} \right\}^{p/2} dx < \infty$$

for some polynomials P_{2^n} of degree 2^n.

Theorem 1 is an L^p analog of a well-known theorem of S. M. Nikol'skiĭ and A. F. Timan. Theorem 2 is a nonperiodic analog of the Littlewood-Paley theorem on binary expansion.

The article also contains similar results for analytic functions in domains "with corners" as well as a study of the behavior of the spaces considered here under the action of Faber-Toeplitz operators.

Bibliography: 27 titles.

Contents

1980 *Mathematics Subject Classification*. Primary 46E35, 30G20; Secondary 41A10, 41A17.

©1983 American Mathematical Society
0081-5438/83 $1.00 + $.25 per page

§0. Introduction

In this article we study the concept of pseudoanalytic continuation of functions of one variable which are elements of Sobolev and Besov classes. By means of pseudoanalytic continuation we give a constructive characterization of these classes in terms of weighted polynomial approximations, and establish their Faber invariance. Constructive characteristics of various function classes have been well studied in the periodic case, i.e. for approximations of periodic functions by trigonometric polynomials. The classical result of Jackson and Bernstein states that a function f defined on the segment $[-\pi, \pi]$ satisfies the Lipschitz condition of order s, $0 < s < 1$, if and only if its best uniform approximations by trigonometric polynomials of degree n decrease as n^{-s} when $n \to \infty$. Later similar characterizations were obtained for many other classes of functions which are smooth "in the mean". First of all, let us state the following constructive characterization of Besov's periodic classes $\overset{*}{B}{}^s_{pq}$ on $[-\pi, \pi]$ [2].

THEOREM A. *A function f defined on $[-\pi, \pi]$ belongs to the class $\overset{*}{B}{}^s_{pq}$, $s > 0$, $1 \leq p$, $q \leq \infty$ if and only if*

$$\left[\sum_{n=1}^{\infty} \frac{1}{n}\left(n^s e_n(f)_p\right)^q\right]^{1/q} < \infty,$$

where

$$e_n(f)_p = \inf_{T_n}\left(\int_{-\pi}^{\pi} |f(x) - T_n(x)|^p\, dx\right)^{1/p}$$

is the best approximation of f in $L^p[-\pi, \pi]$ by trigonometric polynomials of degree n.

The Littlewood-Paley theorem on binary expansion [3] can also be viewed, after a minor modification, as a constructive characterization of another class, namely the Sobolev periodic class $\overset{*}{W}{}^l_p$.

THEOREM B. *A function f defined on $[-\pi, \pi]$ belongs to the class W^l_p, l a positive integer, $1 < p < \infty$, if and only if there exists a sequence $\{T_{2^k}\}$ of trigonometric polynomials of degree 2^k, $k = 0, 1, \ldots$, such that*

$$\int_{-\pi}^{\pi}\left(\sum_{k=0}^{\infty} |f(x) - T_{2^k}(x)|^2 2^{2kl}\right)^{p/2} dx < \infty.$$

The theorem on binary expansion can be obtained from Theorem B if we take for the polynomials T_{2^k} the partial sums of the Fourier series of f.

Investigations of the nonperiodic case, i.e. of approximations of nonperiodic functions by ordinary polynomials, started only after World War II. The basic phenomenon, first discovered by Nikol'skiĭ [4], consists in the necessity of going over from the homogeneous approximations $e_n(f)$ to weighted approximations which take into account the location of the point in the interval. The phenomenon of nonhomogeneity arises from the fact that the endpoints of the interval are singular points of it in the complex plane, while there are no singular points in the periodic case. Indeed, the latter case essentially reduces to the study of functions on the circle which has no such singular points. The following theorem, established by Nikol′skiĭ [4], Timan [5] and Dzyadyk [6], is a classical result in the nonperiodic case:

THEOREM C. *A function f defined on $[-1,1]$ satisfies the Lipschitz condition of order $s>0$ if and only if for some sequence of (algebraic) polynomials $\{P_n\}$ of degree n, $n=1,2,\ldots$,*

$$|f(x)-P_n(x)| \leqslant C\left(n^{-1}\sqrt{1-x^2}+n^{-2}\right)^s, \qquad x\in[-1,1]. \tag{0.1}$$

Theorem C is the nonperiodic analog of the Jackson-Bernstein theorem. A large number of articles (for a survey, see, for example, [7]) have been devoted to obtaining analogs of this theorem in the L^p metric, $1\leqslant p\leqslant\infty$. It seems natural to consider for a function f from $L^p[-1,1]$ the weighted best approximation

$$E_n(f)=E_n(f)_{p,s}=\inf_{P_n}\left(\int_{-1}^{1}\left|\frac{f(x)-P_n(x)}{\left(\sqrt{1-x^2/n}+1/n^2\right)^s}\right|^p dx\right)^{1/p},$$

where the infimum is taken over all (algebraic) polynomials P_n of degree n.

One could expect that the condition

$$E_n(f)_{p,s}\leqslant C<\infty, \qquad n=1,2,\ldots, \tag{0.2}$$

coinciding with (0.1) for $p=\infty$, is (in the spirit of Theorem A) a constructive characterization of the Nikol′skiĭ class $H_p^s[-1,1]$ (for definitions of the classes H_p^s, B_{pq}^s and W_p^l see the monograph [2]; we recall that $H_p^s=B_{p\infty}^s$ and that B_{pp}^s is denoted simply by B_p^s). However, Motornyĭ [8], on the basis of previous work of V. K. Lebed′ and M. K. Potapov, has shown that this conjecture is incorrect, namely that the class of functions satisfying (0.2) does not coincide with H_p^s and that the class H_p^s does not seem to allow a simple constructive characterization.(1)

Although a direct structural description of the class of functions satisfying (0.2) as well as other similar conditions is of independent interest (see [7], [9] and [10]), the problem of finding an exact analog of Theorem C in L^p remains open, and so does the problem of finding a constructive characterization, in the spirit of Theorem A, of the Besov classes B_{pq}^s. (We have just seen that for $q=\infty$ but $p<\infty$ such a characterization is not known.) Furthermore, until now a nonperiodic analog of Theorem B, i.e. a constructive characterization of the nonperiodic Sobolev classes W_p^l, has been missing.

In the present paper we study constructive properties of Besov and Sobolev classes by means of pseudoanalytic continuations [11], [12]. By a pseudoanalytic continuation of a function f, originally given on $[-1,1]$, we shall understand an extension of f to a function F such that F has compact support in the plane and the "discrepancy" in the Cauchy-Riemann equations for F

$$\bar\partial F(z)=\frac{1}{2}\left(\frac{\partial F}{\partial x}+i\frac{\partial F}{\partial y}\right), \qquad z=x+iy,$$

(1) In [8] it was suggested that for a constructive description of H_p^s one use not polynomials, but rather certain piecewise polynomial combinations.

rapidly decreases when approaching the segment $[-1, 1]$ in the plane. The speed of this decrease uniquely characterizes the smoothness of the original function f.

One often succeeds in reformulating standard definitions of classes of smooth functions in terms of estimates of the modulus of the function $\bar{\partial} F$. For instance, we shall show in §3 of this article that $f \in B_p^s[-1, 1]$ if and only if one can extend f to a function F such that(²)

$$\int\int |\bar{\partial} F(x)|^p \rho(z, [-1, 1])^{-p(s-1)-1}\, dx\, dy < \infty. \tag{0.3}$$

Such a description of function classes in terms of pseudoanalytic continuations turns out to be useful in the approximation theory, because completely different constructions of continuations may lead to identical estimates of type (0.3). Thus, for instance, there exists a continuation related to the moduli of smoothness of the function f as well as a continuation of f related to its best approximations. If these two methods of continuation lead to identical estimates of $\bar{\partial} F$, we conclude that the two corresponding function classes coincide, i.e. we have obtained a constructive characterization of the type of Theorems A, B, and C. This approach allows us to treat the classes B_{pq}^s and the class (0.2) ([1], [10], [13]) in the same way.

As is well known, a theorem analogous to Theorem C was established for functions analytic in domains "with corners" [5], [14]. In this case the problem of an L^p-analog of Theorem C has also been open. In the present paper we shall study simultaneously the cases of the segment and of a domain. In particular, this generalization allows us to better understand the geometric nature of results concerning the segment (cf. [4]) and to carry out a continuous transition from the periodic case to the nonperiodic one (see §7).

In §1 of this article we state the necessary preliminary facts on function classes, domains, etc. Note that among various definitions of moduli of smoothness the one most suitable for us turned out to be that of Brudnyĭ [15], expressed in terms of local approximations.

In §2 we present three methods of pseudoanalytic continuation of functions, namely by means of quasiconformal symmetry, by means of local polynomial approximations, and by means of global polynomial approximations.

§3 contains a description of the Besov classes in terms of pseudoanalytic continuation. In particular, it is shown that condition (0.3) describes the class $B_p^s[-1, 1]$, while the classes $B_{pq}^s[-1, 1]$ for $q \neq p$ are described by the more complicated condition

$$\left(\int_0^\infty \frac{dr}{r} \left(\int_{\Lambda_r} |\bar{\partial} F(z)|^p \rho(z, [-1, 1])^{-p(s-1)} |dz| \right)^{q/p} \right)^{1/p} < \infty, \tag{0.4}$$

where $\Lambda_r = \{z, \rho(z, [-1, 1]) = r\}$ is the level curve of the euclidean distance.

In §4 we give a similar description of Sobolev classes. The corresponding estimate turns out to be more complicated than (0.3): $f \in W_p^l[-1, 1]$ if and only if

$$\int_{-1}^1 \left(\iint_{S(x)} |\rho(\zeta, [-1, 1])^{-l} \bar{\partial} F(\zeta)|^2\, d\xi\, d\eta \right)^{p/2} dx < \infty,$$

(²) $\rho(z, [-1, 1]) = \inf_{-1 \leq x \leq 1} |z - x|$.

where $1 < p < \infty$, $l = 1, 2, \ldots$, and $S(x) = \{\zeta, \rho(\zeta, [-1, 1]) \geqslant |\zeta - x|/2\}$ is the Luzin sector.

§5 is devoted to the study of constructive properties of the Besov classes. By analogy with Theorem A one considers, instead of (0.2), a more general condition

$$\left(\sum_{n=1}^{\infty} \frac{1}{n} E_n(f)_{p,s}^q \right)^{1/q} < \infty. \tag{0.5}$$

In §5 it is shown that to this condition there corresponds an estimate

$$\left(\int_0^{\infty} \frac{dr}{r} \left(\int_{L_r} |\bar{\partial} F(z)|^p \rho(z, [-1, 1])^{-p(s-1)} |dz| \right)^{q/p} \right)^{1/q} < \infty, \tag{0.6}$$

which differs from (0.4) in that the level curves of the usual distance are replaced by the level curves of the outer Green's function of the segment $[-1, 1]$:

$$L_r = \left\{ z, |z + \sqrt{z^2 - 1}| = 1 + r \right\}, \qquad r > 0.$$

It is clear that for $p \neq q$ estimates (0.4) and (0.6) are essentially different and cannot be reduced to each other by means of weights or substitutions of variables. This explains why (for $q = \infty$) condition (0.2) does not describe the class H_p^s; and one may expect the same behavior for all classes B_{pq}^s for $q \neq p$. However, for $q = p$ the iterated integrals (0.4) and (0.6) turn into ordinary double integrals which in our case simply coincide. Hence the class $B_p^s = B_{pp}^s$, unlike B_{pq}^s for $p \neq q$, allows a constructive characterization which, according to (0.5), is of the form

$$\left(\sum_{n=1}^{\infty} \frac{1}{n} E_n(f)_{p,s}^p \right)^{1/p} < \infty.$$

For $p = \infty$ such a description becomes Theorem C and is evidently a natural analog of it in L^p.([3])

In §6 we give a constructive characterization of the Sobolev classes, i.e. we give a nonperiodic analog of Theorem B: A function f, defined on $[-1, 1]$, belongs to the class $W_p^l[-1, 1]$, $1 < p < \infty$, $l = 1, 2, \ldots$, if and only if for some sequence $\{P_{2^k}\}$ of algebraic polynomials of degrees 2^k, $k = 0, 1, \ldots$,

$$\int_{-1}^{1} \left(\sum_{k=0}^{\infty} |f(x) - P_{2^k}(x)|^2 \left(2^{-k}\sqrt{1 - x^2} + 4^{-k} \right)^{-2l} \right)^{p/2} dx < \infty.$$

The proof of a constructive characterization of B_p^s, given in §5, consists of two parts. The first part involves relatively simple considerations of pseudoanalytic continuations; the second one deals with rather tedious estimates related to an explicit construction of approximating polynomials. It is shown in §7 that this second part can be significantly simplified by means of Faber-Toeplitz operators. These operators map functions analytic in one region into functions analytic in another region. For instance, a function analytic in the unit disc goes into a function analytic on the segment. The simplest example is provided by the usual Faber

([3]) In the final analysis the difference between B_p^s and H_p^s lies in the fact that the variables $x \in [-1, 1]$ and δ (= the step in the modulus of smoothness) appear asymmetrically in the definition of H_p^s and symmetrically in the definition of B_p^s.

operator [16] whose action consists in replacing all powers z^n in the Taylor series of the function f by the Tchebycheff polynomials $T_n(x)$.

In §7 the action of Faber-Toeplitz operators is related to pseudoanalytic continuations, and it is shown that these operators preserve the smoothness of functions. In particular, in §7 we construct the Faber-Toeplitz operators which realize the canonical isomorphism between the space A_p^s of functions which are analytic in the unit disc and have boundary values from B_{pq}^s and the space $B_p^s[-1,1]$ (we also construct canonical isomorphisms between similar spaces over domains with corners). A similar theorem is established for the Sobolev classes. This isomorphism can be used to transfer results established for the disc to the case of the segment or that of a domain, i.e. it can be used in attempts to obtain a characterization of $B_p^s[-1,1]$ as an immediate consequence of Theorem A. In §7 this method leads to a constructive characterization when $s < 2/p$. The applicability of this simple approach for large values of s is yet to be clarified. This makes it necessary to resort to the "technical" proof from §5.

Everywhere in the sequel we make a significant use of results of [17], which represents a "technical introduction" to the present paper.

The basic results of this paper were briefly announced in [1], and were presented in the seminar of S. M. Nikol'skiĭ and L. D. Kudrjavcev on embedding theorems held at the Steklov Institute of Mathematics of the Academy of Sciences of the USSR.

The author expresses his sincere thanks to N. A. Shirokov, Yu. A. Brudnyĭ and L. V. Potepun with whom he discussed various aspects of the results.

§1. Preliminary facts

1.1. Notation and definitions.

1.1.1. $\mathbb{C}$ is the plane of the complex variable $z = x + iy$ or $\zeta = \xi + i\eta$.

We set

$$\bar{\partial} f(z) \overset{\text{def}}{=} \frac{1}{2}\left(\frac{\partial f}{\partial x} + i\frac{\partial f}{\partial y}\right);$$

$\rho(z, E)$ is the distance from the point z to the set E.

$\mathbb{D} = \{z : |z| < 1\}$ is the unit disc and $|I|$ is the length of the arc I.

1.1.2. Let K be a continuum in $\mathbb{C}$. We shall consider the case $K = [-1, 1]$ and the case when K is the closure of a Jordan domain G with Lipschitz boundary [3], [18]. Such continua will be called *admissible*. In both cases ∂K is a rectifiable curve. For $r > 0$ set

$$\Lambda_r = \{z, \rho(z, K) = r\}.$$

All Λ_r are smooth curves. It is easy to see that for any function F summable in $\mathbb{C} \setminus K$ one has

$$\iint_{\mathbb{C}\setminus K} F(z)\,dx\,dy = \int_0^\infty dr \int_{\Lambda_r} F(z)\chi(z)\,dz, \tag{1.1}$$

where $|\chi| \asymp 1$.

Let $z \in \mathbb{C} \setminus K$. $I(z)$ will denote the arc on ∂K with center at the point $\hat{z}$ (= the closest point to z on ∂K) and of length $\rho(z, \partial K)$. If I is an arc, $I \subset \partial K$, by λI, $\lambda > 0$, we shall denote the arc with the same center as I but of length $\lambda |I|$.

1.1.3. The Luzin sector (see [3]) $S_\gamma(z)$, $z \in \partial K$, $\gamma > 0$, will be defined as follows:

$$S_\gamma(z) = \{\zeta \in \mathbb{C} \setminus K, |\zeta - z| \leqslant (1+\gamma)\rho(\zeta, K)\}.$$

The value of γ will not be essential in the sequel and will usually be omitted.

It is easy to see that for any function F summable in $\mathbb{C} \setminus K$

$$\iint_{\mathbb{C}\setminus K} F(z)\,dx\,dy = \int_{\partial K} |dz| \iint_{S(z)} F(\zeta)\chi(\zeta, z)\frac{d\xi\,d\eta}{\rho(\zeta, \partial K)}, \tag{1.2}$$

where $|\chi| \asymp 1$.

1.1.4. Let $\varphi\colon \mathbb{C}\setminus K \to \mathbb{C}\setminus \overline{\mathbb{D}}$, $\varphi(\infty) = \infty$, $\varphi'(\infty) > 0$, be a conformal mapping and ψ its inverse. The level curves of K are the curves

$$L_r = \{z \in \mathbb{C}\setminus K, |\varphi(z)| = 1 + r\}, \qquad r > 0.$$

It is easy to see that for any function F summable in $\mathbb{C}\setminus K$

$$\iint_{\mathbb{C}\setminus K} F(z)\,dx\,dy = \int_0^\infty dr \int_{L_r} F(z)\chi(z)\frac{dz}{\varphi'(z)}, \tag{1.3}$$

where $|\chi| \asymp 1$.

1.1.5. Let $z \in \mathbb{C}\setminus K$ and let $r > 0$. The conformal shift $z_{\langle r\rangle}$ is defined by $z_{\langle r\rangle} = \psi((1-r)\varphi(z))$. If $z \in \partial K$, then $z_{\langle r\rangle} \in L_r$ and one can show (see [17]) that

$$|z - z_{\langle r\rangle}| \asymp \rho(z, L_r) \asymp r/|\varphi'(z_{\langle r\rangle})|. \tag{1.4}$$

The same estimates remain valid for $z \in L_\sigma$, $\sigma \leqslant r/2$. But if $z \in L_\sigma$, $\sigma > r/2$, then

$$|z - z_{\langle r\rangle}| \asymp r/|\varphi'(z)| \asymp \rho(z, K)\cdot r/\sigma. \tag{1.5}$$

If K is the closure of a Lipschitz domain G, then z^* will denote a point quasiconformally symmetric to z with respect to ∂G [19]. The mapping $z \to z^*$ is continuously differentiable in the plane.

1.1.6. The distances to the level curves exhibit a rather regular behavior [17]: for $z \in \partial K$

$$c_1\left(\frac{r}{\sigma}\right)^\alpha \leqslant \frac{\rho(z, L_r)}{\rho(z, L_\sigma)} \leqslant c_2\left(\frac{r}{\sigma}\right)^\beta, \qquad r > \sigma, \tag{1.6}$$

where $\alpha > 0$ and $0 < \beta < 2$. Furthermore, for $z, \zeta \in \partial K$

$$\frac{\rho(z, L_r)}{\rho(\zeta, L_r)} \leqslant c\left(1 + \frac{|\zeta - z|}{\rho(\zeta, L_r)}\right)^\gamma, \qquad 0 < \gamma < 1. \tag{1.7}$$

Finally, for $z, \zeta \in L_r$

$$\frac{\rho(z, K)}{\rho(\zeta, K)} \leqslant c\left(1 + \frac{|\zeta - z|}{\rho(\zeta, K)}\right)^\delta, \qquad 0 < \delta < 1. \tag{1.8}$$

1.1.7. The domain G is called a *Radon domain* [20] if its boundary is rectfiable and the slope of the tangent line to ∂G is a function of bounded variation such that all jumps of this function are strictly less than π. Radon domains are Lipschitz domains. If G is a Radon domain then for any function $f \in L^p(\partial G)$, $1 < p < \infty$, the function

$$\Phi(z) = \int_{\partial G} f(\zeta)\frac{d\zeta}{\zeta - z}, \qquad z \notin \partial G,$$

belongs both in G and in $\mathbb{C}\setminus\overline{G}$ to the class E^p of V. I. Smirnov (see [21]) and

$$\|\Phi\|_{E^p} \leqslant c\|f\|_{L^p(\partial G)}.$$

The same estimate also holds for the domain L_r^0 bounded by the level curve L_r of the domain G, and the constant c does not depend on r.

Calderón [22] recently established a similar fact for certain Lipschitz domains which are not Radon domains.

1.1.8. In the sequel we shall need functions from the Sobolev classes W_q^1, $q > 1$, in $\mathbb{C}$ and in $\mathbb{C}\setminus K$. It was shown in [18] that these classes possess boundary values on ∂K, and the set of these boundary values was described. On the other hand, if $K = \overline{G}$ is the closure of a Lipschitz domain and $f \in E^q(G)$, then f possesses boundary values almost everywhere on ∂G. It is easy to see that both kinds of boundary values are in agreement. In particular, if a function $f_0 \in E^q(G)$ coincides almost everywhere on ∂G with the boundary value of some function f_1 from $W_q^1(\mathbb{C}\setminus\overline{G})$, then for $z \in \mathbb{C}\setminus\overline{G}$

$$0 = \frac{1}{2\pi i}\int_{\partial G} f_0(\zeta)\frac{d\zeta}{\zeta - z} = \frac{1}{2\pi i}\int_{\partial G} f_1(\zeta)\frac{d\zeta}{\zeta - z} = \lim_{r\to 0}\frac{1}{2\pi i}\int_{\Lambda_r} f_1(\zeta)\frac{d\zeta}{\zeta - z}$$

$$= \lim_{r\to 0}\left[f_1(z) + \frac{1}{\pi}\iint\limits_{\rho(\zeta, G)>r}\bar{\partial} f_1(\zeta)\frac{d\xi\, d\eta}{\zeta - z}\right] = f_1(z) + \frac{1}{\pi}\iint\limits_{\mathbb{C}\setminus\overline{G}}\bar{\partial} f_1(\zeta)\frac{d\xi\, d\eta}{\zeta - z}.$$

Hence,

$$f_0(z) = -\frac{1}{\pi}\iint\limits_{\mathbb{C}\setminus\overline{G}}\bar{\partial} f_1(\zeta)\frac{d\xi\, d\eta}{\zeta - z}$$

almost everywhere on ∂G, and by the Calderón-Zygmund theorem [19] the function F, which equals f_0 in G and f_1 in $\mathbb{C}\setminus\overline{G}$, is an element of $W_q^1(\mathbb{C})$.

1.1.9. The major topic of our later considerations will be the well-known classes $W_p^l[-1, 1]$ and $B_p^s[-1, 1]$ (see [2] and [18]) and their analytic analogs in domains "with corners". The class $W_p^l[-1, 1]$, $1 < p < \infty$, $l > 0$ an integer (see [2] and [18]) consists of functions f, defined on $[-1, 1]$, continuously differentiable through order $(l - 1)$ such that $f^{(l-1)}$ is absolutely continuous and $f^{(l)} \in L^p[-1, 1]$. This class is a Banach space with the norm $\|f\|_{W_p^l} = \|f\|_{L^p} + \|f^{(l)}\|_{L^p}$.

If G is a Lipschitz domain, $1 < p < +\infty$ and l is a positive integer, then we define the class $E_p^l(G)$ as the set of all functions $f \in E^p(G)$ which are analytic in G and such that $f^{(l)} \in E^p(G)$. This class is a Banach space with the norm $\|f\|_{E_p^l} = \|f\|_{E^p} + \|f^{(l)}\|_{E^p}$. If $G = \mathbb{D}$ is the unit disc, then $E_p^l(\mathbb{D})$ can be identified with the set of functions analytic in $\mathbb{D}$ such that $f(e^{it}) \in W_p^l[-1, 1]$. Hence $E_p^l(G)$ can be viewed as an analytic analog of $W_p^l[-1, 1]$.

The spaces $W_p^l[-1, 1]$ and $E_p^l(\mathbb{D})$ were discussed by Hardy and Littlewood (see, for example, [23]). The classes $E_p^l(\mathbb{D})$ play an important role in the modern theory of Hardy function spaces (see, for example, [24]), in particular in the theory of invariant subspaces. They do not bear a special name, but they are one-dimensional analytic variants of well-known Sobolev classes [12], [18]. For the sake of uniformity, in this article we shall call both the spaces $W_p^l[-1, 1]$ and their analytic variants $E_p^l(G)$ (one-dimensional) *Sobolev classes*.

The Besov class $B_p^s[-1, 1]$, $1 \leqslant p \leqslant \infty$, $s > 0$, consists of all functions f on $[-1, 1]$ such that for $m > s$ the integral modulus of smoothness of order m of the function f from L^p, $\omega_m(f, \delta)_p$ satisfies the inequality (see [2] and [18])

$$c(f) = \left(\int_0^1 (\omega_m(f, \delta)_p/\delta^s)^p \frac{d\delta}{\delta}\right)^{1/p} < +\infty.$$

This class is a Banach space with the norm $\|f\|_{B_p^s} = \|f\|_{L^p} + c(f)$. Below we shall consider an analog of B_p^s, consisting of functions analytic in G. Again, for the sake of uniformity we shall call it a Besov class. These analogs, denoted by $A_p^s(G)$, and the spaces $E_p^l(G)$ arise naturally when one attempts to calculate the image of the standard classes W_p^l and B_p^s under the action of Faber-Toeplitz operators (see §7). In the problem of approximation theory we are considering they enable us to carry out a continuous transition from the periodic case to the nonperiodic one.

1.2. Local approximation by polynomials. The classes $A_p^s(G)$ (cf. Brudnyĭ [15]).

1.2.1. Let us recall that K denotes either the segment $[-1, 1]$ or the closure of a Lipschitz domain G. Let $f \in L^1(\partial K)$, and let I be an interval, $I \subset \partial K$.

We shall consider the best approximation of the function $f \in L^p(I)$ by polynomials of degree m:

$$E_m(f, I)_p = \inf\left(\int_I |f - T|^p\right)^{1/p},$$

where the infimum is taken over all polynomials T of degree m, $m = 0, 1, \ldots$. We will be interested in the case where m is fixed. Let us start with an interpolation construction which will yield an "almost best" approximation. Let P_0 be a polynomial such that $\deg P_0 = m$ and

$$\int_0^1 P_0(x)\,dx = 1, \quad \int_0^1 x^k P_0(x)\,dx = 0, \qquad k = 1, 2, \ldots, m.$$

Let J be a Lipschitz arc in the plane with endpoints z_0 and z_1, and let

$$P_J(z) = \frac{1}{z_1 - z_0} P_0\left(\frac{z - z_0}{z_1 - z_0}\right).$$

Then for any polynomial T, $\deg T \leqslant m$,

$$T(z_0) = \int_J P_J(z) T(z)\,dz.$$

Let I be the initial interval, $I \subset \partial K$. Let us partition I by the points $z_0, \ldots, z_{m+1}$ into $m + 1$ intervals $I_0, \ldots, I_m$ of length $|I|/(m + 1)$.

Let $f \in L^1(I)$. Then $\mathcal{P}_I(f)$ will denote the polynomial of degree m which at the points $z_0, \ldots, z_m$ takes on the values

$$\int_{I_0} P_{I_0}(z) f(z)\,dz, \ldots, \int_{I_m} P_{I_m}(z) f(z)\,dz,$$

respectively.

Obviously, $\mathcal{P}_I T = T$ for any polynomial T, $\deg T \leqslant m$, and for any $\lambda > 0$

$$\max_{\rho(z, I) \leqslant \lambda |I|} |\mathcal{P}_I f(z)| \leqslant \frac{c}{|I|} \int_I |f|, \tag{1.9}$$

where c depends on λ, m and K, but not on I or f. If follows immediately from (1.9) that

$$\| f - \mathcal{P}_I f \|_{L^p(I)} \leqslant (c+1) E_m(f, I)_p,$$

and hence the polynomial $\mathcal{P}_I f$ may be used instead of the polynomial of best approximation. Relation (1.9) together with the Lagrange interpolation formula easily yields a number of standard properties of local approximations [15]. For instance, if $I = I_1 \cup I_2$, $|I_1 \cap I_2| > q|I|$, $0 < q < 1$, then

$$E_m(f, I)_p \leqslant c\big(E_m(f, I_1)_p + E_m(f, I_2)_p\big).$$

These estimates imply the inequality ($p' = p/(p-1)$)

$$E_m(f, I)_1 \leqslant |I|^{1/p'} E_m(f, I)_p.$$

1.2.2. Let us go over to the definition of moduli of smoothness. Let $f \in L^p(\partial K)$ and $\delta > 0$. Consider all possible partitions $\partial K = \bigcup_1^N I_j$ of ∂K into intervals such that $\delta/2 \leqslant |I_j| \leqslant \delta$ for all j. Set

$$\omega_m(f, \delta)_p = \sup\left(\sum_{j=1}^{N} E_m(f, I)_p^p\right)^{1/p},$$

where the supremum is taken over all partitions of this type. $\omega_m(f, \delta)_p$ will be called the *modulus of smoothness of f of order m* in $L^p(\partial K)$. Brudnyĭ [15] showed that in the case of the segment or an analytic curve this definition is equivalent to the standard one.

REMARKS. a) $\omega(f, \delta) \leqslant \| f \|_{L^p(\partial K)}$ and $\omega(f, \delta) = 0(1)$, $\delta \to 0$.

b) Let T_δ be a piecewise polynomial function which coincides on I_j with the polynomial $\mathcal{P}_{I_j} f$. Then

$$\omega(f, \delta)_p \asymp \| f - T_\delta \|_{L^p(\partial K)}.$$

1.2.3. Let G be a Lipschitz domain, $1 \leqslant p \leqslant \infty$ and $s > 0$. We define the class $A_p^s(G)$ as the set of all functions $f \in E^p(G)$ such that for $m > s$

$$c(f) = \left(\int_0^1 \left(\omega_m(f, \delta)_p / \delta^s\right)^p \frac{d\delta}{\delta}\right)^{1/p} < \infty. \tag{1.10}$$

$A_p^s(G)$ is a Banach space with the norm $\| f \|_{A_p^s} = \| f \|_{E^p} + c(f)$.

1.3. Maximal functions. Muckenhaupt's condition.

1.3.1. Let $1 < p < \infty$, let Γ be a Lipschitz curve and let $w \geqslant 0$ be a function on Γ. The function w is said to satisfy the Muckenhaupt condition (A_p) if for any arc $I \subset \Gamma$

$$\left(\int_I w\right)\left(\int_I w^{1-p'}\right)^{p-1} \leqslant c|I|^p.$$

A detailed discussion of condition (A_p) and its consequences can be found in [17].

In particular, if $\varphi\colon \mathbf{C} \setminus K \to \mathbf{C} \setminus \overline{\mathbf{D}}$ is conformal and ψ is the inverse mapping, then on ∂K the function $|\varphi'|$ satisfies condition (A_p) with some p, $1 < p < \infty$, and on $\partial \mathbf{D}$ $|\psi'|$ satisfies condition (A_{p_1}) with some p_1, $1 < p_1 < 2$.

1.3.2. Let f be a function from $L^1(\Gamma)$. The Hardy-Littlewood *maximal function* is then defined by

$$Mf(z) = \sup_{I \ni z} \frac{1}{I} \int_{|I|} |f|, \qquad z \in \Gamma$$

(the supremum is taken over all arcs $I \subset \Gamma$ containing z), and the *truncated maximal function* is defined by

$$M_r f(z) = \sup_{I \ni z,\, |I| > r} \frac{1}{I} \int_{|I|} |f|, \qquad z \in \Gamma$$

(the supremum is taken over all arcs $I \subset \Gamma$, $I \ni z$, such that $|I| > r$). It is clear that $|f| \leqslant Mf$ and $M_r f \leqslant Mf$.

The *maximal theorem* [3], [25] states that if w satisfies condition (A_p), $1 < p < \infty$, then

$$\left(\int_\Gamma (Mf)^p w \right)^{1/p} \leqslant c \left(\int_\Gamma |f|^p w \right)^{1/p}.$$

1.3.3. Let $f \in L^1(\Gamma)$, and for a point $z \in \mathbf{C}$ let $\hat{z}$ denote the closest point (on Γ) to z.

LEMMA 1. *For arbitrary $r > |z - \hat{z}|$ and $m > 0$*

$$\frac{1}{r} \int_{|\zeta - z| < r} |f(\zeta)|\,|d\zeta| + \int_{|\zeta - z| > r} \frac{r^m}{|\zeta - z|^{m+1}} |f(\zeta)|\,|d\zeta| \leqslant c M_r f(\hat{z}),$$

where c depends only on Γ and m.

PROOF. The estimate of the first summand is obvious. In the second we partition the contour of integration into parts $I_k = \{\zeta \in \Gamma, 2^k r \leqslant |\zeta - \hat{z}| \leqslant 2^{k+1} r\}$. Then the contribution from the part I_k does not exceed

$$\frac{1}{2^{k(m+1)} r} \int_{I_k} |f| \leqslant c \cdot 2^{-km} M_r f(\hat{z}),$$

since, obviously, $|I_k| \asymp 2^k r$.

1.3.4. Fefferman and Stein [26] established the following strengthened version of the maximal theorem. We shall formulate it directly for Lipschitz curves. Let $f_1, f_2, \ldots$ be a sequence of functions from $L^1(\Gamma)$. Then for $1 < p < \infty$

$$\left(\int_\Gamma \left(\sum_{n=1}^{\infty} |Mf_n|^2 \right)^{p/2} \right)^{1/p} \leqslant c \left(\int_\Gamma \left(\sum_{n=1}^{\infty} |f_n|^2 \right)^{p/2} \right)^{1/p}.$$

1.4. Approximation of the Cauchy kernel. Let us recall that K is either the segment $[-1, 1]$ or the closure of a Lipschitz domain G. We shall need the following result on approximation of the Cauchy kernel by polynomials; its proof may be found in [14]. Let $m > 0$ and let $\xi \in \mathbf{C} \setminus K$. Then for any $n = 0, 1, \ldots$ there is a function $K_n(\zeta, z)$ which is a polynomial in the variable z of degree n such that for $z \in \partial K$:

a) for $|z - \zeta| < \rho(z, L_{1/n})$

$$|K_n(\zeta, z)| \leqslant c/\rho(z, L_{1/n});$$

b) for $|z-\zeta| \geqslant \rho(z, L_{1/n})$

$$\left|\frac{1}{\zeta - z} - K_n(\zeta, z)\right| \leqslant c\frac{|\zeta - \zeta_{\langle 1/n\rangle}|^m}{|\zeta - z|^{m+1}}.$$

The estimates (1.5) easily imply that for $\zeta \in L_r$, $r < 1/n$,

$$|\zeta - \zeta_{\langle 1/n\rangle}| \asymp \rho(\hat{\zeta}, L_{1/n}),$$

where $\hat{\zeta}$ is the point on ∂K closest to ζ, and for $\zeta \in L_r$, $r > 1/n$,

$$|\zeta - \zeta_{\langle 1/n\rangle}| \asymp \rho(\zeta, \partial K)/nr.$$

1.5. Admissible functions. Recall that K denotes either $[-1, 1]$ or the closure of a domain G with Lipschitz boundary. A function f defined on K will be called *admissible* provided $f \in L^1[-1, 1]$ for $K = [-1, 1]$ and $f \in E^1(G)$ for $K = \overline{G}$. We shall identify an admissible function with its boundary value on ∂K, and we shall write $f \in L^p(\partial K)$, etc.

In order to preserve formulations we shall denote by $A_p^s(K)$ the space $B_p^s[-1, 1]$ for $K = [-1, 1]$ and by $A_p^s(G)$ the same space for $K = \overline{G}$. In the same way let $E_p^l(K)$ denote $W_p^l[-1, 1]$ when $K = [-1, 1]$ and $E_p^l(G)$ when $K = \overline{G}$.

§2. Three methods of pseudoanalytic continuation

Let K be an admissible continuum. In §2.1 we introduce the concept of a pseudoanalytic continuation for functions defined on K [11], [12]. In §§2.2–2.4 we describe three methods for constructing pseudoanalytic continuations with special properties.

2.1. The concept of a pseudoanalytic continuation. Let f be an admissible function on K, and let $q > 1$.

THEOREM 1. *The following conditions on a function f are equivalent*:

(i) *There exists a function* F_1 *with compact support*, $F_1 \in W_q^1(\mathbb{C})$, *such that* $F_{1|K} = f$ (*the restriction* $F_{1|K}$ *of f is defined in the sense of boundary values*).

(ii) *There exists a function* $F_2 \in W^1(\mathbb{C}\setminus K)$ *such that* $F_2|_{\partial K} = f|_{\partial K}$.

(iii) *f can be represented by the integral*

$$f(z) = \iint_{\mathbb{C}\setminus K} \lambda(\zeta)\frac{d\xi\, d\eta}{\zeta - z}, \tag{2.1}$$

where $\iint_{\mathbb{C}\setminus K} |\lambda|^q < \infty$.

The proof of Theorem 1 was actually given in §1.1.8. ●

If a function f satisfies conditions (i)–(iii) of Theorem 1, we shall say that f admits a *pseudoanalytic continuation*. Either of the two functions F_1 and F_2 will be called a pseudoanalytic continuation of f and will be denoted by $\mathbf{f}$. When it is necessary to specify explicitly the value of q in Theorem 1 we shall speak about a W_q^1-continuation. Pseudoanalytic continuation is not unique, and below we describe three different methods of constructing such continuations.

2.2. Continuation by symmetry. Let $K = \overline{G}$. It is obvious that for a given function $f \in E^1(G)$ to admit a W_q^1-continuation it is necessary that $f' \in L^q(G)$.

THEOREM 2. *If $f \in E^1(G)$ and $f' \in L^q(G)$ for some $q > 1$, then for each $m = 1, 2, \ldots$ there exists a W_q^1-continuation* $\mathbf{f}$ *such that*

$$|\bar{\partial}\mathbf{f}(z)| \leqslant c|f^{(m)}(z^*)|\rho(z, \partial G)^{m-1}, \qquad z \in \mathbf{C}\setminus\overline{G}, \tag{2.2}$$

where z^* is a point quasiconformally symmetric to z with respect to ∂G.

PROOF. It follows easily from the mean value theorem that, under the conditions of our theorem,

$$\iint_G \left(|f^{(k)}(z)|\rho(z, \partial G)^{k-1}\right)^q dx\,dy < \infty, \qquad k = 1, 2, \ldots.$$

Set

$$F(z) = \sum_{k=1}^{m-1} f^{(k)}(z^*)\frac{(z - z^*)^k}{k!} \tag{2.3}$$

for $z \in \mathbf{C}\setminus\overline{G}$. Then $F \in W_q^1(\mathbf{C}\setminus\overline{G})$. Obviously the boundary value of F on ∂G coincides almost everywhere with f. Finally,

$$\bar{\partial}F(z) = f^{(m)}(z^*)\frac{(z - z^*)^{m-1}}{(m-1)!}\bar{\partial}z^*,$$

which proves the theorem. ●

2.3. Continuation by means of local approximations. Let f be an admissible function on K and m an integer, and let $z \in \mathbf{C}\setminus K$. Set (see §1.2.1)

$$E(f, z) = E_m(f, I(z))_1.$$

THEOREM 3. *For some $q > 1$ let*

$$E(f, z)\rho(z, K)^{-2} \in L^q(\mathbf{C}\setminus K).$$

Then $f \in L^q(\partial K)$, and there exists a W_q^1-continuation of f such that

$$|\bar{\partial}\mathbf{f}(z)| \leqslant cE(f, z)/\rho(z, K)^2, \qquad z \in \mathbf{C}\setminus K. \tag{2.4}$$

PROOF. a. First we shall show that $f \in L^q(\partial K)$. For any arc $I \subset \partial K$, set

$$\Omega(I) = \{z \in \mathbf{C}\setminus K, \rho(z, I) < |I|/10, \rho(z, K)|I|/100\}.$$

Obviously, meas $\Omega(I) \asymp |I|^2$ and $E(f, z) \geqslant E_m(f, I)_1$ for any $z \in \Omega(I)$.

Consider the sequence of partitions of ∂K into arcs

$$\partial K = \bigcup_{k=1}^{2^n} I_k^n, \qquad n = 0, 1, \ldots,$$

where $I_1^0 = \partial K$ and each subsequent partition is obtained from the previous one by halving each of its arcs.

The function T^n is defined on ∂K by setting $T^n(z) = \mathcal{P}_{I_k^n}f(z)$ for $z \in I_k^n$ (the polynomial projection $\mathcal{P}_I$; §1.2.1). Then almost everywhere on ∂K

$$f = T^0 + \sum_{n=0}^{\infty}(T^{n+1} - T^n)$$

and it suffices to verify that

$$\sum_{n=0}^{\infty}\|T^{n+1} - T^n\|_{L^q(\partial K)} < \infty.$$

However, on I_k^{n+1} we obviously have

$$|T^{n+1}-T^n|\leqslant c\cdot 2^n E_m\left(f, 2I_k^{n+1}\right)_1\leqslant c\cdot 2^n E(f,z)$$

for any $z\in\Omega(2I_k^{n+1})$. Hence

$$\int_{I_k^{n+1}}|T^{n+1}-T^n|^q\leqslant c\cdot 2^{n(q-1)}\inf\left\{E(f,z), z\in\Omega\left(2I_k^{n+1}\right)\right\}$$

$$\leqslant c\cdot 2^{-n(q-1)}\iint_{\Omega(2I_k^{n+1})}E(f,z)^q\rho(z,K)^{-2q}\,dx\,dy.$$

Upon introducing the domains

$$\Omega^n=\left\{z, 2^{-n-10}\leqslant\rho(z,K)/|\partial K|\leqslant 2^{-n-3}\right\},$$

we have

$$\|T^{n+1}-T^n\|_{L^q(\partial K)}\leqslant c\cdot 2^{-n(q-1)/q}\left(\iint_{\Omega^n}E(f,z)^q\rho(z,K)^{-2q}\,dx\,dy\right)^{1/q}$$

and, finally,

$$\sum_{n=0}^{\infty}\|T^{n+1}-T^n\|_{L^q(\partial K)}\leqslant C\left(\iint_{\mathbf{C}\setminus K}E(f,z)^q\rho(z,K)^{-2q}\,dx\,dy\right)^{1/q}<\infty.$$

b. Now we construct a continuation of the function f. Consider a Whitney partition of unity (see [3]) in the domain $\mathbf{C}\setminus K$, $1=\Sigma\chi_k$. For each k let z_k be the center of the support of χ_k and let $I_k\subset\partial K$ denote the arc whose center is the point which is closest to z_k and whose length is equal to $\rho(z_k, K)/10$. For $z\in\mathbf{C}\setminus K$ we set

$$F(z)=\sum_k\chi_k(z)\mathcal{P}_{I_k}f(z). \tag{2.5}$$

If T is an arbitrary polynomial, then

$$F(z)=T(z)+\sum_k\chi_k(z)\left[\mathcal{P}_{I_k}f(z)-T(z)\right]$$

and

$$\bar\partial F(z)=\sum_k\left[\mathcal{P}_{I_k}f(z)-T(z)\right]\bar\partial\chi_k(z).$$

Now let us take $T=\mathcal{P}_{I(z)}f$. For all k with $\chi_k(z)\neq 0$ we have $I_k\subset I(z)$, whence

$$|\mathcal{P}_{I_k}f(z)-T(z)|\leqslant cE(f,z)\rho(z,K)^{-1},$$

which proves (2.4). In particular, $\bar\partial F\in L^q(\mathbf{C}\setminus K)$.

The Cauchy-Green formula implies that for $r<\rho(z,K)$

$$F(z)=-\frac{1}{\pi}\iint_{\mathbf{C}\setminus\Lambda_r^0}\bar\partial F(\zeta)\frac{d\xi\,d\eta}{\zeta-z}+\frac{1}{2\pi i}\int_{\Lambda_r}F(\zeta)\frac{d\zeta}{\zeta-z}. \tag{2.6}$$

However, it obviously follows from (2.5) that $|F(z)|\leqslant cMf(\hat z)$, where $\hat z$ is the point on K which is closest to z and f is considered as a function on ∂K; furthermore, it also follows that F converges to f almost everywhere on ∂K. Since $f\in L^q(\partial K)$ and thus also $Mf \in L^q(\partial K)$, we may go over to the limit in (2.6) as $r \to 0$, and obtain

$$\int_{\Lambda_r} F(\zeta)\frac{d\zeta}{\zeta - z} \to \int_{\partial K} f(\zeta)\frac{d\zeta}{\zeta - z} = 0,$$

$$F(z) = -\frac{1}{\pi}\iint_{\mathbf{C}\setminus K} \bar\partial F(\zeta)\frac{d\xi\, d\eta}{\zeta - z}.$$

In view of the Calderón-Zygmund theorem this implies that F is a W_q^1-continuation of f. ●

In the proof of Theorem 3 the polynomials of local best approximation were used for the continuation of f. Conversely, suppose we are given a W_q^1-continuation of f to the whole plane and an arc $I \subset \partial K$.

Let z_0 be the center of I. Consider the polynomial

$$P_I(z) = -\frac{1}{\pi}\iint_{|\zeta - z_0| > 2|I|} \bar\partial \mathbf{f}(\zeta)\left[\frac{1}{\zeta - z_0} + \frac{z - z_0}{(\zeta - z_0)^2} + \cdots + \frac{(z - z_0)^m}{(\zeta - z_0)^{m+1}}\right] d\xi\, d\eta \quad (2.7)$$

of degree m.

In many cases the polynomial P_I, similarly to the polynomial $\mathcal{P}_I f$ from §1, can be used instead of the polynomial of best approximation of f on I. Then for $z \in I$ we obviously have

$$|f(z) - P_I(z)| \leqslant \iint_{|\zeta - z_0| > 2|I|} |\bar\partial \mathbf{f}(\zeta)| \frac{d\xi\, d\eta}{|\zeta - z|} + \iint_{|\zeta - z_0| > 2|I|} |\bar\partial \mathbf{f}(\zeta)| \frac{|I|^{m+1}}{|\zeta - z|^{m+2}} d\xi\, d\eta. \quad (2.8)$$

Estimate (2.8) will be used in the next section.

2.4. Continuation by means of global approximations. Recall that $\varphi: \mathbf{C}\setminus K \to \mathbf{C}\setminus\overline{\mathbf{D}}$ is a conformal mapping. Let a function f on K be admissible, and let $P_1, P_2, \ldots$ be a sequence of polynomials of degrees $1, 2, \ldots$, respectively, converging to f in $L^1(\partial K)$. Let $z \in \mathbf{C}\setminus K$ and let $2^{-n} \leqslant |\varphi(z)| - 1 \leqslant 2^{-n+1}$. Set

$$\lambda(z) = (1/\rho(z, K))\,|P_{2^{n+1}}(z) - P_{2^n}(z)|.$$

THEOREM 4. *Let $\lambda \in L^q(\mathbf{C}\setminus K)$ for some $q > 1$. Then $f \in L^q(\partial K)$, and there exists a W_q^1-continuation of f such that*

$$|\bar\partial \mathbf{f}(z)| \leqslant c\lambda(z), \qquad z \in \mathbf{C}\setminus K.$$

PROOF. Let η be a function from $C^\infty(0, \infty)$, $0 \leqslant \eta \leqslant 1$, $\eta(t) = 1$ for $t \leqslant 1$ and $\eta(t) = 0$ for $t \geqslant 2$. The desired continuation of the function f for $2^{-n} \leqslant |\varphi(z)| - 1 \leqslant 2^{-n+1}$ will be defined by

$$\mathbf{f}(z) = P_{2^n}(z) + \eta[2^n(|\varphi(z)| - 1)](P_{2^{n+1}}(z) - P_{2^n}(z)).$$

Obviously, this function is infinitely differentiable in $\mathbf{C}\setminus K$, and $|\bar\partial \mathbf{f}(z)| \leqslant c\lambda(z)$.

For each N we shall now introduce the function F_N: $F_N(z) = \mathbf{f}(z)$ for $z \in \mathbf{C}\setminus K$, $|\varphi(z)| - 1 > 2^{-N}$, and $F_N(z) = P_{2^{N+1}}(z)$ for $|\varphi(z)| - 1 \leqslant 2^{-N}$ and for $z \in K$. F_N is infinitely differentiable everywhere, it is analytic inside the level curve $L_{2^{-N}}$, and outside the level curve $|\bar{\partial} F_N(z)| \leqslant c\lambda(z)$. The Cauchy-Green formula implies that for every $z \in K$

$$P_{2^{N+1}}(z) = F_N(z) = -\frac{1}{\pi} \iint_{\mathbf{C} \setminus K} \bar{\partial} F_N(\zeta) \frac{d\xi\, d\eta}{\zeta - z},$$

and obviously, in this formula one can pass to the limit as $N \to \infty$. ●

In applications the polynomials P_n will be polynomials of best approximation of f in some metric. Then we shall be able to estimate $\lambda(z)$ by means of inequalities of Bernstein type. Conversely, suppose we are given a W_q^1-continuation of f. In this case the approximating polynomials can be constructed by approximating the Cauchy kernel in the Cauchy-Green integral as was done in §1.4.

§3. Pseudoanalytic continuation of functions from Besov classes

3.1. The Main Theorem. As above, K stands either for the segment $[-1, 1]$ or for the closure of a Lipschitz domain G.

THEOREM 5. *Let $1 \leqslant p \leqslant \infty$ and $s > 0$, and let f be an admissible function on K. Then $f \in A_p^s(K)$ if and only if f admits a pseudoanalytic continuation such that*

$$\iint_{\mathbf{C} \setminus K} |\bar{\partial} \mathbf{f}(z)|^p \rho(z, K)^{-p(s-1)-1}\, dx\, dy < \infty \tag{3.1}$$

for $p < \infty$, and

$$|\bar{\partial} \mathbf{f}(z)| \leqslant c\rho(z, K)^{s-1}, \qquad z \in \mathbf{C} \setminus K \tag{3.1'}$$

for $p = \infty$.

In the proof of Theorem 5 we shall use the following characterization of a pseudoanalytic continuation of f:

$$\sigma_p(\mathbf{f}, r) = \left(\int_{\Lambda_r} |\bar{\partial} \mathbf{f}(z)|^p\, |dz| \right)^{1/p}, \qquad r > 0$$

(for the definition of Λ_r, see §1.1).

Equality (1.1) implies that condition (3.1) (respectively, (3.1′)) is equivalent to

$$\left(\int_0^1 \left(\frac{\sigma_p(\mathbf{f}, r)}{r^{s-1}} \right)^p \frac{dr}{r} \right)^{1/p} < \infty. \tag{3.2}$$

Condition (3.2) will be derived from the sharper estimates (3.3) and (3.4). On the one hand we shall construct, for any m, a contination of f such that

$$\sigma_p(\mathbf{f}, r) \leqslant c\omega_m(f, 20r)_p / r. \tag{3.3}$$

On the other hand, if a contination of f is already given, then

$$\omega_m(f, \delta)_p \leqslant c \int_0^{2\delta} \sigma_p(\mathbf{f}, r) \log \frac{2\delta}{r}\, d\delta + c\delta^{m+1} \int_\delta^\infty \sigma_p(\mathbf{f}, r) \frac{dr}{r^{m+1}}. \tag{3.4}$$

Obviously, (3.3) and (3.4), together with Hardy's inequality [23], allow us to identify (3.2) with the standard definition of Besov classes (1.10).

Moreover, suppose that $1 \leqslant q \leqslant \infty$. Then it follows, e.g., from (3.3) and (3.4) that f belongs to $B^s_{pq}[-1, 1]$ (for the definition, see [2] or [18]) if and only if

$$\left(\int_0^1 \frac{dr}{r}\left(\int_{\Lambda_r} |\bar{\partial}\mathbf{f}(z)|^p \frac{|dz|}{\rho(z, K)^{p(s-1)}}\right)^{q/p}\right)^{1/q} < \infty. \tag{3.5}$$

Such a description certainly remains valid for the classes $A^s_{pq}(G)$, which can be defined by analogy with $A^s_p(G)$. Thus, it remains to prove (3.3) and (3.4).

3.2. Proof of (3.3). We use the construction of a pseudoanalytic continuation from §2.3. Conditions of Theorem 3 are clearly satisfied, and there exists a continuation of f with estimate (2.4):

$$|\bar{\partial}\mathbf{f}(z)| \leqslant \left(c/\rho(z, K)^2\right) E_m(f, I(z))_1,$$

where $I(z)$ is the interval on ∂K with center at $\hat{z}$ and length $\rho(z, K)$. First of all, instead of approximation in L^1, we may consider approximation in L^p:

$$|\bar{\partial}\mathbf{f}(z)| \leqslant \left(c/\rho(z, K)^{1+1/p}\right) E_m(F, I(z))_p.$$

Hence, obviously,

$$\sigma_p(\mathbf{f}, r) \leqslant \frac{c}{r^{1+1/p}}\left(\int_{\Lambda_r} E_m(f, I(z))^p_p\, |dz|\right)^{1/p}.$$

Intervals $I(z)$, $z \in \Lambda_r$, clearly form a covering of ∂K. They are all of the same length r. By a well-known lemma on coverings [27], we can find a finite set of intervals $\{I_k\}_1^N$ such that $|I_k| = 5r$ for all k; for each $z \in \Lambda_r$ the interval $I(z) \subset I_k$ for some k, and each point of ∂K is covered by at most five intervals I_k. It is obvious that for a fixed k the set of those $z \in \Lambda_r$ for which $I(z) \subset I_k$ has length at most $7r$ on Λ_r. Hence

$$\int_{\Lambda_r} E_m(f, I(z))^p_p\, |dz| \leqslant 7r \sum_{k=1}^N E_m(f, I_k)^p_p \leqslant cr\omega_m(f, 10r)^p_p$$

which proves (3.3).

3.3. Proof of (3.4). We use (2.8), which implies that

$$\omega_m(f, \delta)_p \leqslant c\left(\|g\|_{L^p(\partial K)} + \|h\|_{L^p(\partial K)}\right),$$

where

$$g(z) = \iint_{|\zeta - z| < 2\delta} |\bar{\partial}\mathbf{f}(\zeta)| \frac{d\xi\, d\eta}{|\zeta - z|}, \qquad z \in \partial K,$$

$$h(z) = \iint_{|\zeta - z| > \delta} |\bar{\partial}\mathbf{f}(\zeta)| \frac{\delta^{m+1}}{|\zeta - z|^{m+2}} d\xi\, d\eta, \qquad z \in \partial K.$$

Furthermore, let

$$g_r(z) = \int_{\substack{\Lambda_r \\ |\zeta - z| < 2\delta}} |\bar{\partial}\mathbf{f}(\zeta)| \frac{|d\zeta|}{|\zeta - z|},$$

$$h_r(z) = \int_{\substack{\Lambda_r \\ |\zeta - z| > \delta}} |\bar{\partial}\mathbf{f}(\zeta)| \frac{\delta^{m+1}}{|\zeta - z|^{m+2}} |d\zeta|.$$

Then

$$\omega_m(f,\delta)_p \leq c_1\int_0^{2\delta}\|g_r\|_{L^p(\partial K)}\,dr + c_2\int_0^{\infty}\|h_r\|_{L^p(\partial K)}\,dr.$$

We shall show that

$$\begin{aligned}
&\|g_r\|_{L^p(\partial K)} \leq c\sigma_p(\mathbf{f},r)\log(2\delta/r),\\
&\|h_r\|_{L^p(\partial K)} \leq c\sigma_p(\mathbf{f},r), \qquad 0<r<\delta, \\
&\|h_r\|_{L^p(\partial K)} \leq c\sigma_p(\mathbf{f},r)(\delta^{m+1}/r^{m+1}), \qquad r>\delta.
\end{aligned}\tag{3.6}$$

It suffices to verify each of these three estimates only for $p=1$ and $p=\infty$, since for $1<p<\infty$ they follow from the Riesz-Thorin interpolation theorem [3]. For $p=1$

$$\int_{\partial K}|g_r(z)|\,|dz| = \int_{\Lambda_r}|\bar\partial\mathbf{f}(\zeta)|\,|d\zeta|\int_{\substack{\partial K\\ |z-\zeta|<2\delta}}\frac{|dz|}{|\zeta-z|} \leq c\sigma_1(\mathbf{f},r)\log\frac{2\delta}{r}.$$

For $p=\infty$

$$|g_r(z)| \leq \sigma_\infty(\mathbf{f},r)\int_{\substack{\Lambda_r\\ |\zeta-z|<2\delta}}\frac{|d\zeta|}{|\zeta-z|} \leq c\sigma_\infty(\mathbf{f},r)\log\frac{2\delta}{r},$$

which proves (3.6). The two remaining estimates for h_r can be established similarly. This proves estimate (3.4), and Theorem 5 is completely established. ●

§4. Pseudoanalytic continuation of functions from Sobolev classes

In this section we are compelled to consider the case of a domain G separately from the case of a segment $[-1,1]$. The former will now be a Radon domain (see §1.1.7).

4.1. Pseudoanalytic continuation of functions from $E_p^l(G)$.

THEOREM 6. *Let G be a Radon domain, let $1<p<\infty$, and let l be an integer. A function $f\in E^p(G)$ belongs to $E_p^l(G)$ if and only if f admits a pseudoanalytic continuation such that*(4)

$$\left(\int_{\partial G}|dz|\left(\iint_{S(z)}|\rho(\zeta,\partial G)^{-l}\bar\partial\mathbf{f}(\zeta)|^2\,d\xi\,d\eta\right)^{p/2}\right)^{1/p}<\infty. \tag{4.1}$$

PROOF. Let $f\in E_p^l(G)$. The conditions of Theorem 2 from §2 are obviously satisfied, and so there is a pseudoanalytic continuation of f such that

$$|\bar\partial\mathbf{f}(z)| \leq c\rho(z,\partial G)^l|f^{(l+1)}(z^*)|, \qquad z\in\mathbf{C}\setminus\overline{G}$$

(symmetry with respect to ∂G). But now (4.1) follows immediately from the definition of $E_p^l(G)$ by virtue of an inequality of Luzin type for G (see [17]).

Conversely, let there be given a pseudoanalytic continuation of f satisfying (4.1). Suppose for now that $l=1$. It suffices to verify that for any function $g\in L_{p'}^{\prime}(\partial G)$

$$\left|\int_{\partial G}f'(z)g(z)\,dz\right| \leq c\|g\|_{L^{p'}(\partial G)}.$$

(4) $S(z)$ is a Luzin sector (see §1.1.3).

However, by the Cauchy-Green formula

$$\int_{\partial G} f'(z)g(z)\,dz = \frac{1}{\pi}\int_{\partial G} g(z)\,dz \iint_{\mathbf{C}\setminus G} \bar\partial \mathbf{f}(\zeta)\frac{d\xi\,d\eta}{(\zeta - z)^2}$$

$$= \frac{1}{\pi}\iint_{\mathbf{C}\setminus G} \bar\partial \mathbf{f}(\zeta)\,d\xi\,d\eta \int_{\partial G} g(z)\frac{dz}{(z-\zeta)^2}.$$

Set

$$\Phi(\zeta) = \frac{1}{\pi}\int_{\partial G} g(z)\frac{dz}{\zeta - z}, \qquad \zeta \in \mathbf{C}\setminus \bar{G}.$$

Then $\Phi \in E^{p'}(\mathbf{C}\setminus\bar{G})$ and $\|\Phi\|_{E^{p'}(\mathbf{C}\setminus G)} \leqslant c\|g\|_{L^{p'}(\partial G)}$ (see §1). Consequently, upon a double application of the Hölder inequality and (1.2) we obtain

$$\left|\int_{\partial G} f'(z)g(z)\,dz\right| \leqslant \iint_{\mathbf{C}\setminus G} |\Phi'(\zeta)|\,|\bar\partial \mathbf{f}(\zeta)|\,d\xi\,d\eta$$

$$\leqslant c\int_{\partial G}|dz|\iint_{S(z)} |\Phi'(\zeta)|\,|\bar\partial \mathbf{f}(\zeta)|\frac{1}{\rho(\zeta,\partial G)}\,d\xi\,d\eta$$

$$\leqslant c\int_{\partial G}|dz|\left(\iint_{S(z)}|\Phi'(\zeta)|^2\,d\xi\,d\eta\right)^{1/2}\left(\iint_{S(z)}|\rho(\zeta,\partial G)^{-1}\bar\partial \mathbf{f}(\zeta)|^2\,d\xi\,d\eta\right)^{1/2}$$

$$\leqslant c\left[\int_{\partial G}|dz|\left(\iint_{S(z)}|\Phi'(\zeta)|^2\,d\xi\,d\eta\right)^{p'/2}\right]^{1/p'}$$

$$\times\left[\int_{\partial G}|dz|\left(\iint_{S(z)}|\rho(\zeta,\partial G)^{-1}\bar\partial \mathbf{f}(\zeta)|^2\,d\xi\,d\eta\right)^{p/2}\right]^{1/p}.$$

However, by Luzin's inequality [17] for $\mathbf{C}\setminus\bar{G}$, the first factor does not exceed $\|\Phi\|_{E^{p'}(\mathbf{C}\setminus G)}$, and the second factor is finite because of (4.1) (recall that $l = 1$). Thus our theorem is completely established in the case $l = 1$.

Finally, let (4.1) hold but $l > 1$. Then for $z \in \bar{G}$ The Cauchy-Green formula implies

$$f^{(l-1)}(z) = -\frac{(l-1)!}{\pi}\iint_{\mathbf{C}\setminus G} \bar\partial \mathbf{f}(\zeta)\frac{d\xi\,d\eta}{(\zeta - z)^l}. \tag{4.2}$$

We shall use the following elementary identity which holds for any r, $0 < r < |\zeta - z|$, and $t = \sigma + i\tau$:

$$\frac{1}{(\zeta - z)^l} = \frac{l}{\pi}\iint_{|t-\zeta|<r/4}\left(\frac{4}{r}\right)^{2l}\frac{\overline{(\zeta - t)}^{\,l-1}}{t - z}\,d\sigma\,d\tau. \tag{4.3}$$

This identity follows by applying the Cauchy-Green formula to the function of z which is equal to $(\zeta - z)^{-l}$ for $|\zeta - z| > r/4$ and to $(4/r)^{2l}\overline{(\zeta - z)}^{\,l}$ for $|\zeta - z| \leqslant r/4$.

Upon substitution of (4.3) into (4.2) for $r = \rho(\zeta, \partial G)$, we obtain

$$f^{(l-1)}(z) = \iint_{\mathbf{C}\setminus G}\mu(\zeta)\frac{d\xi\,d\eta}{\zeta - z},$$

where

$$\mu(\zeta) = -\frac{16^l l!}{\pi^2} \iint_{\mathbf{C}\setminus G} \rho(t, \partial G)^{-2l} \overline{(t-\zeta)}^{l-1} \chi(\zeta, t) \bar{\partial} \mathbf{f}(t)\, d\sigma\, d\tau,$$

$$\chi(\zeta, t) = \begin{cases} 1, & |\zeta - t| < 1/4\rho(t, \partial G), \\ 0, & |\zeta - t| \geqslant 1/4\rho(t, \partial G). \end{cases}$$

Obviously,

$$|\mu(\zeta)| \leqslant c\rho(\zeta, \partial G)^{-l-1} \iint_{|t-\zeta|<\rho(\zeta, \partial G)/3} |\bar{\partial}\mathbf{f}(t)\, d\sigma\, d\tau,$$

and thus for any $z \in \partial G$

$$\iint_{S(z)} |\rho(\zeta, \partial G)^{-1} \mu(\zeta)|^2\, d\xi\, d\eta \leqslant c \iint_{S(z)} |\rho(\zeta, \partial G)^{-l} \bar{\partial}\mathbf{f}(\zeta)|^2\, d\xi\, d\eta.$$

Hence by Theorem 1 of §2 the function $f^{(l-1)}$ possesses a pseudoanalytic continuation satisfying (4.1) for $l = 1$; and by virtue of the already established part of the theorem we have $f^{(l-1)} \in E_p^1(G)$, i.e. $f \in E_p^l(G)$. ●

4.2. Pseudoanalytic continuation of functions from $W_p^l[-1, 1]$. In the proof of Theorem 6 we used quasisymmetry with respect to ∂G to continue a given function from $E_p^l(G)$. Thus the proof of Theorem 6 cannot be transferred immediately to the case of the segment. However, this proof can be suitably modified and the final result remains valid.

THEOREM 7. *Let* $1 < p < \infty$, *and let* l *be an integer. The function* $f \in L^p[-1, 1]$ *belongs to* $W_p^l[-1, 1]$ *if and only if* f *admits a pseudoanalytic continuation such that*

$$\left(\int_{-1}^{1} dx \left(\iint_{S(x)} |\rho(\zeta, [-1, 1])^{-l} \bar{\partial}\mathbf{f}(\zeta)|^2\, d\xi\, d\eta \right)^{p/2} \right)^{1/p} < \infty. \tag{4.4}$$

PROOF. An inequality of Luzin's type remains valid for the domain $\mathbf{C}\setminus[-1, 1]$ (see [17]). Hence the entire second part of the proof of Theorem 6, i.e. the proof of the relation $f \in W_p^l[-1, 1]$ under condition (4.4), can be carried out without modifications. Thus it remains to prove that each function f from $W_p^l[-1, 1]$ has a pseudoanalytic continuation with property (4.4).

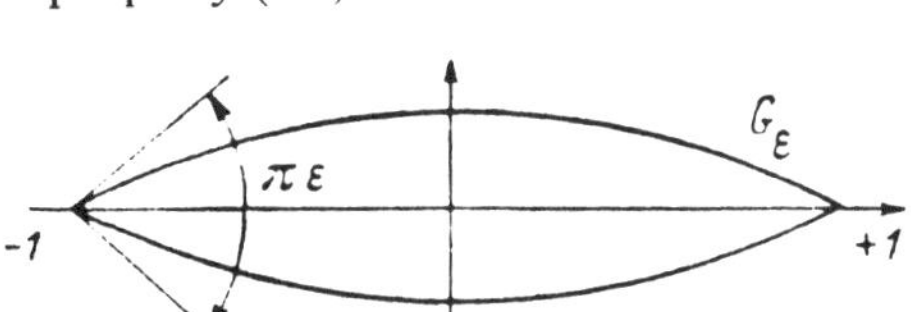

Consider the region G_ε which is a circular lune formed by two circular arcs with endpoints at ± 1. These arcs are symmetric with respect to the real axis, and at the end points they form the angle $\pi\varepsilon$ (see the figure). Consider also the conformal mapping φ_ε: $\mathbf{C}\setminus G_\varepsilon \to \mathbf{C}\setminus[-1, 1]$, $\varphi_\varepsilon(\infty) = \infty$, $\varphi_\varepsilon'(\infty) > 0$, and its inverse ψ_ε. The function φ_ε is real analytic everywhere on ∂G_ε with the exception of the vertices ± 1, and possesses the following properties:

$$\psi_\varepsilon(\pm 1) = \pm 1, \qquad \varphi_\varepsilon(z) = \overline{\varphi_\varepsilon(\bar{z})} = -\varphi_\varepsilon(-z),$$

and near the point $+1$

$$|\varphi_\varepsilon(z)-1| \asymp |z-1|^{2/(2-\varepsilon)}, \tag{4.5}$$

$$|\varphi_\varepsilon'(z)| \asymp |z-1|^{\varepsilon/(2-\varepsilon)},$$

$$|\varphi_\varepsilon^{(k)}(z)| \leqslant c_k |z-1|^{2/(2-\varepsilon)-k}, \quad k=1,3,\ldots. \tag{4.6}$$

Let $f \in W_p^l[-1,1]$. Then $f = f_0 + P$, where P is a polynomial of degree at most $2l$ and $f_0(\pm 1) = f_0'(\pm 1) = \cdots = f_0^{(l-1)}(\pm 1) = 0$.

Define the function g on ∂G_ε by the equality

$$g(z) = \varphi_\varepsilon'(z)^{1/p-l} f_0[\varphi_\varepsilon(z)], \qquad z \in \partial G_\varepsilon.$$

Then a direct calculation shows that g has $(l-1)$ continuous derivatives on ∂G_ε with $g^{(l-1)}$ being absolutely continuous, and $g^{(l)} \in L^p(\partial G)_\varepsilon$. Assume, for instance, that $l=1$. Then

$$g'(z) = (1/p-1)\varphi_\varepsilon'(z)^{1/p} f_0[\varphi_\varepsilon(z)]\varphi_\varepsilon''(z) + \varphi_\varepsilon'(z)^{1/p} f_0'[\varphi_\varepsilon(z)]$$

and the relation $g' \in L^p(\partial G_\varepsilon)$ follows by Hardy's inequality from (4.5), (4.6) and the condition $f' \in L^p[-1,1]$ upon the substitution of the variable $z = \psi_\varepsilon(x)$. Therefore, there exists a decomposition $g = g_+ + g_-$, where $g_+ \in E_p^l(G_\varepsilon)$, $g_- \in E_p^l(\mathbf{C}\setminus G_\varepsilon)$.

By Theorem 2 the function g_+ admits a pseudoanalytic continuation such that

$$|\bar\partial \mathbf{g}_+(z)| \leqslant c\rho(z,\partial G_\varepsilon)^l |g_+^{(l+1)}(z^*)|, \qquad z \in \mathbf{C}\setminus G_\varepsilon,$$

where the quasiconformal symmetry $z \to z^*$ is understood with respect to ∂G_ε. For $\varepsilon < 2/p'$ the function $|\varphi_\varepsilon'|$ satisfies the Muckenhaupt condition (A_p) on ∂G_ε (see § 1.3). Hence for such ε's an inequality of Luzin type with weight $|\varphi_\varepsilon'|$ holds in G_ε (see [17]). This inequality will be applied to g. If we take into account that φ_ε' is an outer function [21] and introduce the substitution $z \to z^*$, we immediately obtain the estimate

$$\int_{\partial G_\varepsilon} |\varphi_\varepsilon'(z)||dz| \left(\iint_{S(z)} |\varphi_\varepsilon'(\zeta)^{-1/p} \rho(\zeta,\partial G_\varepsilon)^{-l} \bar\partial \mathbf{g}_+(\zeta)|^2 \, d\xi\, d\eta \right)^{p/2} < \infty. \tag{4.7}$$

Here $S(z)$ denotes the Luzin sector for G_ε.

It is now clear that the function $\mathbf{g}_0 = \mathbf{g}_- + \mathbf{g}_+$ belongs in $\mathbf{C}\setminus G_\varepsilon$ to $W_p^l(\mathbf{C}\setminus \overline{G}_\varepsilon)$, that its boundary condition on ∂G_ε coincides with g, and that $\bar\partial \mathbf{g}_0 = \bar\partial \mathbf{g}_+$, because $\bar\partial g_- = 0$.

Let us now define the function F in $\mathbf{C}\setminus[-1,1]$ by the formula

$$F(z) = \psi_\varepsilon'(z)^{1/p-l} \mathbf{g}_0[\psi_\varepsilon(z)] + P(z).$$

It is easy to verify that for a sufficiently small ε (not depending, of course, on f) we have $F \in W_q^1(\mathbf{C}\setminus[-1,1])$ for some q, $1 < q < \infty$, and, furthermore, that the boundary value of F on $[-1,1]$ coincides with f. But, obviously,

$$|\bar\partial F(z)| \leqslant c|\psi_\varepsilon'(z)|^{1/p-l+1} |\bar\partial \mathbf{g}_+[\psi_\varepsilon(z)]|.$$

Thus F is a pseudoanalytic continuation of f, and (4.4) follows immediately from (4.7) by means of the substitution $z = \psi_\varepsilon(t)$.

§5. A constructive characterization of Besov classes

5.1. A preliminary discussion. In this section we again need the conformal mapping $\varphi\colon \mathbf{C}\setminus K \to \mathbf{C}\setminus\overline{\mathbf{D}}$, $\varphi(\infty)=\infty$, its level curves $L_r=\{z, |\varphi(z)|=1+r\}$, and the distance $\rho(z, L_r)$ from the point $z\in\partial K$ to L_r.

Let $1\leqslant p\leqslant\infty$ and $s>0$. For any admissible function f on K we set

$$E_n(f)=E_n(f)_{p,s}=\inf\left(\int_{\partial K}|\rho(z, L_{1/n})^{-s}(f(z)-P_n(z))|^p\,|dz|\right)^{1/p}, \quad (5.1)$$

where the infimum is taken over all polynomials P_n of degree $n=1,2,\ldots$. If $K=[-1,1]$ then $\rho(x, L_r)\asymp r\sqrt{1-x^2}+r^2$, and (5.1) turns into

$$E_n(f)=\inf\left(\int_{-1}^{1}\left|\frac{f(x)-P_n(x)}{\left(\sqrt{1-x^2}/n+1/n^2\right)^s}\right|^p dx\right)^{1/p}$$

(cf. [7] and [8]).

It is well known that for $p=\infty$ the Lipschitz condition

$$\omega_m(f,\delta)_p\leqslant c\delta^s, \qquad \delta>0, \quad (5.2)$$

is equivalent to the condition

$$E_n(f)_{p,s}\leqslant c<\infty, \quad (5.3)$$

i.e. for $p=\infty$ condition (5.3) is a constructive characterization of the Lipschitz class of order s (see [4], [6] and [28] for the case of the segment, and [5] and [14] for the case of a domain).

Attempts to generalize this result to other values of p met with unexpected difficulties. In [8] it was shown that for $p<\infty$ conditions (5.2) and (5.3) are not equivalent. The class of functions satisfying (5.2) (in the case of the segment it is the Nikol'skiĭ class $H_p^s[-1,1]$) apparently does not allow a constructive characterization by means of ordinary polynomial approximations (in [8] it was suggested to use piecewise polynomial aggregates for a constructive description of H_p^s). What is the cause of this phenomenon?

Let $\{P_n\}$ be a sequence of polynomials of best approximation of f in the sense of (5.1). In Theorem 4 we constructed a pseudoanalytic continuation of f such that

$$|\bar\partial \mathbf{f}(z)|\leqslant (c/\rho(z,K))\,|P_{2^{n+1}}(z)-P_{2^n}(z)|$$

for $z\in\mathbf{C}\setminus K$ and $2^{-n}\leqslant|\varphi(z)|-1\leqslant 2^{-n+1}$.

Consider the function Φ:

$$\Phi(z)=\frac{P_{2^{n+1}}(z)-P_{2^n}(z)}{\varphi(z)^{2^{n+2}}}\varphi'(z_{\langle 2^{-n}\rangle})^s, \qquad z\in\mathbf{C}\setminus K$$

(the conformal shift is defined by $z_{\langle r\rangle}\overset{\text{def}}{=}\psi((1+r)\varphi(z))$; see §1.1). The function Φ is analytic in $\mathbf{C}\setminus K$, and $\Phi(\infty)=0$. The maximum principle then implies that for each $r>0$

$$\left(\int_{L_r}|\Phi(z)|^p\,|dz|\right)^{1/p}\leqslant\left(\int_{\partial K}|\Phi(z)|^p\,|dz|\right)^{1/p}.$$

If $z \in \partial K$, then, according to (1.4),

$$|\Phi(z)| \asymp |P_{2^{n+1}}(z) - P_{2^n}(z)| \cdot 2^{-ns}\rho(z, L_{2^{-n}})^{-s};$$

$$\left(\int_{\partial K} |\Phi(z)|^p |dz|\right)^{1/p} \leq c \cdot 2^{-ns} E_n(f)_{p,s}.$$

On the other hand, when $2^{-n} \leq r \leq 2^{-n+1}$ and $z \in L_r$, (1.5) implies that

$$|\Phi(z)| \asymp |P_{2^{n+1}}(z) - P_{2^n}(z)| \cdot 2^{-ns}\rho(z, \partial K)^{-s},$$

$$\left(\int_{L_r} |\Phi(z)|^p |dz|\right)^{1/p} \asymp c \cdot 2^{-sn}\left(\int_{L_r} |\rho(z, \partial K)^{-s}(P_{2^{n+1}}(z) - P_{2^n}(z))|^p |dz|\right)^{1/p}.$$

Hence, finally,

$$S_p(\mathbf{f}, r) \stackrel{\text{def}}{=} \left(\int_{L_r} |\bar{\partial}\mathbf{f}(z)\rho(z, \partial K)^{s-1}|^p |dz|\right)^{1/p} \leq cE_{2^n}(f)_{p,s}. \tag{5.4}$$

In other words, a knowledge of $E_n(f)$ allows us to estimate the quantity $S_p(\mathbf{f}, r)$, i.e. the L^p-norm of $\bar{\partial}\mathbf{f}/\rho^{s-1}$ on the L_r-level curve of the conformal mapping. At the same time, by (3.3), a knowledge of the modulus of continuity $\omega_m(f, \delta)_p$ leads to an estimate of $\sigma_p(\mathbf{f}, r)$, i.e. of the L^p-norm of the same function $\bar{\partial}\mathbf{f}/\rho^{s-1}$ on the Λ_r-level curve of the usual distance. This enables us to understand the disagreement between (5.2) and (5.3). Indeed, they are related to estimates on essentially different families of curves. And we can also understand the positive answer when $p = \infty$: In this case all estimates are point estimates, and do not depend on any curves.

Now let us consider conditions more general then (5.2) and (5.3), namely

$$\left(\sum_{n=1}^{\infty} \frac{1}{n} E_n(f)_{p,s}^q\right)^{1/q} < \infty, \tag{5.5}$$

$$\left(\int_2^1 \left(\frac{\omega_m(f, \delta)_p}{\delta^s}\right)^q \frac{d\delta}{\delta}\right)^{1/q} < \infty, \tag{5.6}$$

where $1 \leq q \leq \infty$.

Obviously, in view of (3.3) and (5.4) conditions (5.5) and (5.6) lead respectively to estimates of $\bar{\partial}\mathbf{f}$ in certain mixed $L^{(p,q)}$-norms, i.e. to iterated integrals taken first over L_r or over Λ_r, and then over r from 0 to 1. When $q \neq p$ these estimates cannot be equivalent. However, when $q = p$ the mixed norm and the iterated integral turn respectively into a usual L^p-norm and into the double integral over the area of $\mathbf{C} \setminus K$. Therefore we may expect that, unlike the class $B^s_{pq}[-1, 1]$ (whose definition is (5.6) when $q \neq p$), the class $B^s_p[-1, 1]$ (whose definition is (5.6) when $q = p$) does allow a constructive characterization of type (5.5). The theorem of §5.2 will confirm this expectation.

5.2. The main theorem.

THEOREM 8. *Let* $1 \leq p \leq \infty$ *and* $s > 0$, *and let* f *be an admissible function on* K. *Then* $f \in A^s_p(K)$ *if and only if*

$$\left(\sum_{n=1}^{\infty} \frac{1}{n} E_n(f)_{p,s}^p\right)^{1/p} < \infty. \tag{5.7}$$

PROOF. We shall show that (5.7) is equivalent to estimate (3.1) of the pseudoanalytic continuation **f**. Then the statement of our theorem will become a consequence of Theorem 5 from §3.

Let (5.7) hold. Since a weight is now present, the best approximations $E_n(f)$ do not necessarily decrease as n increases. However, it is not difficult to conclude from (1.6) that

$$0 < c_1(\lambda) < E_{\lambda n}(f)/E_n(f) \leqslant c_2(\lambda) < \infty, \qquad \lambda > 0.$$

Hence (5.7) is equivalent to the condition

$$\left(\sum_{n=0}^{\infty} E_{2^n}(f)_{p,s}^p\right)^{1/p} < \infty. \tag{5.8}$$

In view of (5.8) and (5.4) the conditions of Theorem 4 of §2 are clearly satisfied, and there exists a pseudoanalytic continuation of f such that, by (5.4),

$$S_p(\mathbf{f}, r) \leqslant cE_{2^n}(f), \qquad 2^{-n} \leqslant r \leqslant 2^{-n+1}$$

and

$$\left(\int_0^1 S_p(\mathbf{f}, r)^p \frac{dr}{r}\right)^{1/p} \leqslant c\left(\sum_{n=0}^{\infty} E_{2^n}(f)^p\right)^{1/p} < \infty, \tag{5.9}$$

which according to (1.3) and (1.4) is equivalent to (3.1) or (3.1′).

Conversely, let there exist a pseudoanalytic continuation of f with estimates (5.9). Condition (5.7) is an obvious consequence of the sharper estimate

$$E_n(f)_{p,s} \leqslant c_1 \int_0^{1/n} S_p(\mathbf{f}, r)(nr)^{\varepsilon s} \frac{dr}{r} + c_2 \int_{1/n}^{\infty} S_p(\mathbf{f}, r) \frac{1}{(nr)^m} \frac{dr}{r}, \tag{5.10}$$

where $\varepsilon > 0$ and m is arbitrarily large. Thus it remains to prove (5.10). We shall use the approximation of the Cauchy kernel by polynomials described in §1.4. Set, in the notation of that subsection,

$$P_n(z) = -\frac{1}{\pi} \iint_{\mathbf{C} \setminus K} \bar{\partial}\mathbf{f}(\zeta) K_n(\zeta, z)\, d\xi\, d\eta.$$

Then, in view of §1.4, for $z \in \partial K$ and $r = |\varphi(\zeta)| - 1$ we have

$$|f(z) - P_n(z)| \leqslant U(z) + V(z) + W_1(z) + W_2(z),$$

$$U(z) = \iint_{|\zeta - z| < \rho(z, L_{1/n})} |\bar{\partial}\mathbf{f}(\zeta)| \frac{d\xi\, d\eta}{|\zeta - z|}, \tag{5.11}$$

$$V(z) = \iint_{|\zeta - z| < \rho(z, L_{1/n})} |\bar{\partial}\mathbf{f}(\zeta)|\, |K_n(\zeta, z)|\, d\xi\, d\eta, \tag{5.12}$$

$$W_1(z) = \iint_{\substack{|\zeta - z| < \rho(z, L_{1/n}) \\ r < 1/n}} |\partial\mathbf{f}(\zeta)| \frac{\rho(\zeta, L_{1/n})^m}{|\zeta - z|^{m+1}} d\xi\, d\eta, \tag{5.13}$$

$$W_2(z) = \iint_{r > 1/n} |\partial\mathbf{f}(\zeta)| \frac{\rho(\zeta, L_{2/n})^m}{|\zeta - z|^{m+1}} d\xi\, d\eta. \tag{5.14}$$

First of all, for $|\zeta - z| < \rho(z, L_{1/n})$

$$|K_n(\zeta, z)| \leqslant \frac{c}{\rho(z, L_{1/n})} \leqslant \frac{c}{|\zeta - z|}.$$

Hence $V(z) \leqslant cU(z)$. According to (1.4), (1.6) and (1.7), for the same ζ we have

$$\rho(\zeta, K) \leqslant c(nr)^{\varepsilon}\rho(z, L_{1/n}), \qquad \varepsilon > 0,$$

and, in view of (1.3),

$$U(z)\rho(z, L_{1/n})^{-s} \leqslant c\int_0^{1/n}(nr)^{\varepsilon s}\frac{dr}{r} \iint\limits_{\substack{L_r \\ |\zeta - z| < \rho(z, L_{1/n})}} |\bar{\partial}\mathbf{f}(\zeta)|\,\rho(\zeta, K)^{1-s}\frac{|d\zeta|}{|\zeta - z|}.$$

Set

$$g_r(z) = \iint\limits_{\substack{L_r \\ |\zeta_r - z| < \rho(z, L_{1/n})}} |\partial\mathbf{f}(\zeta)|\,\rho(\zeta, K)^{1-s}\frac{|d\zeta|}{|\zeta - z|}.$$

It is easy to verify (e.g., for $p = 1$ and $p = \infty$ and then by the Riesz-Thorin theorem) that

$$\|g_r\|_{L^p(\partial K)} \leqslant cS_p(\mathbf{f}, r)\log(2/nr),$$

and so, by Minkowski's inequality,

$$\left(\int_{\partial K} |U(z)\rho(z, L_{1/n})^{-s}|^p\,|dz|\right)^{1/p} \leqslant c\int_0^{1/n} S_p(\mathbf{f}, r)(nr)^{\varepsilon s/2}\frac{dr}{r}. \tag{5.15}$$

Certainly, this estimate also holds for $V(z)$.

Let us go over to the estimate of $W_1(z)$. Obviously,

$$W_1(z)\rho(z, L_{1/n})^{-s} \leqslant c\int_0^{1/n}\frac{dr}{r}\int_{L_r, |\zeta - z| > \rho(z, L_{1/n})} |\rho(\zeta, K)^{1-s}\partial\mathbf{f}(\zeta)|\,\rho(\zeta, L_{2/n})^m$$
$$\times \rho(\zeta, K)^s\rho(z, L_{1/n})^{-s}\frac{|d\zeta|}{|\zeta - z|^{m+1}}.$$

According to (1.7) and (1.8) we have

$$\rho(\zeta, L_{2/n}) \leqslant c\cdot|\zeta - z|^{\gamma}\rho(z, L_{1/n})^{1-\gamma}, \qquad 0 < \gamma < 1,$$
$$\rho(\zeta, K) \leqslant c(nr)^{\varepsilon}\rho(z, L_{1/n})^{1-\gamma}|\zeta - z|^{\gamma}.$$

Hence

$$W_1(z)\rho(z, L_{1/n})^{-s}$$
$$\leqslant c\int_0^{1/n}(nr)^{\varepsilon}\frac{dr}{r}\int\limits_{\substack{L_r \\ |\zeta - z| > \rho(z, L_{1/n})}} |\rho(\zeta, K)^{1-s}\bar{\partial}\mathbf{f}(\zeta)|\,\frac{\rho(z, L_{1/n})^q}{|\zeta - z|^{q+1}}\,|d\zeta|,$$
$$q = m(1 - \gamma).$$

We introduce the function

$$h_r(z) = \int\limits_{\substack{L_r \\ |\zeta - z| > \rho(z, L_{1/n})}} |\rho(\zeta, K)^{1-s}\bar{\partial}\mathbf{f}(\zeta)|\,\frac{\rho(z, L_{1/n})^q}{|\zeta - z|^{q+1}}\,|d\zeta|,$$

and we shall show that $\|h_r\|_{L^p(\partial K)} \leqslant cS_p(\mathbf{f}, r)$. For $p = \infty$ this follows from the inequality

$$\int_{\substack{L_r \\ |\zeta - z| > a}} \frac{a^q}{|\zeta - z|^{q+1}} |d\zeta| \leqslant c < \infty. \tag{5.16}$$

For $p = 1$ we get

$$\int_{\partial K, |z - \zeta| > \rho(z, L_{1/n})} \frac{\rho(z, L_{1/n})^q}{|\zeta - z|^{q+1}} |dz| \leqslant c < \infty,$$

which, in view of (1.7), reduces to an inequality of type (5.16) with another q. Finally, for the proof of (5.16) we decompose the integral into a sum of integrals over the portions

$$Q_k = \{\zeta \in L_r, 2^k a \leqslant |\zeta - z| \leqslant 2^{k+1} a\}.$$

However, within Q_k we have $|\zeta - z| \asymp 2^k a$, and the length of Q_k does not exceed $c \cdot 2^k a$. Inequality (5.16) is established. By the Riesz-Thorin theorem this gives us the desired estimate of h_r.

Now, by Minkowski's inequality we obtain

$$\left(\int_{\partial K} |W_1(z)\rho(z, L_{1/n})^{-s}|^p |dz|\right)^{1/p} \leqslant c\int_0^{1/n} S_p(\mathbf{f}, r)(nr)^{\varepsilon s} \frac{dr}{r}. \tag{5.17}$$

Finally,

$$W_2(z)\rho(z, L_{1/n})^{-s} \leqslant c\int_{1/n}^{\infty} \frac{1}{(nr)^m} \frac{dr}{r}$$
$$\times \int_{L_r} |\bar\partial \mathbf{f}(\zeta)| \rho(\zeta, K)^{m+1} \rho(z, L_{1/n})^{-s} \frac{|d\zeta|}{|\zeta - z|^{m+1}}.$$

In view of (1.6)–(1.8) we have

$$\rho(\zeta, K)^{m+s} \rho(z, L_{1/n})^{-s} |z - \zeta|^{-m-1} \leqslant c\rho(\zeta, K)^q |\zeta - z|^{-q-1} (nr)^\alpha,$$

where $q = m + s\delta/(1 - \delta)$ (α and δ from (1.6) and (1.8)).

Now the estimate of $W_2(z)$ can be completed in the same way as that of $W_1(z)$, and we obtain (replacing $m - \alpha$ by m, since m was arbitrary) that

$$\left(\int_{\partial K} |W_2(z)\rho(z, L_{1/n})^{-s}|^p |dz|\right)^{1/p} \leqslant c\int_{1/n}^{\infty} S_p(\mathbf{f}, r)\frac{1}{(nr)^m} \frac{dr}{r}. \tag{5.18}$$

Estimates (5.15), (5.17) and (5.18) together with the obvious inequality

$$E_n(f) \leqslant \left(\int_{\partial K} |\rho(z, L_{1/n})^{-s}(f(z) - P_n(z))|^p |dz|\right)^{1/p}$$

complete the proof of (5.10) and thus also of the entire theorem. ●

§6. A constructive characterization of Sobolev classes

6.1. The main theorem. In this section K will denote either the segment $[-1, 1]$ or the closure of a Radon domain G. The constructive characterization of Besov classes given in §5 was modeled on the corresponding theorem on Lipschitz classes on

$[-1, 1]$. The model for the main theorem of this section is the theorem on binary expansion in harmonic analysis [3].

THEOREM 9. *Let* $1 < p < \infty$, *let* l *be an integer and let* f *be an admissible function on* K. *Then* $f \in E_p^l(K)$ *if and only if there exists a sequence of polynomials* $\{P_{2^n}\}_0^\infty$ *such that* $\deg P_{2^n} \leqslant 2^n$, $n = 0, 1, \ldots$, *and*

$$\int_{\partial K} |dz| \left(\sum_{n=0}^{\infty} |f(z) - P_{2^n}(z)|^2 \rho(z, L_{2^{-n}})^{-2l} \right)^{p/2} < \infty. \tag{6.1}$$

PROOF. We shall use the characterization of Sobolev classes given in Theorems 6 and 7, i.e. $f \in E_p^l(K)$ if and only if f has a pseudoanalytic continuation such that

$$\int_{\partial K} |dz| \left(\iint_{S(z)} |\rho(\zeta, K)^{-l} \bar\partial \mathbf{f}(\zeta)|^2 \, d\xi \, d\eta \right)^{p/2} < \infty, \tag{6.2}$$

where $S(z)$ is the Luzin sector in $\mathbf{C} \setminus K$ with vertex at z.

6.2. Proof of the sufficiency. Let (6.1) hold. Then Theorem 4 gives a pseudoanalytic continuation of f such that

$$|\bar\partial \mathbf{f}(z)| \leqslant (c/\rho(z, K)) \, |P_{2^{n+1}}(z) - P_{2^n}(z)|$$

for $r \in \mathbf{C} \setminus K$ and $2^{-n} \leqslant |\varphi(z)| - 1 \leqslant 2^{-n+1}$.

In what follows we shall need the partition of the sector $S(z)$ into the parts

$$S_n(z) = \{\zeta \in S(z), 2^{-n} \leqslant |\varphi(z)| - 1 \leqslant 2^{-n+1}\}.$$

Let us mention some of the properties of $S_n(z)$:

a) $\rho(\zeta, K) \asymp \rho(z, L_{2^{-n}})$ for $\zeta \in S_n(z)$.

b) If $|S_n(z)|$ is the area of $S_n(z)$, then $|S_n(z)| \asymp \rho(z, L_{2^{-n}})^2$.

Set

$$a_n(z) = |P_{2^{n+1}}(z) - P_{2^n}(z)| \, \rho(z, L_{2^{-n}})^{-l},$$

$$b_n(z) = \left(\iint_{S_n(z)} |\bar\partial \mathbf{f}(\zeta)|^2 \rho(\zeta, K)^{-2l} \, d\xi \, d\eta \right)^{1/2}, \qquad z \in \partial K, n = 0, 1, \ldots. \tag{6.3}$$

LEMMA 2. *Let* $Mf(z)$ *denote the maximal function of* f *on* ∂K (*see* §1.3). *Then* $b_n(z) \leqslant cMa_n(z)$.

If condition (6.3) is satisfied, then in view of the maximal theorem of Fefferman and Stein (see §1.3) we obtain

$$\int_{\partial K} \left(\sum_{n=0}^{\infty} b_n(z)^2 \right)^{p/2} |dz| \leqslant c \int_{\partial K} \left(\sum_{n=0}^{\infty} a_n(z)^2 \right)^{p/2} |dz|.$$

However, in view of (6.1), the right-hand side is obviously finite (because $|P_{2^{n+1}} - P_{2^n}| \leqslant |P_{2^{n+1}} - f| + |P_{2^n} - f|$), and

$$\sum_{n=0}^{\infty} b_n(z)^2 = \iint_{S(z)} |\rho(\zeta, K)^{-l} \bar\partial \mathbf{f}(\zeta)|^2 \, d\xi \, d\eta,$$

which proves the sufficiency part of Theorem 9. Thus it remains to prove Lemma 2.

We consider a function g analytic in $\mathbb{C}\setminus K$:

$$g(z)=\frac{P_{2^{n+1}}(z)-P_{2^n}(z)}{\varphi(z)^{2^{n+1}}}\varphi'(z_{\langle 2^{-n}\rangle})^l,\qquad z\in\mathbb{C}\setminus K.$$

If $\zeta\in\partial K$, then, as in §5,

$$|g(\zeta)|\asymp 2^{-nl}|P_{2^{n+1}}(\zeta)-P_{2^n}(\zeta)|\rho(\zeta,L_{2^{-n}})^{-l}.$$

If $\zeta\in S_n(z)$, then

$$|g(\zeta)|\asymp 2^{-nl}|P_{2^{n+1}}(\zeta)-P_{2^n}(\zeta)|\rho(\zeta,K)^{-l}.$$

Applying to g in $\mathbb{C}\setminus K$ the maximal estimate from [17], we obtain

$$|g(\zeta)|\leqslant\frac{c_1}{\rho(\zeta,K)}\int_{I(\zeta)}|g(\tau)||d\tau|+c_2\int_{\partial K\setminus I(\zeta)}|g(\tau)|\frac{\rho(\zeta,K)^m}{|\tau-\zeta|^{m+1}}|d\tau|;$$

it should be recalled that $I(\zeta)$ denotes the arc on ∂K with center $\hat{\zeta}$ (which is the closest point on ∂K to the point ζ) and length $\rho(\zeta,K)$, and m is arbitrarily large. This estimate immediately yields

$$\begin{aligned}b_n(z)\leqslant{}&\frac{c_1}{\rho(z,L_{2^{-n}})}\int_{|\zeta-z|<\rho(z,L_{2^{-n}})}a_n(\zeta)|d\zeta|\\&+c_2\int_{|\zeta-z|>\rho(z,L_{2^{-n}})}a_n(\zeta)\frac{\rho(z,L_{2^{-n}})^m}{|\zeta-z|^{m+1}}|d\zeta|.\end{aligned}$$

However, the right-hand side of this inequality obviously does not exceed $cMa_n(z)$ (or even $cM_{\rho(z,L_{2^{-n}})}a_n(z)$). Thus Lemma 2, and with it also the sufficiency part of Theorem 9, are established.

6.3. Proof of the necessity. We shall use again the approximation construction of §1.4. As in §5, we set

$$P_{2^n}(z)=-\frac{1}{\pi}\iint_{\mathbb{C}\setminus K}\bar{\partial}\mathbf{f}(\zeta)K_{2^n}(\zeta,z)\,d\xi\,d\eta.$$

Then again

$$|f(z)-P_{2^n}(z)|\leqslant U(z)+V(z)+W_1(z)+W_2(z),$$

where U, V, W_1 and W_2 are defined by (5.11)–(5.14).

It was shown in §5 that $V(z)\leqslant cU(z)$. Hence when assuming the validity of condition (6.2) we must estimate the contribution of V, W_1 and W_2 to (6.1). First, let us estimate $U(z)$. Let us apply the formula of iterated integration to (5.11), and recall that $S(\tau)=\bigcup_k S_k(\tau)$ and, furthermore, that $\zeta\in S_k(\tau)$ can satisfy the condition

$$|\zeta-z|<\rho(z,L_{2^{-n}})$$

only when $|\tau-z|<\lambda_1\rho(z,L_{2^{-n}})$ and $k>\lambda_2 n$, where $\lambda_1,\lambda_2>0$ do not depend on z or n. Hence

$$U(z)\leqslant\int_{|\tau-z|<\lambda_1\rho(z,L_{2^{-n}})}|d\tau|\sum_{k>\lambda_2 n}\iint_{S_k(\tau)}|\bar{\partial}\mathbf{f}(\zeta)|\frac{1}{|\zeta-z|}\frac{1}{\rho(\zeta,K)}d\xi\,d\eta.$$

However, $|\zeta - z| \asymp |\tau - z| + \rho(\zeta, K) \geqslant \rho(\zeta, K)$, and for $\zeta \in S_k(\tau)$ we have $\rho(\zeta, K) \asymp \rho(\tau, L_{2^{-k}})$. Applying to each integral over $S_k(\tau)$ the Schwarz inequality, we obtain

$$\rho(z, L_{2^{-n}})^{-l} U(z) \leqslant \sum_{k > \lambda_2 n} \int_{|\tau - z| < \lambda_1 \rho(z, L_{2^{-n}})} b_k(\tau) \frac{\rho(\tau, L_{2^{-k}})^{l-1}}{\rho(z, L_{2^{-n}})^l} |d\tau| .$$

Furthermore, according to (1.6)

$$\rho(\tau, L_{2^{-k}}) \leqslant 2^{-\alpha(k-n)} \rho(\tau, L_{2^{-n}}) \asymp 2^{-\alpha(k-n)} \rho(z, L_{2^{-n}}),$$

and thus also

$$\begin{aligned} \rho(z, L_{2^{-n}})^{-l} U(z) &\leqslant \sum_{k > \lambda_2 n} c \frac{2^{-\varepsilon(k-n)}}{\rho(z, L_{2^{-n}})} \int_{|\tau - z| < \lambda_1 \rho(z, L_{2^{-n}})} b_k(\tau) |d\tau| \\ &\leqslant c \sum_{k > \lambda_2 n} 2^{-\varepsilon(k-n)} M b_k(z). \end{aligned} \tag{6.4}$$

Let us go over to the estimate of $W_1(z)$. The same computations must be carried out for the case where $k > n$ but $|\tau - z| > \lambda_2 \rho(z, L_{2^{-n}})$:

$$W_1(z) \leqslant \int_{|\tau - z| > \lambda_2 \rho(z, L_{2^{-n}})} |d\tau| \sum_{k > n} \iint_{S_k(\tau)} |\bar{\partial} \mathbf{f}(\zeta)| \frac{1}{\rho(\zeta, K)} \frac{\rho(\zeta, L_{2^{-n-1}})^m}{|\zeta - z|^{m+1}} d\xi\, d\eta.$$

Now $|\zeta - z| \asymp |\tau - z| + \rho(\zeta, K)$, $\rho(\zeta, L_{2^{-n-1}}) \asymp \rho(z, L_{2^{-n}})$ and

$$\rho(z, L_{2^{-n}})^{-l} W_1(z) \leqslant c \sum_{k > n} \int_{|\tau - z| > \rho(z, L_{2^{-n}})} b_k(\tau) \frac{\rho(\tau, L_{2^{-k}})^l \rho(\tau, L_{2^{-n}})^m}{|\tau - z|^{m+1} \rho(z, L_{2^{-n}})^l} |d\tau| .$$

Moreover, $\rho(\tau, L_{2^{-k}}) \leqslant 2^{-\alpha(k-n)} \rho(\tau, L_{2^{-n}})$ and, by (1.7),

$$\frac{\rho(\tau, L_{2^{-n}})}{\rho(z, L_{2^{-n}})} \leqslant c \left(\frac{|\tau - z|}{\rho(z, L_{2^{-n}})} \right)^\gamma, \qquad 0 < \gamma < 1.$$

Therefore,

$$\rho(z, L_{2^{-n}})^{-l} W_1(z) \leqslant c \sum_{k > n} 2^{-\varepsilon(k-n)} \int_{|\tau - z| > \rho(z, L_{2^{-n}})} b_k(\tau) \frac{\rho(z, L_{2^{-n}})^{m(1-\gamma) - \gamma l}}{|\tau - z|^{m(1-\gamma) - \gamma l + 1}} |d\tau|$$

and for m sufficiently large, more precisely for $m > (\gamma/(1-\gamma))l$,

$$\rho(z, L_{2^{-n}})^{-l} W_1(z) \leqslant c \sum_{k > n} 2^{-\varepsilon(k-n)} M b_k(z). \tag{6.5}$$

Thus it remains to estimate $W_2(z)$. We proceed in the same way as before but with $k \leqslant n$:

$$W_2(z) \leqslant \int_{\partial K} |d\tau| \sum_{k=0}^{n} \iint_{S_k(\tau)} 2^{(k-n)m} \frac{\rho(\zeta, K)^{m-1}}{|\zeta - z|^{m+1}} |\bar{\partial} \mathbf{f}(\zeta)|\, d\xi\, d\eta,$$

$$W_2(z) \leqslant \sum_{k=0}^{n} \frac{c}{2^{(n-k)m}} \int_{\partial K} b_k(\tau) \frac{\rho(\tau, L_{2^{-k}})^{m+l}}{\rho(\tau, L_{2^{-k}}) + |\tau - z|^{m+1}} |d\tau| .$$

As above it is now easy to verify that for m large enough (again for $m > \gamma l/(1-\gamma)$)

$$W_2(z) \leqslant \sum_{k=0}^{n} \frac{c}{2^{(n-k)m}} \rho(z, L_{2^{-k}})^l M b_k(z).$$

However, $\rho(z, L_{2^{-k}}) \leqslant 2^{(n-k)\beta}\rho(z, L_{2^{-n}})$, and, finally,

$$\rho(z, L_{2^{-n}})^{-l} W_2(z) \leqslant \sum_{k=0}^{n} \frac{c}{2^{(n-k)\varepsilon}} Mb_k(z), \qquad \varepsilon = m - \beta l. \tag{6.6}$$

Estimates (6.4)–(6.6) together imply the inequality

$$\rho(z, L_{2^{-n}})^{-l} |f(z) - P_{2^n}(z)| \leqslant c_1 \sum_{k=0}^{n} 2^{(k-n)\varepsilon} Mb_k(z) + c_2 \sum_{k=n+1}^{\infty} 2^{(n-k)\varepsilon} Mb_k(z).$$

It now remains to apply the following elementary statement:

LEMMA 3. *Let* $\{x_n\}_0^\infty$ *and* $\{y_n\}_0^\infty$ *be sequences of positive terms such that*

$$y_n \leqslant \sum_{k=0}^{n} q^{n-k} x_k + \sum_{k=n+1}^{\infty} q^{k-n} x_k, \qquad 0 < q < 1, n = 0, 1, \ldots.$$

Then

$$\sum_{n=0}^{\infty} y_n^2 \leqslant \frac{2}{1-q} \sum_{n=0}^{\infty} x_n^2.$$

PROOF. Indeed, it is obvious that for all n

$$\begin{gathered} y_n \leqslant \sup_k \left(\sum_{k=0}^{n} q^{n-k} + \sum_{k=n+1}^{\infty} q^{k-n} \right) \sup_k x_k \leqslant \sup x_k \frac{2}{1-q}, \\ \sup_n y_n \leqslant \frac{2}{1-q} \sup x_n \end{gathered} \tag{6.7}$$

and

$$\sum_{n=0}^{\infty} y_n \leqslant \sum_{n=0}^{\infty} x_k \left(\sum_{n=k}^{\infty} q^{n-k} + \sum_{n=0}^{k-1} q^{k-n} \right) \leqslant \frac{2}{1-q} \sum_{n=0}^{\infty} x_n. \tag{6.8}$$

Now the estimate of $\Sigma_0^\infty y_n^2$ follows from (6.7), (6.8) and the Riesz-Thorin theorem.

Applying Lemma 3 to the sequences $\{\rho(z, L_{2^{-k}})^{-l} |f(z) - P_{2^n}(z)|\}$ and $\{Mb_n(z)\}$, we obtain

$$\sum_{n=0}^{k} |f(z) - P_{2^n}(z)|^2 \rho(z, L_{2^{-n}})^{-2l} \leqslant c \sum_{n=0}^{\infty} Mb_n(z)^2,$$

and condition (6.1) follows from (6.2) and the Fefferman-Stein maximal theorem. ●

§7. The Faber-Toeplitz operators

7.1. General aspects. Let G be a Lipschitz domain and f a function defined on its boundary. If $f \in L^2(\partial G)$, one can define the Cauchy type integral of f, which is a function f_+ analytic in G, by

$$f_+(z) = \frac{1}{2\pi i} \int_{\partial G} f(\zeta) \frac{d\zeta}{\zeta - z}, \qquad z \in G.$$

Let G_1 and G_2 be two Lipschitz domains, and let $\varphi: \mathbf{C} \setminus G_2 \to \mathbf{C} \setminus G_1$, $\varphi(\infty) = \infty$, $\varphi'(\infty) > 0$ be a conformal mapping. The *Faber operator* (or Faber transformation; see [16]) T maps a function f analytic in G_1 in a function Tf analytic in G_2 by the formula $Tf = (f \circ \varphi)_+$, i.e.

$$Tf(z) = \frac{1}{2\pi i} \int_{\partial G_2} f[\varphi(\zeta)] \frac{d\zeta}{\zeta - z}, \qquad z \in G_2.$$

Finally, given a function w analytic in $\mathbf{C}\setminus G_2$, we shall define the *Faber-Toeplitz operator* T_w with weight w (see [16]) by the formula $T_w f = [w\cdot(f\circ\varphi)]_+$, i.e.

$$T_w f(z) = \frac{1}{2\pi i}\int_{\partial G_2} f[\varphi(\zeta)]\,w(\zeta)\frac{d\zeta}{\zeta - z}, \qquad z\in G_2.$$

For $w\equiv 1$ we obtain $T_2 = T$.

Suppose that for the original function f we are given its pseudoanalytic continuation from $W_q^1(\mathbf{C}\setminus G_1)$. Then obviously

$$f_+(z) = -\frac{1}{\pi}\iint_{\mathbf{C}\setminus G_1} \bar\partial \mathbf{f}(\zeta)\frac{d\xi\,d\eta}{\zeta - z}, \qquad z\in G_1,$$

and correspondingly

$$T_w f(z) = -\frac{1}{\pi}\iint_{\mathbf{C}\setminus G_2} \mu(\zeta)\frac{d\xi\,d\eta}{\zeta - z}, \qquad z\in G_2,$$

where

$$\mu(\zeta) = \bar\partial_\zeta(\mathbf{f}[\varphi(\zeta)]\,w(\zeta)) = w(\zeta)\,\overline{\varphi'(\zeta)}\,\bar\partial\mathbf{f}[\varphi(\zeta)].$$

In this section we study the behavior of Besov and Sobolev classes under the action of the operator T_w. We shall find those weights w for which T_w implements an isomorphism of these spaces over different domains. However, there are two reasons why we shall not be able to use the simple way of introducing the operators T_w as indicated above. First of all, we shall have to consider the case of the segment $[-1,1]$ where the "naive" definition of the Cauchy type integral loses its meaning. Second, even for a smooth function f, the function $w\cdot(f\circ\varphi)$ is not necessarily summable over ∂G_2, and we will have to consider "generalized" Cauchy type integrals in order to handle this situation. Let us start with Cauchy type integrals.

Let K denote, as above, either the segment $[-1,1]$ or the closure of a Lipschitz domain G. Let $\partial(\mathbf{C}\setminus K)$ be the set of boundary ends of the region $\mathbf{C}\setminus K$. If K is the closure of G, then $\partial(\mathbf{C}\setminus K)$ coincides with ∂G. But then $K=[-1,1]$, $\partial(\mathbf{C}\setminus K)$ is the set of all points of $[-1,1]$ in two copies: to each x, $1\leqslant x\leqslant 1$, there correspond two points of $\partial(\mathbf{C}\setminus K)$, namely $x+i0$ and $x-i0$.

Let f be a measurable function on $\partial(\mathbf{C}\setminus K)$ (hence for $K=[-1,1]$ it is a "two-valued" function, but the values $f(\pm 1)$ are defined uniquely). We shall say that *f admits a Cauchy type integral*, and write $f\in\mathfrak{C}(K)$, provided there exist functions h and Φ in $\mathbf{C}\setminus K$ such that

a) $h\in E^\delta(\mathbf{C}\setminus K)$ for some $\delta>0$,

b) $\Phi\in W_q^1(\mathbf{C}\setminus K)$ for some $q>1$, and

c) the boundary value of $h+\Phi$ on $\partial(\mathbf{C}\setminus K)$ coincides with f almost everywhere, $h+\Phi|_{\partial(\mathbf{C}\setminus K)} = f$.(5)

If $f\in\mathfrak{C}(K)$, then by the Cauchy type integral of f on K we shall understand a function f_+ which is defined on K by the formula

$$f_+(z) = -\frac{1}{\pi}\iint_{\mathbf{C}\setminus K} \bar\partial\Phi(\zeta)\frac{d\xi\,d\eta}{\zeta - z}. \tag{7.1}$$

(5) h takes on boundary values in the sense of E^δ [21], while Φ does it in the sense of embedding theorems [18]. At any rate, $\Phi+h|_{\partial(\mathbf{C}\setminus K)}$ is almost everywhere the angular limit of $\Phi+h$.

LEMMA 4. *The function f_+ defined by* (7.1) *is independent of the choice of h and Φ.*

PROOF. It suffices to show that $f_+ = 0$ if $f = 0$. If $f = 0$, then $\Phi|_{\partial(\mathbf{C}\setminus K)} = -h|_{\partial(\mathbf{C}\setminus K)}$ and, by a theorem of V. I. Smirnov (see [21]), $h \in E^q(\mathbf{C}\setminus K)$. On the other hand, the function (7.1) belongs to W^1_q in the entire plane, and it suffices to show that its boundary value on $\partial(\mathbf{C}\setminus K)$ is zero.

Let $z \in \mathbf{C}\setminus K$. Then

$$\begin{aligned}
-\frac{1}{\pi}\iint\limits_{\mathbf{C}\setminus K} \bar\partial\Phi(\zeta)\frac{d\xi\,d\eta}{\zeta - z} &= \lim_{r\to 0}\left(-\frac{1}{\pi}\iint\limits_{\rho(\zeta,K)>r} \partial\Phi(\zeta)\frac{d\xi\,d\eta}{\zeta - z}\right)\\
&= \lim_{r\to 0}\left(\Phi(z) - \frac{1}{2\pi i}\int_{\Lambda_r}\Phi(\zeta)\frac{d\zeta}{\zeta - z}\right)\\
&= \Phi(z) + h(z) - \frac{1}{2\pi i}\lim_{r\to 0}\int_{\Lambda_r}[\Phi(\zeta) + h(\zeta)]\frac{d\zeta}{\zeta - z}\\
&= \Phi(z) + h(z),
\end{aligned}$$

since $\Phi + h|_{\partial(\mathbf{C}\setminus K)} = 0$ in the sense of L^q, and $z \in \mathbf{C}\setminus K$. Now it is clear that this function tends to 0 as $\rho(z, K) \to 0$. ●

If $K = \bar G$ and $f \in L^1(G)$, then clearly our definition coincides with the usual one. If $K = [-1, 1]$ and $f \in L^1(\partial(\mathbf{C}\setminus K))$, then it is easy to verify that

$$\begin{aligned}
f_+(x) &= \frac{1}{2}[f(x + i0) + f(x - i0)]\\
&\quad + \frac{1}{2\pi}\text{p.v.}\int_{-1}^{1}[f(t + i0) - f(t - i0)]\frac{dt}{t - x}, \qquad x \in [-1, 1].
\end{aligned}$$

Now we can define the Faber-Toeplitz operators. Let K_1 and K_2 be two admissible continua. Let $\varphi\colon \mathbf{C}\setminus K_2 \to \mathbf{C}\setminus K_1$, $\varphi(\infty) = \infty$, $\varphi'(\infty) > 0$ be a conformal mapping, and let w be a function analytic in $\mathbf{C}\setminus K_2$, with $w \in E^\delta(\mathbf{C}\setminus K_2)$ for some $\delta > 0$.

We shall say that the operator T_w is applicable to an admissible function f on K_1 provided that on $\partial(\mathbf{C}\setminus K_2)$ the function $w(f\circ\varphi)$ belongs to the class $\mathfrak{L}(K_2)$. In this case we set

$$T_w f = [w(f\circ\varphi)]_+ \quad \text{on } K_2$$

in the sense of the above definition.

Observe that by definition the function $T_w f$ always has a pseudoanalytic continuation into $\mathbf{C}\setminus K_2$ (e.g., by (7.1)). Assume that f has a pseudoanalytic continuation into $\mathbf{C}\setminus K_1$ such that

$$w\varphi'\bar\partial\mathbf{f}[\varphi] \in L^q(\mathbf{C}\setminus K_2)$$

with some $q > 1$. Then the function

$$\Phi(z) = -\frac{1}{\pi}\iint\limits_{\mathbf{C}\setminus K_2} w(\zeta)\,\overline{\varphi'(\zeta)}\,\bar\partial\mathbf{f}[\varphi(\zeta)]\frac{d\xi\,d\eta}{\zeta - z} \tag{7.2}$$

belongs to W^1_q. The function $w\cdot(\mathbf{f}\circ\varphi) - \Phi$ is clearly analytic in $\mathbf{C}\setminus K_2$ and, in view of the conditions on w, is of class $E^\delta(\mathbf{C}\setminus K_2)$ for some $\delta > 0$. Therefore the operator T_w is applicable to f, and the function (7.2) coincides with $T_w f$.

Furthermore, let K_3 be a third admissible continuum, and let there be given a commutative diagram of conformal mappings

$$\begin{array}{ccc} \mathbb{C}\setminus K_1 & \overset{\varphi_1}{\longleftarrow} & \mathbb{C}\setminus K_2 \\ \uparrow & & \uparrow \varphi_2 \\ \llcorner & \overset{\varphi}{\longrightarrow} & \mathbb{C}\setminus K_3 \end{array}$$

Moreover, let w_1 and w_2 be admissible weights in $\mathbb{C}\setminus K_2$ and $\mathbb{C}\setminus K_3$, respectively. Set $w = w_2 \cdot (w_1 \circ \varphi_2)$. Then w is an admissible weight in $\mathbb{C}\setminus K_3$.

LEMMA 5. *Assume that the operator T_{w_1} is applicable to the function f on K_1 and that the operator T_{w_2} is applicable to the function $T_{w_1} f$ on K_2. Then the operator T_w is applicable to the function f, where $T_w f = T_{w_2}(T_{w_1} f)$.*

PROOF. By definition, on $\partial(\mathbb{C}\setminus K_2)$ we have

$$w_i(f \circ \varphi_1) = h + \Phi,$$

where $h \in E^{\delta}(\mathbb{C}\setminus K_2)$, $\Phi \in W_q^1$ and $\Phi_+ = \Phi$. Then

$$w(f \circ \varphi) = w_2(h \circ \varphi_2) + w_2(\Phi \circ \varphi_2),$$

however, by our hypothesis $w_2(\Phi \circ \varphi_2) \in \mathfrak{L}(K_3)$, and obviously $w_2(h \circ \varphi_2) \in E^{\delta_1}(\mathbb{C}\setminus K_3)$ for some $\delta_1 > 0$. Therefore

$$w \cdot (f \circ \varphi) \in \mathfrak{L}(K_3) \qquad \text{and} \qquad [w(f \circ \varphi)]_+ = [w_2(\Phi \circ \varphi_2)]_+ . \; \bullet$$

7.2. The invariance of Besov and Sobolev classes. Now we shall see that, as a rule, the Faber-Toeplitz operator preserves the smoothness of functions, and that for a suitable w this operator yields an isomorphism of Besov and Sobolev classes in various domains.

It is obvious from (7.2) that the function $T_w f$ admits a pseudoanalytic continuation into $\mathbb{C}\setminus K_2$, and

$$|\bar{\partial}(T_w f(z))| \leqslant |w(z)\overline{\varphi'(z)}|\,|\bar{\partial}\mathbf{f}[\varphi(z)]|, \qquad z \in \mathbb{C}\setminus K_2. \tag{7.3}$$

Hence $\bar{\partial}(T_w f)$ can be estimated in terms of $\bar{\partial}\mathbf{f}$, and the smoothness of f must affect the smoothness of $T_w f$.

THEOREM 10. *Let $1 \leqslant p \leqslant \infty$ and $s > 0$. Then for $w(z) = \varphi'(z)^{1/p-s}$, $z \in \mathbb{C}\setminus K_2$, the Faber-Toeplitz operator T_w maps $A_p^s(K_1)$ continuously onto $A_p^s(K_2)$.*

PROOF. Estimate (7.3) together with the substitution $z = \varphi(\zeta)$ enables us to extract from condition (3.1) or (3.1′) on f a similar condition on $T_w f$. Hence T_w maps $A_p^s(K_1)$ into $A_p^s(K_2)$. However, the same argument applies to the inverse mapping of $\mathbb{C}\setminus K_1$ onto $\mathbb{C}\setminus K_2$, and by Lemma 5 the corresponding Faber-Toeplitz operator is the inverse of T_w. $\bullet$

COROLLARY. *For any Lipschitz domain G, the Banach spaces $A_p^s(G)$ and $B_p^s[-1, 1]$ are canonically isomorphic.*

Observe that the weight w in Theorem 10 depends on s. Therefore for different values of s we obtain different isomorphism operators. The problem of the existence of one single operator which would, for instance, map $B_p^s[-1, 1]$ onto $A_p^s(G)$ for all

$s > 0$, i.e. the problem of the isomorphism of the scales $\{A_p^s(G)\}_{s>0}$ for various G, remains open.

Let us turn to Sobolev classes. In this case we must assume that G is necessarily a Radon domain when $K = \bar{G}$. The following exact analog of Theorem 10 holds.

THEOREM 11. *Let $1 < p < \infty$, and let l be an integer. Then for $w(z) = \varphi'(z)^{1/p-l}$, $z \in \mathbf{C}\setminus K_2$, the Faber-Toeplitz operator T_w maps $E_p^l(K_1)$ continuously onto $E_p^l(K_2)$.*

PROOF. We again use (7.3). According to Theorem 2, if $f \in E_p^l(G)$, then f has a pseudoanalytic continuation such that

$$|\bar{\partial}\mathbf{f}(z)| \leqslant c\rho(z, \partial G)^l |f^{(l+1)}(z^*)|, \qquad z \in \mathbf{C}\setminus\bar{G}. \tag{7.4}$$

Let μ be an outer function [21] on $\mathbf{C}\setminus\bar{G}$ such that $|\mu|$ satisfies on ∂G the Muckenhaupt condition (A_p) (see §1.3).

The following Luzin type inequality with weight was established in [17]:

$$\int_{\partial G} |\mu(z)||dz| \left(\iint_{S(z)} |f^{(l+1)}(\zeta^*)\mu(\zeta)^{-1/p}|^2 \, d\xi \, d\eta \right)^{p/2} \leqslant c\|f\|^p_{E_p^l(G)}. \tag{7.5}$$

Suppose that the domains G_1 and G_2 are such that the derivative φ' of the conformal mapping $\varphi\colon \mathbf{C}\setminus G_2 \to \mathbf{C}\setminus G_1$, $\varphi(\infty) = \infty$, $\varphi'(\infty) > 0$, satisfies condition (A_p) on ∂G_2. Then, upon the substitution $z = \varphi(\zeta)$ for $\mu = \varphi'$, conditions (7.3)–(7.5) immediately yield condition (4.1) for $T_w f$. Hence the operator T_w maps $E_p^l(G_1)$ into $E_p^l(G_2)$. If $K_2 = [-1, 1]$, the whole argument can be repeated without modifications. If $K_1 = [-1, 1]$, then the inequality for $\partial\mathbf{f}$ with weight $|\varphi'|$ follows from the construction of the pseudoanalytic continuation in the proof of Theorem 7 (it can be obtained from an inequality of type (7.5) applied to the domain G_ε introduced in the proof of Theorem 7). To complete the proof it suffices to remove condition (A_p) on φ'. However, it was shown in [17] that one can always find a finite sequence of domains with piecewise analytic boundaries $G_0, G_1, \ldots, G_N$, such that the derivatives of all conformal mappings $\varphi_j\colon \mathbf{C}\setminus G_j \to \mathbf{C}\setminus G_{j-1}$, $\varphi_j(\infty) = \infty$, $\varphi_j'(\infty) > 0, j = 1, \ldots, N$ and $\varphi_{-1}\colon \mathbf{C}\setminus G_0 \to \mathbf{C}\setminus K_1$, $\varphi_{N+1}\colon \mathbf{C}\setminus K_2 \to \mathbf{C}\setminus G_N$ satisfy condition (A_p) on the corresponding contours. Theorem 5 now follows from Lemma 5 and the already established part of this theorem. ●

7.3. The Faber-Toeplitz operators and approximation theory. We shall show that for sufficiently small values of $s > 0$ the "direct" theorem on approximation of functions from $B_p^s[-1, 1]$ can be obtained by means of Faber-Toeplitz operators, i.e. without using the results of §1.4 on approximations of the Cauchy kernel.

Let $\varphi\colon \mathbf{C}\setminus K \to \mathbf{C}\setminus\bar{\mathbf{D}}$ again be a conformal mapping of $\mathbf{C}\setminus K$ onto the exterior of the disk with $\varphi(\infty) = \infty$ and $\varphi'(\infty) > 0$. Set $w = (\varphi')^{1/p-s}$ and consider the operator T_w which, by Theorem 10, maps $A_p^s(\mathbf{D})$ onto $A_p^s(K)$ ($B_p^s[-1, 1]$ for $K = [-1, 1]$).

If P_n is a polynomial of degree n, then obviously so is $T_w P_n$.

LEMMA 6. *If the function $|\varphi'^{ps}|$ satisfies condition (A_p) on ∂K, then for any polynomial P_n of degree n*

$$\left(\int_{\partial K} |\rho(z, L_{1/n})^{-s} T_w P_n(z)|^p |dz| \right)^{1/p} \leqslant cn^s \left(\int_{\partial\mathbf{D}} |P_n|^p \right)^{1/p},$$

where c does not depend on n.

PROOF. First of all, Bernstein's inequality implies that

$$\left(\int_{\partial\mathbb{D}} |P_n|^p\right)^{1/p} \geqslant e^{-1}\left(\int_{|\zeta|=1+1/n} |P_n(\zeta)|^p |d\zeta|\right)^{1/p}.$$

Upon substituting $\zeta = \varphi(z)$ we obtain

$$\left(\int_{L_{1/n}} |\varphi'(z)^s w(z) P_n[\varphi(z)]|^p |dz|\right)^{1/p} \leqslant c\left(\int_{\partial\mathbb{D}} |P_n|^p\right)^{1/p}.$$

However, $|\varphi'|^{ps}$ satisfies the Muckenhaupt condition (A_p) on ∂K and thus also (according to [17]) on $L_{1/n}$. Hence we can take the Cauchy type integral, and obtain

$$\left(\int_{L_{1/n}} |\varphi'(z)^s T_w P_n(z)|^p |dz|\right)^{1/p} \leqslant c\left(\int_{\partial\mathbb{D}} |P_n|^p\right)^{1/p}.$$

On $L_{1/n}$ we have $|\varphi'(z)| \asymp (1/n\rho(z, K))$, and therefore

$$\left(\int_{L_{1/n}} |\rho(z, K)^{-s} T_w P_n(z)|^p |dz|\right)^{1/p} \leqslant cn^s\left(\int_{\partial\mathbb{D}} |P_n|^p\right)^{1/p}.$$

Finally, in view of the maximal estimate [17] for $T_w P_n$ in the interior of the level curve $L_{1/n}$ we obtain

$$\left(\int_{\partial K} |\rho(z, L_{1/n})^{-s} T_w P_n(z)|^p |dz|\right)^{1/p} \leqslant cn^s\left(\int_{\partial\mathbb{D}} |P_n|^p\right)^{1/p}. \bullet$$

Now we can give a new proof of the necessity part of Theorem 8 from §5:

THEOREM 12. *If $f \in A_p^s(K)$ and the function $|\varphi'^{ps}|$ satisfies the Muckenhaupt condition (A_p) on ∂K, then*

$$\left(\sum_{n=1}^{\infty} \frac{1}{n} E_n(f)_{p,s}^p\right)^{1/p} < \infty.$$

PROOF. By Theorem 10, $f = T_w g \cdot g \in A_p^s(\mathbb{D})$. Let P_n be a polynomial of best approximation to g in $H^p(\mathbb{D})$ (i.e. in the metric of $L^p(\partial\mathbb{D})$). Then, as is well known (see [2]),

$$\left(\sum_{k=0}^{\infty} \left(2^{ks} \| P_{2^{k+1}} - P_{2^k} \|_{L^p(\partial\mathbb{D})}\right)^p\right)^{1/p} < \infty.$$

However,

$$f = T_w g = T_{w_0} P_0 + \sum_{k=0}^{\infty} T_w (P_{2^{k+1}} - P_{2^k}).$$

Applying Lemma 2 to each term of this series, we easily obtain

$$\left(\sum_{k=0}^{\infty} E_{2^k}(f)_{p,s}^p\right)^{1/p} < \infty,$$

which is equivalent to the statement of the theorem.

REMARKS. a) The conditions of Theorem 12 are satisfied for any K provided s is small enough; for a convex set K it suffices to assume that $s < 2/p$ [17].

b) In case of uniform approximations a similar approach turned out to be applicable for all values of s [10], [13]. However, it is not known whether in our case one can also give a similar "direct" proof for arbitrary values of s.

BIBLIOGRAPHY

1. E. M. Dyn'kin, *On a constructive characterization of Sobolev and Besov classes*, Dokl. Akad. Nauk SSSR **233** (1977), no. 5, 773–775; English transl. in Soviet Math. Dokl. **18** (1977).

2. S. M. Nikol'skiĭ, *Approximation of functions of several variables and embedding theorems*, "Nauka", Moscow, 1969; English transl., Springer-Verlag, 1974.

3. Elias M. Stein, *Singular integrals and differentiability properties of functions*, Princeton Univ. Press, Princeton, N. J., 1970.

4. S. M. Nikol'skiĭ, *On the best approximation of functions satisfying Lipschitz's condition by polynomials*, Izv. Akad. Nauk SSSR Ser. Mat. **10** (1946), 295–322. (Russian; English summary)

5. A. F. Timan, *The approximation of functions satisfying a Lipschitz condition by ordinary polynomials*, Dokl. Akad. Nauk SSSR **77** (1951), 969–972. (Russian)

6. V. K. Dzyadyk, *Constructive characterization of functions satisfying the condition* Lip α $(0 < \alpha < 1)$ *on a finite segment of the real axis*, Izv. Akad. Nauk SSSR Ser. Mat. **20** (1956), 623–642. (Russian)

7. M. K. Potapov, *On structural characterizations of classes of functions with a given degree of best approximations*, Trudy Mat. Inst. Steklov. **134** (1975), 260–277; English transl. in Proc. Steklov Inst. Math. **134** (1975).

8. V. P. Motornyĭ, *Approximation of functions by algebraic polynomials in the L_p metric*, Izv. Akad. Nauk SSSR Ser. Mat. **35** (1971), 874–899; English transl. in Math. USSR Izv. **5** (1971).

9. J.-E. Andersson, *On the degree of weighted polynomial approximation of holomorphic functions*, Anal. Math. **2** (1976), 163–171.

10. E. M. Dyn'kin, *The rate of polynomial approximation in $E^p(G)$*, Dokl. Akad. Nauk SSSR **231** (1976), 529–531; English transl. in Soviet Math. Dokl. **17** (1976).

11. ______, *Smooth functions on plane sets*, Dokl. Akad. Nauk SSSR **208** (1973), 25–27; English transl. in Soviet Math. Dokl. **14** (1973).

12. ______, *Extensions and integral representations of smooth functions of a complex variable*, Author's Summary of Candidate's Dissertation, Leningrad, 1973. (Russian)

13. ______, *Uniform approximation of functions in Jordan domains*, Sibirsk. Mat. Zh. **18** (1977), 775–786; English transl. in Siberian Math. J. **18** (1977).

14. V. K. Dzyadyk, *On the theory of approximation of functions on closed sets of the complex plane* (*with reference to a problem of S. M. Nikol'skiĭ*), Trudy Mat. Inst. Steklov. **134** (1975), 63–114; English transl. in Proc. Steklov Inst. Math. **134** (1975).

15. Ju. A. Brudnyĭ, *Spaces defined by means of local approximations*, Trudy Moskov. Mat. Obshch. **24** (1971), 69–132; English transl. in Trans. Moscow Math. Soc. **24** (1971).

16. V. I. Smirnov and N. A. Lebedev, *The constructive theory of functions of a complex variable*, "Nauka", Moscow, 1964; English transl., *Functions of a complex variable: Constructive theory*, M.I.T. Press, Cambridge, Mass., 1968.

17. E. M. Dyn'kin, *Estimates for analytic functions in Jordan domains*, Zap. Nauchn. Sem. Leningrad. Otdel. Mat. Inst. Steklov. (LOMI) **73** (1977), 70–90; English transl. in J. Soviet Math. (to appear).

18. O. V. Besov, V. P. Il'in and S. M. Nikol'skiĭ, *Integral representations of functions and imbedding theorems*, "Nauka", Moscow, 1975; English transl., Vols. 1, 2, Wiley, 1979.

19. Lars V. Ahlfors, *Lectures in quasiconformal mappings*, Van Nostrand, Princeton, N. J., 1966.

20. I. I. Danilyuk, *Nonregular boundary-value problems in the plane*, "Nauka", Moscow, 1975. (Russian)

21. I. I. Privalov, *Boundary properties of analytic functions*, 2nd ed., GITTL, Moscow, 1950; German transl., VEB Deutscher Verlag Wiss., Berlin, 1956.

22. A. Calderón, *On the Cauchy integral on Lipschitz curves and related operators*, Preprint, 1977.*

23. G. H. Hardy and J. E. Littlewood, *Some properties of fractional integrals*. I, Math. Z. **27** (1927/28), 565–606.

24. Peter L. Duren, *Theory of H^p-spaces*, Academic Press, 1970.

25. R. R. Coifman and C. Fefferman, *Weighted norm inequalities for maximal functions and singular integrals*, Studia Math. **51** (1974), 241–250.

26. C. Fefferman and E. M. Stein, *Some maximal inequalities*, Amer. J. Math. **93** (1971), 107–115.

27. Miguel de Guzmán, *A covering lemma with applications to differentiability of measures and singular integral operators*, Studia Math. **34** (1970), 299–317 (errata insert).

Translated by M. A. DOSTAL

---

* *Editor's note*. See also his note in Proc. Nat. Acad. Sci. U.S.A. **74** (1977), 1324–1327.

Труды Матем. Инст. Стеклов.
155(1981)

Proceedings of the Steklov
Institute of Mathematics
1983, Issue 1

THE FOURIER COEFFICIENTS OF THE BOUNDARY VALUES OF FUNCTIONS ANALYTIC IN THE DISK AND IN THE BIDISK

UDC 513.8 + 519.4

S. V. KISLYAKOV

ABSTRACT. 1. Let $a = \{a_n\} \in l^2(\mathbb{Z}_+)$; then there exists a function f on the circle $\mathbb{T}$ such that $\hat{f}(n) = 0$ for $n < 0$, $|\hat{f}(n)| \geqslant |a_n|$ for $n \geqslant 0$, and $\Sigma_{n \geqslant 0} \hat{f}(n) z^n$ converges uniformly in the closed unit disk.

2. If

$$a = \{a_{mn}\} \in l^2(\mathbb{Z}_+ \times \mathbb{Z}_+),$$

then there is an f in $C(\mathbb{T}^2)$ such that $|\hat{f}(m, n)| \geqslant |a_{mn}|$ for $m, n \geqslant 0$ and $\hat{f}(m, n) = 0$ if $m < 0$ or $n < 0$.

In both cases f can be chosen so that its norm in the corresponding space does not exceed $K \| a \|_{l^2}$, where K is an absolute constant.

Bibliography: 18 titles.

Contents

§0. Introduction

How can the Fourier coefficients $\hat{f}(n)$ of a continuous function f behave? It was clear from the classical work of Carleman [1], Banach [2] and Paley [3] that, roughly speaking, the sequence $\{|\hat{f}(n)|\}$ possesses virtually no additional properties beyond the property $\Sigma_{n \in \mathbb{Z}} |\hat{f}(n)|^2 < \infty$. This same principle holds also for certain other classes of functions on the circle that are smaller than $C(\mathbb{T})$. We present two typical propositions which hold for many such classes and may serve as a precise mathematical expression of this principle.

Let X be some Banach space of functions on the circle (where usually $X \subset C(\mathbb{T})$ and $\| f \|_X \leqslant \| f \|_{C(\mathbb{T})}$ for all f in X); let M be either $\mathbb{Z}$ or $\mathbb{Z}_+$ (the second case arises when all functions in X satisfy $\hat{f}(n) = 0$ for $n < 0$).

1980 *Mathematics Subject Classification.* Primary 42B99, 42B30, 42A16, 42B05.

©1983 American Mathematical Society
0081-5438/83 $1.00 + $.25 per page

PROPOSITION A. *Let $\{a_n\}_{n\in M}$ be a numerical sequence such that for all f in X*

$$\sum_{n\in M} |a_n \hat{f}(n)| \leqslant K\|f\|_X$$

(K is independent of f). Then $(\sum_{n\in M}|a_n|^2)^{1/2} \leqslant LK$, where L is determined by the space X alone.

PROPOSITION B. *Suppose the set $N \subset M$ is "sufficiently sparse" (this often means that N is a set of type Λ_2, for whose definition see §1). Then the operator T from X into $l^2(N)$ given by the formula $Tf = \{\hat{f}(n)\}_{n\in N}$ maps X onto $l^2(N)$.*

The work of S. A. Vinogradov (see [4]–[6]) contains many results of this type. One of his deepest theorems says that Propositions A and B hold for the space U whose elements are the restrictions to the circle of functions representable by a power series which converges uniformly in the unit disk.

It is not hard to see that Propositions A and B are very different in character: while the second gives more than satisfactory information about the behavior of Fourier coefficients with *sparse* indices, the first leaves open many of the natural questions about how "massive" a part of $l^2(M)$ is occupied by the totality of sequences $\{\hat{f}(n)\}_{n\in M}$, $f \in X$. Quite recently de Leeuw, Katznelson, and Kahane [7] obtained the following result, which fills in this gap in the case where $X = C(\mathbb{T})$, and answers a question of Sidon [8]: If $\sum_{n\in\mathbf{Z}}|a_n|^2 < \infty$, then there exists a function f in $C(\mathbb{T})$ such that $\hat{f}(n) \geqslant |a_n|$ for all integers n and $\|f\|_{C(\mathbb{T})} \leqslant K(\sum_{n\in\mathbf{Z}}|a_n|^2)^{1/2}$ (K is a constant independent of the sequence $\{a_n\}$).

The purpose of this article is to extend the theorem of de Leeuw, Katznelson and Kahane to other classes of functions different from $C(\mathbb{T})$, specifically to the class of functions representable by a uniformly convergent power series (as in Vinogradov's theorem) and to the class $C_A(\mathbb{T}^2)$ (whose definition is given in §1).

In addition it is proved here that a theorem of type B holds for the class $C_A(\mathbb{T}^2)$. It should be noted that up to now it was unknown whether the analogues of Propositions A and B hold for the space $C_A(\mathbb{T}^n)$ with $n \geqslant 2$. Unfortunately I still know nothing sufficiently satisfactory about the case $n > 2$.

Many ideas of de Leeuw, Katznelson, and Kahane will be used in our proofs. The abstract scheme presented below is essentially contained in their work [7] (although not formulated there quite so explicitly). Additional difficulties (in comparison with [7]) arise in verifying that such and such a specific class of functions fits into this abstract scheme. For example, the proof of this fact for the space U relies on a result of Vinogradov on the existence of an estimate of weak type for functionals on the space U, which in turn is deduced from Carleson's theorem on almost everywhere convergence of Fourier series (for more details see [5] and [6]).

I am grateful to A. B. Aleksandrov for many useful discussions (their result can be seen at the beginning of §5). In addition I am deeply indebted to S. V. Khrushchev for permission to use a variant of an abstract scheme of his.

§1. Notation and terminology

The closed unit disk will be denoted by $\mathbb{D}$ and its boundary by $\mathbb{T}$. We shall denote normalized Lebesgue measure on the n-dimensional torus $\mathbb{T}^n$ by m_n. If f is a measurable function on $\mathbb{T}^n$, its norm in the space $L^p(m_n)$, $0 < p \leqslant \infty$,(¹) is denoted by $\|f\|_p$. $\mathbb{Z}$ is the group of integers, and $\mathbb{Z}_+$ is defined as $\{m \in \mathbb{Z}: m \geqslant 0\}$. We shall identify the group $\mathbb{Z}^n$ in the usual manner with the dual of the group $\mathbb{T}^n$. If f is a function (or a measure) on $\mathbb{T}^n$, then its Fourier transform (defined on $\mathbb{Z}^n$) is denoted by $\hat{f}$. If $\hat{f}$ is different from zero only on a finite subset of $\mathbb{Z}^n$, then $\hat{f}$ will be called a trigonometric polynomial in n variables. The symbol z denotes the identity mapping of $\mathbb{T}$ into itself, and $z_1, \ldots, z_n$ are the coordinate functions on the n-dimensional torus $\mathbb{T}^n$. $C_A(\mathbb{T}^n)$ is defined as $\{f: f \in C(\mathbb{T}^n);\ \hat{f}(l) = 0 \text{ if } l \notin (\mathbb{Z}_+)^n\}$, and $H^p(\mathbb{T}^n)$ is the closure of the set of functions $C_A(\mathbb{T}^n)$ in the space $L^p(m_n)$ (when $p = \infty$, the closure is taken in the weak topology $\sigma(L^\infty(m_n), L^1(m_n))$). The space $C_A(\mathbb{T}^n)$ consists precisely of the restrictions to $\mathbb{T}^n$ of functions continuous in $\mathbb{D}^n$ and analytic inside $\mathbb{D}^n$.

The space $H^p(\mathbb{T})$ is denoted simply by H^p; $H^p_0 = \{f: f \in H^p;\ \int f\,dm_1 = 0\}$. We shall call the orthogonal projection P from $L^2(m_1)$ onto $\overline{H^2}$ the *Riesz projection* (the bar denotes complex conjugation), although this notation differs somewhat from the usual. It is well known (see, for example, [9] or [10]) that P is an operator of weak type 1-1 (this, by definition, means that

$$m_1\{\tau: |(Pf)(\tau)| > t\} \leqslant C\|f\|_1/t, \qquad f \in L^2(m_1),\ t > 0,$$

where C is independent of f and t). The space conjugate to $H^2(\mathbb{T}^n)$ will be identified with $\overline{H^2(\mathbb{T}^n)}$ via the duality

$$\langle f, g\rangle = \int_{\mathbb{T}^n} fg\,dm_n, \qquad f \in H^2(\mathbb{T}^n),\ g \in \overline{H^2(\mathbb{T}^n)}\,.$$

We shall denote by U the set of those functions f in $L^1(\mathbb{T})$ for which $\hat{f}(n) = 0$ when $n < 0$ and $\sum_{n\geqslant 0}\hat{f}(n)z^n$ converges uniformly (to f) on $\mathbb{T}$. We shall regard U as a Banach space, providing it with the norm $\|\cdot\|_U$,

$$\|f\|_U = \sup_{k\geqslant 0}\sup_{\zeta\in\mathbb{T}}\left|\sum_{n=k}^{\infty}\hat{f}(n)\zeta^n\right|.$$

In §4 we shall need the concept of sets of type Λ_2. We recall the definition. Suppose $0 < p < \infty$, $N \subset \mathbb{Z}^n$ and $X = \{f: f \in L^p(m_n);\ \hat{f}(l) = 0 \text{ for } l \notin N\}$. N is called a *set of type* Λ_p if the norms of all spaces L^r, $0 < r \leqslant p$, are equivalent on X.

§2. The abstract scheme

In this section we present some arguments which are part of the proofs both of the de Leeuw-Katznelson-Kahane result and of our results.

Let $\mathcal{H}$ be a Hilbert space and X a Banach space continuously embedded in $\mathcal{H}$ (this, by definition, means that X is contained as a linear subspace in $\mathcal{H}$ and $\|f\|_{\mathcal{H}} \leqslant \|f\|_X$ for all f in X). Let E be an orthonormal system in $\mathcal{H}$.(²) We are interested in the question of whether every nonnegative square-summable family defined on E is majorized for some f in X by the family $\{|(f, e)|\}_{e\in E}$.

(¹) By abuse of language we shall use the term "norm" for $p < 1$ also.

(²) E is regarded as a subset in $\mathcal{H}$ (and not as a family of elements of $\mathcal{H}$).

Suppose ω is a nonincreasing positive function on $[0, +\infty)$ with $\lim_{t\to\infty}\omega(t) = 0$. We shall say that the triple $(X, \mathcal{H}, E)$ *possesses property* $(\mathfrak{D}_\omega)$ if for some $\alpha > 0$ the following holds: for any numerical family $\{a_e\}_{e\in E}$ such that $\Sigma_{e\in E}|a_e|^2 = 1$ and any number $\eta \geqslant \alpha$, there exist vectors $g \in X$, $h \in \mathcal{H}$, and a family $\{\varepsilon_e\}_{e\in E}$, $\varepsilon_e = \pm 1$, for which

$$\sum_{e\in E} \varepsilon_e a_e e = g + h, \qquad \|g\|_X \leqslant \eta,\ \|h\|_{\mathcal{H}} \leqslant \omega(\eta).$$

For us functions ω satisfying the following condition will be important:

There exists a sequence of numbers

$$\{c_n\}_{n\geqslant 0} \quad \textit{such that } c_1 > 0,$$

$$c_{n+1} > c_n, \quad c_n \to \infty \quad \textit{and} \quad \sum_{n\in\mathbb{Z}_+} c_{n+1}\omega(c_n)/c_n < \infty. \tag{$*$}$$

It is not difficult to see that for $\varepsilon > 0$ the function $x \mapsto (\log(1+x))^{-1-\varepsilon}$ satisfies this condition (set $c_n = 2^n$), while the function $x \mapsto (\log(1+x))^{-1+\varepsilon}$ does not satisfy it. In all applications, however, the function ω will decrease significantly faster than $(\log(1+x))^{-1-\varepsilon}$.

THEOREM 1. *Suppose the function ω satisfies condition $(*)$ and the triple $(X, \mathcal{H}, E)$ possesses property $(\mathfrak{D}_\omega)$. Then there exists a number K such that for every square-summable family of nonnegative numbers $\{a_e\}_{e\in E}$ there exists a vector G in X for which $|(G, e)| \geqslant a_e$, $e \in E$, and $\|G\|_X \leqslant K(\Sigma_{e\in E} a_e^2)^{1/2}$.*

REMARK. In the definition of property $(\mathfrak{D}_\omega)$ the condition $\Sigma_{e\in E}|a_e|^2 = 1$ was imposed on the sequence $\{a_e\}$, which causes some inconvenience in applications. It is nevertheless easy to see that if the triple $(X, \mathcal{H}, E)$ possesses property $(\mathfrak{D}_\omega)$, then the following holds. Suppose $(\Sigma_{e\in E}|a_e|^2)^{1/2} \leqslant \varepsilon$, $\varepsilon > 0$, and suppose $\eta \geqslant \varepsilon\alpha$. Then for some choice of numbers $\varepsilon_e = \pm 1$ the vector $f = \Sigma_{e\in E}\varepsilon_e a_e e$ can be represented in the form $f = g + h$, where $\|g\|_X \leqslant \eta$ and $\|h\|_{\mathcal{H}} \leqslant \varepsilon\omega(\eta/\varepsilon)$.

PROOF OF THEOREM 1. Since ω satisfies condition $(*)$, we can find an increasing sequence $\{c_j\}_{j\geqslant 0}$ such that $c_j \geqslant \alpha$ for all j (α is the constant in the definition of property $(\mathfrak{D}_\omega)$) and

$$\sum_{j\geqslant 0} \frac{c_{j+1}}{c_j}\omega(c_j) < \frac{1}{2}. \tag{1}$$

We shall construct two numerical sequences $\{\varepsilon_j\}_{j\geqslant 0}$ and $\{\eta_j\}_{j\geqslant 0}$; using them we shall carry out an inductive construction leading in the end to the vector G we are seeking. These sequences are given by the relations

$$\varepsilon_0 = (2c_0)^{-1}, \qquad \eta_0 = 0;$$

$$\varepsilon_{j+1} = \frac{\omega(c_j)}{c_j}, \qquad \eta_{j+1} = \varepsilon_j c_j \qquad (j \geqslant 0).$$

If $j \geqslant 0$, then $\eta_{j+2} = \varepsilon_{j+1}c_{j+1} = (c_{j+1}/c_j)\omega(c_j)$; therefore condition (1) and the obvious equality $\eta_1 = 1/2$ give

$$\eta_0 + \eta_1 + \eta_2 + \cdots < \tfrac{1}{2} + \tfrac{1}{2} = 1.$$

In addition, $\lim_j \varepsilon_j = 0$ (since $\sum c_{j+1}\varepsilon_j = \sum(c_{j+1}/c_j)\omega(c_j) < \infty$) and $\eta_{j+1} \geqslant \varepsilon_j \alpha$ (since $c_j \geqslant \alpha$).

We now take an arbitrary family of nonnegative numbers $\{a_e\}_{e\in E}$ such that $\sum_{e\in E} a_e^2 = \varepsilon_0$. We shall find a vector G in X for which $\|G\|_X \leqslant \sum_{j\geqslant 0}\eta_j$ and $|(G,e)| \geqslant (1-\sum_{j\geqslant 0}\eta_j)a_e$ for any e in E (then obviously the theorem will be proved). For this we construct by induction three sequences of vectors F_j, G_j, and H_j, $j \geqslant 0$ (F_j, $H_j \in \mathcal{H}$; $G_j \in X$) such that $G_0 = \mathbf{0}$, $F_j = G_j + H_j$ for all j and the following relations hold:

(a) $|(F_j, e)| \geqslant a_e(1-\eta_0-\eta_1-\cdots-\eta_j)$, $e \in E, j \geqslant 0$;

(b) $\|G_j - G_{j-1}\|_X \leqslant \eta_j, j \geqslant 1$;

(c) $\|H_j\|_{\mathcal{H}} \leqslant \varepsilon_j$; and

(d) H_j can be represented in the form $H_j = g_j + h_j$, where

$$\|g_j\|_X \leqslant \eta_{j+1}, \qquad \|h_j\|_{\mathcal{H}} \leqslant \varepsilon_j\omega(\eta_{j+1}/\varepsilon_j).$$

If such sequences are already constructed, set $G = \lim_j G_j$ (the limit in the sense of convergence in X exists by virtue of (b)). Then $\|G\|_X \leqslant \sum_{j\geqslant 0}\eta_j$ (again by (b) and the fact that $G_0 = \mathbf{0}$). From (c) and the equality $F_j = G_j + H_j$ it follows that $\lim_j F_j = G$ in $\mathcal{H}$; together with (a) this gives $|(G,e)| \geqslant a_e(1-\sum_{j\geqslant 0}\eta_j)$, $e \in E$, i.e. the vector G is the one sought.

Thus it remains to carry out the inductive construction. First suppose $j = 0$. Set $G_0 = \mathbf{0}$ and $F_0 = H_0 = \sum_{e\in E}\varepsilon_e a_e e$, where the numbers $\varepsilon_e = \pm 1$ are chosen so that (d) holds for $j = 0$ (the possibility of such a choice follows from the remark made before the proof of the theorem). Then (a)–(d) obviously hold for $j = 0$. Now suppose that for some j we have already constructed vectors F_j, G_j, and H_j such that (a)–(d) hold. By (d) we have $H_j = g_j + h_j$, where $\|g_j\|_X \leqslant \eta_{j+1}$ and $\|h_j\|_{\mathcal{H}} \leqslant \varepsilon_j\omega(\eta_{j+1}/\varepsilon_j)$. Set $G_{j+1} = G_j + g_j$; then obviously $\|G_{j+1} - G_j\|_X \leqslant \eta_{j+1}$, i.e. relation (b) holds with j replaced by $j+1$.

To construct H_{j+1} we examine two sets:

$$P = \{e\colon e \in E, |(G_{j+1}, e)| \geqslant a_e(1-\eta_0-\cdots-\eta_{j+1})\} \quad \text{and} \quad Q = E\setminus P.$$

Since by the inductive hypothesis $|(F_j, e)| \geqslant a_e(1-\eta_0-\cdots-\eta_j)$ for all e in E, for each e in Q we have

$$|(F_j, e) - (G_{j+1}, e)| \geqslant a_e\eta_{j+1}.$$

Since $F_j - G_{j+1} = h_j$, the last inequality gives

$$\Big(\sum_{e\in Q} a_e^2\Big)^{1/2} \leqslant \frac{1}{\eta_{j+1}}\|h_j\|_{\mathcal{H}}.$$

Set $H_{j+1} = \sum_{e\in Q}\varepsilon_e \cdot 2a_e(1-\eta_0-\cdots-\eta_{j+1})e$, where the numbers $\varepsilon_e = \pm 1$ will be chosen in a special way, and let $F_{j+1} = G_{j+1} + H_{j+1}$. The following inequalities hold regardless of the choice of the numbers ε_e:

$$\|H_{j+1}\|_{\mathcal{H}} \leqslant 2(1-\eta_1)\frac{1}{\eta_{j+1}}\|h_j\|_{\mathcal{H}} = \frac{1}{\eta_{j+1}}\|h_j\|_{\mathcal{H}} \leqslant \frac{\varepsilon_j}{\eta_{j+1}}\omega\left(\frac{\eta_{j+1}}{\varepsilon_j}\right) = \frac{\omega(c_j)}{c_j} = \varepsilon_{j+1};$$

$$|(F_{j+1}, e)| = |(H_{j+1} + G_{j+1}, e)| \geqslant a_e(1-\eta_0-\cdots-\eta_{j+1}), \qquad e \in E.$$

(In verifying the second inequality, it is necessary to consider separately two cases: $e \in P$ and $e \in Q$.) Therefore it remains to choose the numbers ε_e so that (d) holds (with j replaced by $j+1$), and this is possible by the remark before the proof of the theorem and the fact that $\eta_{j+2} \geqslant \alpha\varepsilon_{j+1}$ for all j. ●

§3. Uniformly convergent power series

The application of the scheme expounded in the preceding section to the triple $(X, \mathcal{H}, E)$, where $X = C(\mathbb{T})$, $\mathcal{H} = L^2(m_1)$, and $E = \{z^n: n \in \mathbb{Z}\}$, was established by de Leeuw, Katznelson and Kahane using the (well-known) lemma on series of random variables given below. This lemma is also quite useful for studying other triples $(X, \mathcal{H}, E)$. For the sake of completeness its proof is presented here.

LEMMA 1. *Let (S, μ) be a space with probability measure, and E a countable collection of uniformly bounded measurable functions on S: $|e(s)| \leqslant M$ if $s \in S$ and $e \in E$. Then for any numerical family $\{a_e\}_{e\in E}$ such that $\Sigma_{e\in E}|a_e|^2 = 1$ and any $\eta > 0$, there exist numbers $\varepsilon_e = \pm 1$ such that the function $f = \Sigma_{e\in E} a_e\varepsilon_e e$ satisfies*

$$\int_{\{s:\,|f(s)|>\eta\}} (|f| - \eta)^2\, d\mu \leqslant 64M^2 \exp\left(-\frac{\eta^2}{16M^2}\right).$$

PROOF. Let $\{r_e\}_{e\in E}$ be a family of independent identically distributed random variables on the interval [0, 1] with Lebesgue measure, each taking on only the values +1 and −1 with probability 1/2. It is well known (see, for example, [10, Appendix D]) that if the b_e are complex numbers, $\Sigma_{e\in E}|b_e|^2 \leqslant c^2$, and $\lambda(\tau) = \operatorname{meas}\{t: t \in [0,1], |\Sigma_{e\in E} b_e r_e(t)| > \tau\}$, then $\lambda(\tau) \leqslant 4\exp(-\tau^2/16c^2)$.

With each t in [0, 1] we associate a function f_t, given by $f_t(s) = \Sigma_{e\in E} r_e(t) a_e e(s)$.(3) The lemma will be proved if we establish that

$$I \stackrel{\text{def}}{=} \int_0^1 \left(\int_{\{s:\,|f_t(s)|>\eta\}} (|f_t| - \eta)^2\, d\mu \right) dt \leqslant 64M^2 \exp\left(-\frac{\eta^2}{16M^2}\right).$$

Changing the order of integration and writing the inner integral as the integral of the distribution function $\lambda_s(\tau) = \operatorname{meas}\{t: |f_t(s)| > \tau\}$, we obtain

$$I = \int_S \left(\int_{\{t:\,|f_t(s)|>\eta\}} (|f_t(s)| - \eta)^2\, dt \right) d\mu(s) = \int_S \left(2\int_\eta^\infty \tau\lambda_s(\tau)\, d\tau \right) d\mu(s).$$

But since $|e(s)| \leqslant M$ on S for all e in E, and $\Sigma |a_e|^2 = 1$, we have

$$\lambda_s(\tau) \leqslant 4\exp(-\tau^2/16M^2)$$

for all s in S. Therefore

$$I \leqslant 2\int_S \int_\eta^\infty 4\tau \exp\left(-\frac{\tau^2}{16M^2}\right) d\tau\, d\mu = 64M^2 \exp\left(-\frac{\eta^2}{16M^2}\right). \quad ●$$

From this lemma it follows immediately that if $X = C(S)$ and $\mathcal{H} = L^2(S, \mu)$ (here S is a compact space and μ is a probability measure on S) while E is an orthonormal

(3) For almost all t this series converges almost everywhere and the function $f_t(s)$ is measurable jointly in the variables s and t.

system in $\mathcal{H}$ such that $|e(s)| \leqslant M$ for $e \in E$ and $s \in S$, then the triple $(X, \mathcal{H}, E)$ satisfies condition $(\mathfrak{D}_\omega)$ with

$$\omega(\eta) = 9M \exp(-\eta^2/32M^2).$$

Indeed, suppose $\Sigma_{e\in E} |a_e|^2 = 1$. Choose signs ε_e as in Lemma 1, and denote by $\tilde{g}$ the truncation of the function $f = \Sigma\, \varepsilon_e a_e e$ at η and $-\eta$. For g choose a function in $C(S)$ such that $\|g\|_{C(S)} \leqslant \eta$ and $\|g - \tilde{g}\|_{L^2(\mu)} \leqslant M \exp(-\eta^2/32M^2)$, and for h choose the function $f - g$.

Applying Theorem 1 now, we will obtain the following slight strengthening of the theorem of de Leeuw, Katznelson, and Kahane:

For any bounded (by the number M) orthonormal system E in $L^2(S, \mu)$ and any family $a = \{a_e\}_{e\in E}$ from $l^2(E)$, there exists a function f in $C(S)$ such that $|(f, e)| \geqslant |a_e|$ for all e in E; moreover, $\|f\|_{C(S)} \leqslant K \|a\|_{l^2(E)}$, where the constant K is determined only by the number M.

We now turn to the space U of uniformly convergent power series. We set $E = \{z^n : n \in \mathbb{Z}_+\}$.

LEMMA 2. *Let $\omega(t) = \exp(-\sqrt{t})$. Then the triple $(U, L^2(m_1), E)$ possesses property $(\mathfrak{D}_\omega)$.*

The result spoken of in the Introduction is obtained immediately from Theorem 1 and Lemma 2.

THEOREM 2. *If $\Sigma_{n\geqslant 0} |a_n|^2 < \infty$, there exists a function f in U such that $|\hat{f}(n)| \geqslant |a_n|$ for all n (and, moreover, one may assume that $\|f\|_U \leqslant K(\Sigma_{n\geqslant 0} |a_n|^2)^{1/2}$, where K is an absolute constant).*

PROOF OF LEMMA 2. Let $\{a_n\}_{n\geqslant 0}$ be a sequence of complex numbers with $\Sigma_{n\geqslant 0} |a_n|^2 = 1$; let $\eta > 0$. By Lemma 1 we can choose signs $\varepsilon_n = \pm 1$ so that for the function $f \stackrel{\text{def}}{=} \Sigma_{n\geqslant 0}\, \varepsilon_n a_n z^n$ we have

$$\int_{\{\zeta:\, |f(\zeta)| > (r\eta)^{1/2}\}} \left(|f| - (r\eta)^{1/2}\right)^2 dm_1 \leqslant 64 \exp\left(-\frac{r\eta}{16}\right). \tag{2}$$

(We shall choose the number r later.)

We shall show that this function f can be represented in the form $f = g + h$, where $\|g\|_U \leqslant \eta$ and $\|h\|_2 \leqslant \exp(-\eta^{1/2})$ (if $\eta \geqslant \alpha$, where α is a suitably chosen absolute constant). The latter can be reformulated as follows: $f \in V_1 + V_2$, where V_1 is the ball of radius η with center at the origin in U and V_2 is the ball of radius $\exp(-\eta^{1/2})$ with center at the origin in H^2 (we remark that in our situation the function h must automatically lie in H^2 and not just in $L^2(m_1)$, as is required in the definition of property $(\mathfrak{D}_\omega)$). Since $V_1 + V_2$ is an absolutely convex subset of H^2 having non-empty interior, to prove the relation $f \in V_1 + V_2$ it is sufficient to show that if $\Phi \in (H^2)^*$ and $|\Phi(x)| \leqslant 1$ for all x in $V_1 + V_2$, then $|\Phi(f)| \leqslant c < 1$.

Thus, let Φ be a functional on H^2 such that $|\Phi(x)| \leqslant 1$, $x \in V_1 + V_2$. There exists a function F in $\overline{H^2}$ for which

$$\int F g \, dm_1 = \Phi(g), \qquad g \in H^2.$$

Since $|\Phi(x)| \leqslant 1$ for all x in V_1 and all x in V_2, we have

$$\|F\|_{U^*} \stackrel{\text{def}}{=} \sup\left\{\left|\int Fg\,dm_1\right| : g \in U, \|g\|_U \leqslant 1\right\} \leqslant \frac{1}{\eta},$$

$$\|F\|_2 \leqslant \exp(\eta^{1/2}).$$

Our immediate task is to show that the number $\|F\|_1$ is "not too large". For this we shall need the following result of Vinogradov [5], [6]:

Suppose $\Psi \in U^*$, and consider the function b on $\mathbb{T}$,

$$b(\zeta) = \lim_{r \to 1-} \Psi\left(\frac{1}{1 - r\bar{\zeta}z}\right)$$

(this limit exists for almost all $\zeta \in \mathbb{T}$). Then

$$m_1\{\zeta : \zeta \in \mathbb{T}, |b(\zeta)| \geqslant t\} \leqslant \frac{K\|\Psi\|_{U^*}}{t}$$

for all $t > 0$, where K is an absolute constant.

This result is highly nontrivial; it suffices to say that its proof makes use of Carleson's theorem on convergence almost everywhere of Fourier series of L^2 functions.

Let us return to our situation. Let Ψ be a functional on U acting like Φ: $\Psi(g) = \int gF\,dm_1$, $g \in U$. Since $F \in \overline{H^2}$, the function b constructed as described above from this functional Ψ coincides with F (since it is easy to see that for all $r < 1$ the Fourier coefficients of the functions $\zeta \mapsto F(r\zeta)$ and $\zeta \mapsto \Psi(1/(1 - r\bar{\zeta}z))$ coincide). Thus, setting

$$\lambda(t) = m_1\{\zeta : \zeta \in \mathbb{T}, |F(\zeta)| > t\},$$

we obtain $\lambda(t) \leqslant K\|F\|_{U^*}/t \leqslant K/\eta t$, $t > 0$, where K is the constant from Vinogradov's theorem.

It is now easy to estimate the number $\|F\|_1$ using the following lemma.

LEMMA 3. *Let (S, μ) be a space with probability measure, and let $G \in L^2(\mu)$, $\|G\|_{L^2(\mu)} \leqslant R$, and $\lambda(t) \leqslant \delta/t$, $t > 0$, where $\lambda(t) = \mu\{s : |G(s)| > t\}$. If $\delta \leqslant R$, then $\|G\|_1 \leqslant 2\delta(1 + \log(R/\delta))$.*

We shall prove this lemma later, but now let us finish the proof of Lemma 2. Supposing η sufficiently large so that $K/\eta \leqslant \exp \eta^{1/2}$, apply Lemma 3 to the function F (with $R = \exp \eta^{1/2}$ and $\delta = K/\eta$). We obtain

$$\|F\|_1 \leqslant \frac{2K}{\eta}\left(1 + \log \frac{\eta \exp \eta^{1/2}}{K}\right).$$

Using this inequality and (2), we estimate the number $\Phi(f)$. Let

$$e = \left\{\zeta : \zeta \in \mathbb{T}, |f(\zeta)| \geqslant (r\eta)^{1/2}\right\}, \quad f_1 = \chi_e \cdot \left(f - (r\eta)^{1/2}\frac{f}{|f|}\right), \quad f_2 = f - f_1.$$

We have

$$|\Phi(f)| = \left|\int fF\,dm_1\right| \leqslant \int_{\mathbb{T}} |f_2F|\,dm_1 + \int_e |f_1F|\,dm_1$$

$$\leqslant (r\eta)^{1/2}\|F\|_1\left(\int_e \left(|f| - (r\eta)^{1/2}\right)^2 dm_1\right)^{1/2}\|F\|_2$$

$$\leqslant r^{1/2}\frac{2K}{\eta^{1/2}}\left(1 + \log\frac{\eta}{K} + \eta^{1/2}\right) + 8\exp\left(-\frac{r\eta}{32}\right)\exp\eta^{1/2}.$$

If r is chosen sufficiently small at the outset, then the last expression is strictly less than unity for all $\eta \geqslant \alpha$, where α is some absolute constant. ●

PROOF OF LEMMA 3. By hypothesis $\lambda(t) \leqslant \delta/t$; from the inequality $\|G\|_{L^2(\mu)} \leqslant R$ it follows easily that $\lambda(t) \leqslant R^2/t^2$; in addition $\lambda(t) \leqslant 1$ (since μ is a probability measure). Fix the numbers a and b, $0 < a \leqslant b$. We have

$$\|F\|_1 = \int_0^\infty \lambda(t)\,dt = \int_0^a \lambda(t)\,dt + \int_a^b \lambda(t)\,dt + \int_b^\infty \lambda(t)\,dt$$

$$\leqslant \int_0^b 1\,dt + \int_a^b \frac{\delta}{t}\,dt + \int_b^\infty \frac{R^2}{t^2}\,dt = a + \delta\log\frac{b}{a} + \frac{R^2}{b^2}.$$

Setting $a = \delta$ and $b = R^2/\delta$ (this is possible if $\delta \leqslant R$, since in this case $\delta \leqslant R^2/\delta$), we obtain what is required. ●

Lemma 2 can be given an abstract form which will be used in the next section. Let (S, μ) be a probability space, E an orthonormal system in $L^2(\mu)$, and X a Banach space continuously embedded in $L^2(\mu)$. Denote by i the injection of X into $L^2(\mu)$ and by H the closure of the set iX in $L^2(\mu)$. Suppose that $E \subset H$ and $|e(s)| \leqslant M$ if $e \in E$ and $s \in S$. Moreover, suppose two nonnegative functions φ and Ω are given; the domain of the first of these is the interval $(0, +\infty)$, and that of the second is a subset of the plane containing the set $[u, +\infty) \times (0, v]$ for some $u, v > 0$.

THEOREM 3. (a) *Suppose the inequality* $\|F\|_{L^1(\mu)} \leqslant \Omega(R, \varepsilon)$ *holds every time that* $F \in \overline{H}$ *(the bar denotes complex conjugation),* $\|F\|_{L^2(\mu)} \leqslant R$, $\|i^*F\|_{X^*} \leqslant \varepsilon$ *and the pair* (R, ε) *lies in the domain of the function* Ω. *If* ω *is a positive function decreasing to zero on* $[0, \pm\infty)$ *such that for all sufficiently large* η(4)

$$\varphi(\eta)^{1/2}\Omega\left(\omega(\eta)^{-1}, \eta^{-1}\right) + 8M\exp\left(-\varphi(\eta)/32M^2\right)\omega(\eta)^{-1} \leqslant c < 1, \qquad (3)$$

then the triple $(X, L^2(\mu), E)$ *possesses property* $(\mathfrak{D}_\omega)$.

(b) *Suppose that there exists a number* K *for which* $F \in \overline{H}$ *implies*

$$\mu\{s : |F(s)| > t\} \leqslant K\|i^*F\|_{X^*}/t, \qquad t > 0.$$

Then the function Ω, *defined on the set* $\{(s, t) : Kt \leqslant s\}$ *by the equality* $\Omega(R, \varepsilon) = 2K\varepsilon(1 + \log(R/K\varepsilon))$, *satisfies the hypothesis of* (a). *If* $\varphi(\eta) = (r\eta)^{1/2}$ *and* $\omega(\eta) = \exp(\eta^{-1/2})$, *then for a suitable choice of* r *inequality* (3) *will hold for all sufficiently large* η.

---

(4) The left-hand side of (3) makes sense for large η because of the condition imposed on the domain of Ω.

The proof repeats (with obvious modifications) the proof of Lemma 2.

(a) According to Lemma 1 we can choose the numbers $\varepsilon_e = \pm 1$ such that for $f \overset{\text{def}}{=} \sum_{n \geqslant 0} \varepsilon_n a_n z^n$ we have

$$\int_{\{s:\, |f(s)| > (\varphi(\eta))^{1/2}\}} \left(|f| - \varphi(\eta)^{1/2}\right)^2 d\mu \leqslant 64M^2 \exp\left(-\frac{\varphi(\eta)}{16M^2}\right).$$

Repeating the estimate from the end of the proof of Lemma 2, we conclude that if $\Phi \in H^*$, $\|\Phi\|_{H^*} \leqslant \omega(\eta)^{-1}$ and $\|i^*\Phi\|_{X^*} \leqslant \eta^{-1}$, then $|\Phi(f)|$ does not exceed the number on the left-hand side of (3); hence, as in Lemma 2 property $(\mathfrak{D}_\omega)$ follows from the triple $(X, L^2(\mu), E)$.

(b) The fact that the function $\Omega(R, \varepsilon) = 2K\varepsilon(1 + \log(R/K\varepsilon))$ satisfies the hypotheses of (a) follows from Lemma 3. The remaining assertion of (b) is immediately verified. ●

REMARK 1. The method of proving Lemma 2 allows us to obtain the following: *There exist constants A and B such that, for any $\eta > 0$, every function f in H^∞ satisfying the condition $\|f\|_\infty = 1$ can be represented in the form $f = g + h$, where $g \in U$, $h \in H^2$, $\|g\|_U \leqslant \eta$ and $\|h\|_2 \leqslant A\exp(-\eta/B)$.* In fact it is sufficient to show that such a decomposition is possible with $A = 1$ and some B for all η in the interval $(\alpha, +\infty)$, where α is some number independent of f (then for $\eta \leqslant \alpha$ it will be possible to set $g = 0$ and $h = f$, and the inequality $\|h\|_2 \leqslant A\exp(-\eta/B)$ can be guaranteed by increasing the constant A). For this it is sufficient to choose B and α such that the relations

$$F \in \overline{H^2}, \qquad \|F\|_2 \leqslant \exp\frac{\eta}{B}, \qquad \|F\|_{U^*} \leqslant \frac{1}{\eta}, \qquad \eta > \alpha,$$

$$f \in H^\infty, \qquad \|f\|_\infty = 1$$

imply the inequality $|\int fF\,dm_1| \leqslant c < 1$. But by Vinogradov's theorem and Lemma 3

$$\left|\int fF\,dm_1\right| \leqslant \|F\|_1 \leqslant \frac{2K}{\eta}\left(1 + \log\frac{\eta\exp(\eta/B)}{K}\right),$$

if $K/\eta \leqslant \exp(\eta/B)$; therefore such a choice of constants B and α is possible.

This fact is related to Men′shov's theorem on correction (for more details, see the article by Khrushchev [11]).([5])

REMARK 2. Since $U \subset C_A$ and $\|f\|_{C_A} \leqslant \|f\|_U$ for all f in U, it of course follows from Lemma 2 that the triple $(C_A, L^2(\mu), E)$, with $E = \{z^n\colon n \geqslant 0\}$, possesses property $(\mathfrak{D}_\omega)$ with $\omega(t) = \exp(-\sqrt{t})$. However, this fact is much simpler than Lemma 2, since it follows immediately from the classical theorem of Kolmogorov and Smirnov that hypothesis (b) of Theorem 3 holds for this triple. Using other reasoning (connected with multiplying by a suitable outer function in a manner similar to that in the following section), we can show that this triple possesses property $(\mathfrak{D}_\omega)$ with $\omega(t) = A\exp(-t^2/B)$, where A and B are suitably chosen.

§4. The space $C_A(\mathbb{T}^2)$

Attempts to extend the results of the preceding section to the space $C_A(\mathbb{T}^n)$, $n \geqslant 2$, encounter certain difficulties, the main one of which is that for $n \geqslant 2$ the orthogonal projection from $L^2(\mathbb{T}^n)$ to $H^2(\mathbb{T}^n)$, in contrast to the case $n = 1$, is no longer of

([5]) See also the later, but earlier-published, article of the present author [12].

weak type 1-1. For $n = 2$ it is nevertheless possible to surmount these difficulties. Almost nothing is known to me about the case $n \geqslant 3$; the methods used for $n = 2$ apparently are no longer helpful here (see §5).

In this section the following assertions will be proved.

THEOREM 4. *Suppose given a double sequence* $\{a_{mn}\}_{m,n\geqslant 0}$, *with* $a_{mn} \geqslant 0$ *and* $\sum_{m,n\geqslant 0} a_{mn}^2 = 1$. *Then there exists a function f in* $C_A(\mathbb{T}^2)$ *such that* $|\hat{f}(m, n)| \geqslant a_{mn}$ *for all* $m, n \geqslant 0$ *and, moreover,* $\|f\|_\infty \leqslant a$, *where the number a is independent of the sequence* $\{a_{mn}\}$.

THEOREM 5. *Suppose* $F \subset \mathbb{Z}_+ \times \mathbb{Z}_+$. *The following assertions are equivalent*:

1) *F is a set of type* Λ_2.

2) *The operator T mapping* $C_A(\mathbb{T}^2)$ *into* $l^2(F)$ *by the formula* $Tf = \{\hat{f}(\nu)\}_{\nu\in F}$ *is surjective*.

The analogue of Theorem 5 for the space $C_A(\mathbb{T})$ is well known (see, for example, [4]). The proof of the implication $1 \Rightarrow 2$ in this theorem for sets of one special type was published by the author in [13]; it differs from the one given here, although it is based on essentially the same idea.

Both theorems will be obtained from the following lemma.

LEMMA 4. *There exist constants u, v, and C with* $u, v > 0$ *and* $C \geqslant 1$, *such that for any numbers* ε, $0 < \varepsilon \leqslant v$, *and* $R \geqslant u$ *and any function F in* $\overline{H^2(\mathbb{T}^2)}$ *satisfying the inequalities*

$$\|F\|_{C_A(\mathbb{T}^2)^*} \overset{\text{def}}{=} \sup\left\{\left|\int Fg\,dm_2\right| : g \in C_A(\mathbb{T}^2),\ \|g\|_\infty \leqslant 1\right\} \leqslant \varepsilon$$

and $\|F\|_2 \leqslant R$, *the following relation holds*:

$$\|F\|_1 = \int |F|\,dm_2 \leqslant 2C\varepsilon^{1/3}R\left(1 + \log\frac{1}{C\varepsilon^{1/3}}\right).$$

We begin by deducing Theorems 4 and 5 from Lemma 4, which we shall prove later.

PROOF OF THEOREM 4. Apply assertion (a) of Theorem 3 to the triple $(C_A(\mathbb{T}^2), L^2(m_2), E)$, where $E = \{z_1^m z_2^n\colon m, n \geqslant 0\}$. For this it is necessary to choose suitable functions Ω, φ, and ω. Lemma 4 shows that it is possible to set

$$\Omega(R, \varepsilon) = 2C\varepsilon^{1/3}R(1 + \log(1/C\varepsilon^{1/3})).$$

Let $\varphi(\eta) = 64\log(1 + \eta)$ and $\omega(\eta) = \eta^{-(1/3)}(\log(1 + \eta))^2$. Then the left-hand side of (3) in Theorem 3 (in which we set $M = 1$) assumes the following form:

$$16C(\log(1 + \eta))^{-3/2}(1 + \log(\eta^{1/3}/C)) + 8(1 + \eta)^{-2}\eta^{1/3}(\log(1 + \eta))^{-2}.$$

This last expression tends to zero as $\eta \to \infty$; therefore by Theorem 3 the triple $(C_A(\mathbb{T}^2), L^2(m_2), E)$ possesses property $(\mathfrak{D}_\omega)$ with $\omega(\eta) = \eta^{-1/3}(\log(1 + \eta))^2$, and it remains only to apply Theorem 1.([6]) ●

([6]) Thus ω decreases no slower than $\eta^{-\alpha}$ for any $\alpha < 1/3$. I have tried to find the "best possible" α which is obtainable from the proof given here. This explains why the formulas are somewhat cumbersome. I do not know if one can take $\alpha = 1/3$.

PROOF OF THEOREM 5. The operator T is surjective if and only if $\|T^*G\| \geqslant K\|G\|$ for all G and some K independent of G. Therefore assertion 2) of Theorem 5 can be reformulated as follows.

2′) *There exists a number K such that $\|p\|_{C_A(\mathbb{T}^2)^*} \geqslant K\|p\|_2$ for any trigonometric polynomial p on $\mathbb{T}^2$ satisfying the relation* $\operatorname{supp}\hat{p} \subset -F$.

The implication 2′) $\Rightarrow$ 1) is obtained quite simply: Suppose $\operatorname{supp}\hat{p} \subset -F$; then $\|p\|_1 \geqslant \|p\|_{C_A(\mathbb{T}^2)^*} \geqslant K\|p\|_2$.

We shall show that 2′) follows from 1). By definition of sets of type Λ_2 there exists a constant a such that $\|p\|_2 \leqslant a\|p\|_1$ for all trigonometric polynomials p such that $\operatorname{supp}\hat{p} \subset -F$. Let u and v be the numbers from the hypothesis of Lemma 4; consider a trigonometric polynomial p such that $\operatorname{supp}\hat{p} \subset -F$ and $\|p\|_2 = u$. Set $\varepsilon = \|p\|_{C_A(\mathbb{T}^2)^*}$. If $\varepsilon \leqslant v$, then, by Lemma 4,

$$u = \|p\|_2 \leqslant a\|p\|_1 \leqslant 2aC\varepsilon^{1/3}\left(1 + \log(1/C\varepsilon^{1/3})\right) \leqslant B\varepsilon^{1/6}$$

for some constant B depending only on the set F. Thus if $\operatorname{supp}\hat{p} \subset -F$ and $\|p\|_2 = u$, then $\|p\|_{C_A(\mathbb{T}^2)^*} \geqslant \min(v, (u/B)^6)$, and this is assertion 2′). ●

To prove Lemma 4 we shall need several new objects. Consider in $L^2(\mathbb{T}^2)$ four closed subspaces:

$$X_1 \stackrel{\text{def}}{=} \operatorname{span}\{z_1^m z_2^n : m \geqslant 0\}, \qquad X_2 \stackrel{\text{def}}{=} \operatorname{span}\{z_1^m z_2^n : n \geqslant 0\},$$

$$X_1^0 \stackrel{\text{def}}{=} \operatorname{span}\{z_1^m z_2^n : m > 0\}, \qquad X_2^0 \stackrel{\text{def}}{=} \operatorname{span}\{z_1^m z_2^n : n \geqslant 0\}.$$

Let P_1 and P_2 be the orthogonal projections whose kernels coincide respectively with X_1^0 and X_2^0. The action of P_j reduces simply to the application of the Riesz projection on the variable z_j ($j = 1, 2$); therefore P_1 and P_2 are of weak type 1-1:

$$m_2\{t : t \in \mathbb{T}^2, |P_j f| > t\} \leqslant A\|f\|_1/t,$$

if $f \in L^2(\mathbb{T}^2)$ and $t > 0, j = 1, 2$.

Consider also operators Γ_1 and Γ_2, given by

$$(\Gamma_1 f)(z_1, z_2) = \frac{1}{i}\left((f - 2P_1 f)(z_1, z_2) + \int_{\mathbb{T}} f(\zeta, z_2)\, dm_1(\zeta)\right),$$

$$(\Gamma_2 f)(z_1, z_2) = \frac{1}{i}\left((f - 2P_2 f)(z_1, z_2) + \int_{\mathbb{T}} f(z_1, \zeta)\, dm_1(\zeta)\right),$$

$$f \in L^2(\mathbb{T}^2).$$

The action of each of these two operators reduces to the application of the usual "one-dimensional" operation of harmonic conjugation with respect to the first and second variables respectively. Γ_1 and Γ_2, like P_1 and P_2, are of weak type 1-1 (and we shall assume that the constants in the inequalities of weak type for these operators also do not exceed A).

It is clear that many facts from the ordinary (one-dimensional) theory of Hardy classes have analogues for the spaces X_1 and X_2. For example, for any function f in $\operatorname{Re} L^2(\mathbb{T}^2)$ the function $\Gamma_j f$ is real and $f + i\Gamma_j f \in X_j$ ($j = 1, 2$); if in addition the equality $\int f(\zeta, \cdot)\, dm(\zeta) = \mathbf{0}$ holds, then $f + i\Gamma_1 f \in X_1^0$ (and an analogous assertion holds for Γ_2). If f is a continuous real function bounded away from zero on $\mathbb{T}^2$, then the function $g = \exp(\log f + i\Gamma_j(\log f))$ lies in X_j and $|g| = |f|$ almost everywhere, etc.

PROOF OF LEMMA 4. Suppose F satisfies the hypotheses of this lemma. Set $V = -(\mathbb{Z}_+ \times \mathbb{Z}_+)$ and $W = (\mathbb{Z} \times \mathbb{Z})\backslash V$. By hypothesis, $\operatorname{supp} \hat{F} \subset V$. Since $C_A(\mathbb{T}^2)^* = C(\mathbb{T}^2)^*/C_A(\mathbb{T}^2)^\perp$, there exists a measure μ on $\mathbb{T}^2$ such that $\operatorname{supp} \hat{\mu} \subset W$ and $\|F \cdot m_2 + \mu\|_{C(\mathbb{T}^2)^*} = \|F\|_{C_A(\mathbb{T}^2)^*} \leqslant \varepsilon$. By considering the convolution of the measure $F \cdot m_2 + \mu$ with the Fejér kernels we see that Lemma 4 will be proved if the following is verified:

Let q and r be the trigonometric polynomials on $\mathbb{T}^2$ with $\operatorname{supp} \hat{q} \subset V$, $\operatorname{supp} \hat{r} \subset W$, $\|q + r\|_1 \leqslant \varepsilon$ *and* $\|q\|_2 \leqslant R$; *then*

$$\|q\|_1 \leqslant 2C\varepsilon^{1/3}R\big(1 + \log(1/C\varepsilon^{1/3})\big),$$

where C is an absolute constant.

Set $s = P_2 r$; then $s \in X_1^0$ (since $\operatorname{supp} \hat{r} \subset W$). Since $P_2 q = q$, for all $t > 0$

$$\lambda(t) \overset{\text{def}}{=} m_2\{\tau \colon \tau \in \mathbb{T}^2, |q(\tau) + s(\tau)| \geqslant t\} \leqslant A\varepsilon/t. \tag{4}$$

Our next task is to replace s by another function s_1 from X_1^0 in such a way that the number $\|q + s_1\|_1$ is "small if ε is small" (after this we shall be able to complete the proof using Lemma 3).

To make this exchange we define the function h on $\mathbb{T}^2$ by the equality

$$h(\tau) = \begin{cases} 1, & \text{if } |q(\tau) + s(\tau)| \leqslant 1, \\ |q(\tau) + s(\tau)|^{-1} & \text{if } |q(\tau) + s(\tau)| \geqslant 1, \end{cases}$$

and set $H = \exp(\log h + i\Gamma_1(\log h))$. It is obvious that $H \in X_1$ and therefore the function s_1 given by $s_1 = sH$ lies in X_1^0 (since s is a trigonometric polynomial in X_1^0). We shall show that if $R \geqslant 1$ and $\varepsilon \leqslant 1$, then

$$\|q + s_1\|_1 \leqslant DR\varepsilon^{1/3}, \tag{5}$$

where D is an absolute constant.

For the time being we suppose that this inequality is already proved and finish the proof of Lemma 4 using it. Making use of the fact that $P_1(q + s_1) = q$ and the operator P_1 is of weak type 1-1, we obtain

$$m_2\{\tau \colon \tau \in \mathbb{T}^2, |q(\tau)| > t\} \leqslant ADR\varepsilon^{1/3}/t, \qquad t > 0.$$

Since $\|q\|_2 \leqslant R$, the following is a consequence of Lemma 3: *if $\varepsilon \leqslant 1$ and $R \geqslant 1$ are such that $ADR\varepsilon^{1/3} \leqslant R$ (i.e., if $R \geqslant 1$ and $\varepsilon \leqslant (AD)^{-3}$), then*

$$\|q\|_1 \leqslant 2ADR\varepsilon^{1/3}\big(1 + \log(R/ADR\varepsilon^{1/3})\big) = 2CR\varepsilon^{1/3}\big(1 + \log(1/C\varepsilon^{1/3})\big),$$

where $C = AD$. But this is the assertion of Lemma 4.

It remains to prove (5). We have

$$\|q + s_1\|_1 = \|q + sH\|_1 \leqslant \|q - qH\|_1 + \|H \cdot (q + s)\|_1.$$

We first estimate the second term on the right in this last inequality. Suppose $E = \{\tau \colon \tau \in \mathbb{T}^2, |q(\tau) + s(\tau)| \leqslant 1\}$. By (4), $m_2(\mathbb{T}^2 \backslash E) \leqslant A\varepsilon$. By the construction of the function H the relations $|H| \leqslant 1$ and $|H||q + s| \leqslant 1$ hold almost everywhere. Therefore if λ is the function defined by (4), then (under the assumption

$A\varepsilon \leqslant 1$) we have

$$\begin{aligned}\int_{\mathbb{T}^2} |H||q+s|\,dm_2 &\leqslant \int_E |q+s|\,dm_2 \leqslant \int_{\mathbb{T}^2\setminus E} dm_2 + \int_0^1 \lambda(t)\,dt + A\varepsilon \\ &= A\varepsilon + \int_0^{A\varepsilon} \lambda(t)\,dt + \int_{A\varepsilon}^1 \lambda(t)\,dt \leqslant A\varepsilon + \int_0^{A\varepsilon} dt + \int_{A\varepsilon}^1 \frac{A\varepsilon}{t}\,dt \\ &= A\varepsilon + A\varepsilon + A\varepsilon \log\frac{1}{A\varepsilon} \leqslant A_1\sqrt{\varepsilon},\end{aligned}$$

where A_1 is an absolute constant. And if $A\varepsilon \geqslant 1$, then the inequality $\|H(q+s)\|_1 \leqslant A\sqrt{\varepsilon}$ is obvious.

We now turn to the estimation of the first term, i.e. the number $\|q - qH\|_1$. First of all,

$$\int_{\mathbb{T}^2} |\log h|\,dm_2 = -\int_1^\infty \log t\,d\lambda(t) = \int_1^\infty \frac{\lambda(t)}{t}\,dt \leqslant A\varepsilon\int_1^\infty \frac{dt}{t^2} = A\varepsilon.$$

Therefore $m_2\{\tau\colon |(\Gamma_1(\log h))(\tau)| > \tau\} \leqslant A^2\varepsilon/t$ for all $t > 0$. Hence

$$\begin{aligned}&m_2\{\tau\colon |(\log h + i\Gamma_1(\log h))(\tau)| > t\} \\ &\quad\leqslant m_2\{\tau\colon |\log h(\tau)| > t/2\} + m_2\{\tau\colon |(\Gamma_1(\log h))(\tau)| > t/2\} \\ &\quad\leqslant 2A\varepsilon/t + 2A^2\varepsilon/t \leqslant 4A^2\varepsilon/t.\end{aligned}$$

Let $e = \{\tau\colon |\log h + i(\Gamma_1(\log h))(\tau)| \leqslant 2A\varepsilon^{1/3}\}$. Putting $t = 2A\varepsilon^{1/3}$ in the preceding inequality, we see that $m_2(\mathbb{T}^2\setminus e) \leqslant 2A\varepsilon^{2/3}$. Let B be a constant such that $|\exp(\zeta) - 1| \leqslant B|\zeta|$ for all ζ satisfying $|\zeta| \leqslant 2A$. We have

$$\begin{aligned}\int |1-H|^2\,dm_2 &= \int |1-H|^2\,dm_2 + \int_{\mathbb{T}^2\setminus e} |1-H|^2\,dm_2 \\ &\leqslant \int_e B\cdot 4A^2\varepsilon^{2/3}\,dm_2 + 4m_2(\mathbb{T}^2\setminus e) \leqslant 4A^2B\varepsilon^{2/3} + 4A\varepsilon^{2/3},\end{aligned}$$

whence it follows that $\|1 - H\|_2 \leqslant B_1\varepsilon^{1/3}$ (B_1 is an absolute constant). Finally, since $\|q\|_2 \leqslant R$, we obtain the following estimate for the first term:

$$\|q - qH\|_1 \leqslant \|q_2\|\,\|1 - H\|_2 \leqslant RB_1\varepsilon^{1/3}.$$

Last of all, $\|q + s_1\|_1 \leqslant RB_1\varepsilon^{1/3} + A_1\varepsilon^{1/2} \leqslant DR\varepsilon^{1/3}$, where D is an absolute constant; and this is (5). ●

§5. Concluding remarks

It is easy to extend Theorems 4 and 5 to subspaces of the space $C(\mathbb{T}^2)$ consisting of functions whose Fourier coefficients are zero outside a fixed sector with vertex at the origin (in place of the operators P_1, P_2, Γ_1 and Γ_2 in the proof of Lemma 4 it is necessary to use the analogous operators constructed from suitably chosen Dirichlet algebras on $\mathbb{T}^2$; see [14], Chapter 2, §3 and Chapter 7). However it is not clear to what extent these theorems may be carried over to the spaces $C_A(\mathbb{T}^n)$, $n \geqslant 3$.

The argument of the preceding section does not go through for the following reason. If we attempt to prove the analogue of Lemma 4 for the three-dimensional torus we come to the following problem: Let $V = \{(l, m, n) \in \mathbb{Z}^3\colon l, m, n \leqslant 0\}$, $W = \mathbb{Z}^3\setminus V$, q and r trigonometric polynomials whose Fourier transforms are concentrated respectively on the sets V and W, and $\|q\|_2 \leqslant R$ and $\|q+r\|_{C_A(\mathbb{T}^3)^*} \leqslant \varepsilon$. It is required to obtain an estimate of the form $\|q_1\| \leqslant \Omega(R, \varepsilon)$, where the function Ω is sufficiently good (for example it must not tend to zero too slowly if ε tends to 0). Let P be the orthogonal projection of $L^2(\mathbb{T}^3)$ onto the subspace span $\{z_1^l z_2^m z_3^n : n \leqslant 0\}$, and (as in the proof of Lemma 3) let $s = Pr$. Then, of course,

$$m_3\{\tau : |q(\tau)+s(\tau)| > t\} \leqslant A\varepsilon/t;$$

nevertheless, as pointed out by A. B. Aleksandrov, in contrast to the case $n = 2$ this inequality no longer carries any information: for any trigonometric polynomial p such that $\operatorname{supp}\hat{p} \subset V$, and any $\delta > 0$, there exists a trigonometric polynomial s such that

$$\operatorname{supp}\hat{s} \subset \{(l, m, n) \in \mathbb{Z}^3 : n \leqslant 0\} \setminus V$$

and $m_3\{\tau : |p(\tau)+s(\tau)| > t\} \leqslant \delta/t$, $t > 0$.

The best thing known to me about the spaces $C_A(\mathbb{T}^n)$, $n \geqslant 3$, is the following theorem.

THEOREM 6. *There exists a number C (independent of n) such that if*

$$\sum_{k \in (\mathbb{Z}_+)^n} |a_k||\hat{f}(k)| \leqslant K\|f\|_\infty, \qquad f \in C_A(\mathbb{T}^n),$$

then for all natural numbers N

$$\left(\sum_{k:\, \max\{k_1,\ldots,k_n\} \leqslant N} |a_k|^2\right)^{1/2} \leqslant CK(n\log(Nn))^{1/2}.$$

The proof uses certain facts and concepts of the theory of p-absolutely summing operators, for which we refer the reader to the article [15] or the survey [16], and the bibliography there found. We shall retain the notation of [16] (in particular, $\pi_p(T)$ is the norm of the operator T in the ideal of p-absolutely summing operators).

Let $\mathcal{P}_{n,N}$ be the space of trigonometric polynomials of the form

$$\sum_{0 \leqslant k_j \leqslant N} a_{k_1,\ldots,k_n} z_1^{k_1} \cdots z_n^{k_n}$$

endowed with the metric of $C_A(\mathbb{T}^n)$. We shall need two (evidently well-known) lemmas.

LEMMA 5. *The Banach-Mazur distance between the space $\mathcal{P}_{n,N}$ and some subspace of $l^\infty_{(100Nn)^n}$ does not exceed ten.*

PROOF. Partition the circle $\mathbb{T}$ into $100Nn$ congruent arcs: $e_1, \ldots, e_{100Nn}$, and let Y be the linear span of the characteristic functions of the sets $A_{i_1,\ldots,i_n} \subset \mathbb{T}^n$, $A_{i_1,\ldots,i_n} = e_{i_1} \times \cdots \times e_{i_n}$, $1 \leqslant i_1, \ldots, i_n \leqslant 100Nn$, and let Y be endowed with the metric form $L^\infty(\mathbb{T}^n)$. Then Y is isometric to $l^\infty_{(100Nn)^n}$. Let s_j be the midpoint of the arc e_j, and $p_{i_1,\ldots,i_n} = (s_{i_1}, \ldots, s_{i_n}) \in \mathbb{T}^n$. Define an operator R from $\mathcal{P}_{n,N}$ into Y by the equality

$$Rf = \sum_{1 \leqslant i_1,\ldots,i_n \leqslant 100Nn} f(p_{i_1,\ldots,i_n})\chi_{A_{i_1,\ldots,i_n}}.$$

Then $\|R\| \leqslant 1$. On the other hand, from Bernstein's inequality it follows that $\|Rf\| \geqslant 10^{-1}\|f\|$, $f \in \mathcal{P}_{n,N}$, and consequently the Banach-Mazur distance between $\mathcal{P}_{n,N}$ and $R(\mathcal{P}_{n,N})$ does not exceed ten. ●

LEMMA 6. *Let k be a natural number, X a subspace of the space l_k^∞, and T a linear operator from X into l^1. Then $\pi_2(T) \leqslant d(\log k)^{1/2}\|T\|$, where d is independent of k, X, and T.*

PROOF. Let K be a discrete space consisting of k points, and μ the measure on K taking on the value k^{-1} on each of its singleton subsets. Let I be the injection of the space $C(K)$ (which is nothing more than l_k^∞) into $L^{\log k}(K, \mu)$. Since $\pi_{\log k}(I) = 1$ (see [16], the example on p. 121), while $\|I^{-1}\| \leqslant \exp(1)$ (this follows from Hölder's inequality), we have $\pi_{\log k}(\mathrm{id}_X) \leqslant \pi_{\log k}(\mathrm{id}_{l_k^\infty}) = \pi_{\log k}(I^{-1}I) \leqslant \|I^{-1}\|\pi_{\log k}(I) \leqslant \exp(1)$. Applying now Proposition 3 of [17] and taking account of the fact that the constant K_q in this proposition does not exceed $C\sqrt{q}$, we obtain

$$\begin{aligned}\pi_2(T) &\leqslant C\sqrt{\log k}\,\pi_{\log k}(T \circ \mathrm{id}_X)\\ &\leqslant C\sqrt{\log k}\,\|T\|\,\pi_{\log k}(\mathrm{id}_X) \leqslant d\sqrt{\log k}\,\|T\|. \quad \bullet\end{aligned}$$

Returning to the proof of Theorem 6, denote by T the operator from $\mathcal{P}_{n,N}$ into l^1 given by

$$Tf = \{\hat{f}(k)a_k\}_{k\in(\mathbf{Z}_+)^n}.$$

By hypothesis $\|T\| \leqslant K$; consequently, Lemmas 5 and 6 give the result that $\pi_2(T) \leqslant 10d(\log((100Nn)^n))^{1/2}K$. By the Grothendieck-Pietsch theorem on operators which commute with group operations (see [18] or [16], p. 123, Theorem 2) we have

$$\|Tf\|_{l^1} \leqslant \pi_2(T)\|f\|_2, \qquad f \in \mathcal{P}_{n,N},$$

i.e.

$$\sum_{0\leqslant k_j\leqslant N} |a_{k_1\cdots k_n}||\hat{f}(k_1,\ldots,k_n)| \leqslant 10d(n\log(100Nn))^{1/2}\cdot K\left(\int_{\mathbf{T}^n}|f|^2\,dm_n\right)^{1/2}$$

for all f in $\mathcal{P}_{n,N}$. The required result follows immediately. $\bullet$

In conclusion I would like to formulate a problem which again relates to the "one-dimensional" situation. Propositions of types A and B from the Introduction hold for many spaces consisting of functions whose supports lie in a fixed subset of positive measure (see, for example, [6]). It would be interesting to find out whether assertions similar to thc theorem of de Leeuw, Katznelson and Kahane hold for these spaces. The simplest of these questions is the following:

Let $e \subset \mathbf{T}$, $m_1 e > 0$, and let $a = \{a_n\}_{n\in\mathbf{Z}} \in l^2$. Is there a function f in $L^2(\mathbf{T})$ which vanishes outside e and is such that $|\hat{f}(n)| \geqslant |a_n|$ for all n?

A positive answer to this question would evidently allow us to prove even stronger assertions (as was done for the spaces considered in this work).

BIBLIOGRAPHY

1. Torsten Carleman, *Über die Fourierkoeffizienten einer stetigen Funktion*, Acta Math. **41** (1916/18), 377–384.

2. S. Banach, *Über einige Eigenschaften der lacunären trigonometrischen Reihen*, Studia Math. **2** (1930), 207–220.

3. R. E. A. C. Paley, *A note on power series*, J. London Math. Soc. **7** (1932), 122–130.

4. S. A. Vinogradov, *Interpolation theorems of Banach-Rudin-Carleson and norms of imbedding operators for certain classes of analytic functions*, Zap. Naučn. Sem. Leningrad. Otdel. Mat. Inst. Steklov. (LOMI) **19** (1970), 6–54; English transl. in Sem. Math. V. A. Steklov Math. Inst. Leningrad **19** (1970).

5. ______, *Convergence almost everywhere of Fourier series of functions in L^2 and the behavior of the coefficients of uniformly convergent Fourier series*, Dokl. Akad. Nauk SSSR **230** (1976), 508–511; English transl. in Soviet Math. Dokl. **17** (1976).

6. ______, *A strengthened form of Kolmogorov's theorem on the conjugate function and interpolation properties of uniformly convergent power series*, this issue.

7. Karel de Leeuw, Yitzhak Katznelson and Jean-Pierre Kahane, *Sur les coefficients de Fourier des fonctions continues*, C. R. Acad. Sci. Paris Sér. A–B **285** (1977), A1001–A1003.

8. S. Sidon, *Einige Sätze und Fragestellungen über Fourier-Koeffizienten*, Math. Z. **34** (1931/32), 477–480.

9. A. Zygmund, *Trigonometric series*, 2nd rev. ed., Vols. I, II, Cambridge Univ. Press, 1959.

10. Elias M. Stein, *Singular integrals and differentiability properties of functions*, Princeton Univ. Press, Princeton, N.J., 1970.

11. S. V. Khrushchev, *Men'šov's correction theorem and Gaussian processes*, this issue.

12. S. V. Kislyakov, *Quantitative aspects of correction theorems*, Zap. Naučn. Sem. Leningrad. Otdel Mat. Inst. Steklov. (LOMI) **92** (1979), 182–191; English transl. in J. Soviet Math. (to appear).

13. ______, *In the space of continuously differentiable functions on the torus there exists no unconditional local structure*, Preprint R-1-77, Leningrad, Otdel. Mat. Inst. Steklov. (LOMI), Leningrad, 1977. (Russian)

14. Theodore W. Gamelin, *Uniform algebras*, Prentice-Hall, Englewood Cliffs, N.J., 1969.

15. J. Lindenstrauss and A. Pełczyński, *Absolute summing operators in p spaces and their applications*, Studia Math. **29** (1968), 275–326.

16. S. V. Kislyakov, *p-absolutely summing operators*, Geometry of Linear Spaces and Operator Theory (B. S. Mityagin, editor), Yaroslav. Gos. Univ., Yaroslavl, 1977, pp. 114–174. (Russian)

17. Bernard Maurey, *Espaces de cotype (p, q) et théorèmes de relèvement*, C. R. Acad. Sci. Paris Sér. A–B **275** (1972), A785–A788.

18. A. Pełczyński, *p-integral operators commuting with group representations and examples of quasi-p-integral operators which are not p-integral*, Studia Math. **33** (1969), 63–70.

Translated by R. L. COOKE

Труды Матем. Инст. Стеклов.
155(1981)

Proceedings of the Steklov
Institute of Mathematics
1983, Issue 1

POSITIVE PROJECTIONS ON SYMMETRIC *KB*-SPACES

UDC 517.948:513.8 + 519.4

V. G. KULAKOVA

ABSTRACT. This article deals with the arbitrary positive projections on symmetric *KB*-spaces (Definition VII.6.1 of B. Z. Vulikh's *Introduction to the theory of partially ordered spaces*, English transl., Noordhoff, Groningen, 1957) (in particular, in Orlicz spaces when the Δ_2-condition is satisfied) and their connection with conditional expectation operators. The question of the structure of closed linear sublattices in symmetric *KB*-spaces is also considered.

Bibliography: 5 titles.

We consider arbitrary positive projections on symmetric *KB*-spaces* (in particular, Orlicz spaces if the Δ_2-condition holds) and their connection with conditional expectation (CE) operators. We hold to the terminology and notation in the theory of partially ordered spaces in [1]. Historically, the necessity of studying positive projections was prompted both by the needs of functional analysis itself and by the fact that such operators arise in the theory of turbulence and statistical mechanics [2]. It is well known that the restriction to L^p of a CE operator is a positive projection. It is natural to try to invert this assertion, i.e., to ask for conditions under which any positive projection is the CE with respect to some σ-subalgebra of measurable sets. The problem has been solved for projections with unit norm in L^p-spaces in a number of papers (for example, [3]). Here no assumptions are made about the norm.

§1. Linear sublattices

Before passing to the solution of the main problem of describing positive projections in terms of conditional expectations, we consider the structure of closed linear sublattices of symmetric *KB*-spaces.

DEFINITION 1. Let $(\Omega, \mathfrak{A}, P)$ be the underlying probability space. The *conditional expectation* (CE) with respect to a σ-subalgebra $\mathfrak{B}$ of $\mathfrak{A}$ and to P is defined to be the operator $E_P^{\mathfrak{B}}$ such that

1) $E_P^{\mathfrak{B}}(L^1(\Omega, \mathfrak{A}, P)) \subset L^1(\Omega, \mathfrak{B}, P \mid \mathfrak{B})$;

2) $\int_B E_P^{\mathfrak{B}} x\, dP \mid \mathfrak{B} = \int_B x\, dP$

for any $x \in L^1(\Omega, \mathfrak{A}, P)$, and $B \in \mathfrak{B}$.

DEFINITION 2. Let $(\Omega, \mathfrak{A}, P)$ be the underlying probability space, and $\mathfrak{B}$ a σ-subalgebra of $\mathfrak{A}$. Then k_P is called a *weight function* for $\mathfrak{B}$ with respect to P if

---

1980 *Mathematics Subject Classification*. Primary 46B30, 47B55; Secondary 76F99, 46E30.

* *Editor's note*. A *KB*-space is an order-complete Banach lattice that is σ-order continuous and monotonically σ-complete.

©1983 American Mathematical Society
0081-5438/83 $1.00 + $.25 per page

$k_P \geqslant \mathbf{0}$, $E_P^{\mathfrak{B}}(k_P) = \chi_{S(k_P)}$ and $S(k_P) \in \mathfrak{B}$ (here $S(k_P)$ is the support of k_P, and $\chi_{S(k_P)}$ is the characteristic function of $S(k_P)$).

THEOREM 1. *Let $(\Omega, \mathfrak{A}, P)$ be the underlying probability space, X a symmetric KB-space of measurable functions on Ω, and Y a nontrivial closed linear sublattice of X. Then there exist a unique σ-subalgebra $\mathfrak{B}$ of $\mathfrak{A}$ and a unique weight function k for $\mathfrak{B}$ with respect to P such that $Y = kX(\mathfrak{B}) \cap X$, where $X(\mathfrak{B}) = \{f: f \in X; f$ is measurable with respect to $\mathfrak{B}\}$. Moreover, there exists a $\mathfrak{B}$-measurable partition of the measure space $\Omega = \bigcup_0^\infty \Omega_n$, $\Omega_n \cap \Omega_m = \varnothing$ $(n \neq m)$, $\Omega_n \in \mathfrak{B}$ for any n, such that $Y_n = k_n X_n(\mathfrak{B})$, on each set Ω_n, where*[1]

$$k_n = k\chi_{\Omega_n} \in Y; \qquad k_n X_n(\mathfrak{B}) \subset X_n;$$
$$Y_n = \{f\chi_{\Omega_n}: f \in Y\}; \qquad X_n(\mathfrak{B}) = \{f\chi_{\Omega_n}: f \in X(\mathfrak{B})\};$$
$$X_n = \{f\chi_{\Omega_n}: f \in X\}.$$

PROOF. We construct a sequence of real-valued functions $\{f_n\}_{n\geqslant 0}$ in Y as follows. Let f_0 be any function in Y such that $P(S(f_0)) > 0$, where $S(f_0)$ is the support of f_0. Suppose that $f_0, \dots, f_{N-1}$ have been constructed, and choose f_N in such a way that $P[S(f_N) - \bigcup_{n=0}^{N-1} S(f_n)] > 1/N$. If this is impossible, set $f_N = \mathbf{0}$.

Let

$$h = \sum_{\substack{n \\ \|f_n\|_X \neq 0}} \left(2^n \|f_n\|_X\right)^{-1} |f_n| \geqslant \mathbf{0}$$

(here $\|\cdot\|_X$ denotes the norm in X).

It is easy to see that $S(f) \subset S(h)$ to within a set of measure zero for each $f \in Y$.

Let clos Y be the closure of Y in the norm of $L^1(\Omega, \mathfrak{A}, P)$. It is not hard to see that clos Y is a sublattice of L^1.

Let us form the class of sets

$$\mathfrak{B}_h = \{A: A \in \mathfrak{A}; h\chi_A \in \operatorname{clos} Y\}.$$

Since clos Y is a sublattice of L^1 and $h \in \operatorname{clos} Y$, it follows that $\mathfrak{B}_h$ is a σ-subalgebra of $\mathfrak{A}$.

We consider two set functions on $\mathfrak{B}_h$. For any $A \in \mathfrak{B}_h$ let

$$\mu_h(A) = \int_A \chi_{S(h)}\, dP = P(S(h) \cap A); \qquad \tilde{\mu}_h(A) = \int_A h\, dP.$$

Then μ_k and $\tilde{\mu}_k$ are mutually absolutely continuous measures on $\mathfrak{B}_h$, and, consequently, by the Radon-Nikodým theorem, there exists a nonnegative $\mathfrak{B}_h$-measurable function g such that $\mu_h(A) = \int_A gh\, dP$ for any $A \in \mathfrak{B}_h$.

Let $\{g_n\}_{n\geqslant 0}$ be a sequence of nonnegative $\mathfrak{B}_h$-measurable simple functions such that $g_n \uparrow g$. We form the sequence $\{g_n h\}_{n\geqslant 0}$, $\{g_n h\}_{n\geqslant 0} \subset \operatorname{clos} Y$. Then $g_n h \to gh$ in the L^1-norm and, hence, $gh \in \operatorname{clos} Y$.

Let $k = gh$. This function k has the properties that $k \in \operatorname{clos} Y$, $k \geqslant \mathbf{0}$ and $\int_A \chi_{S(h)}\, dP = \int_A k\, dP$ for any $A \in \mathfrak{B}_h$, i.e., $E^{\mathfrak{B}_h}(k) = \chi_{S(h)}$, where $E^{\mathfrak{B}_h}$ is the CE operator in $L^1(\Omega, \mathfrak{A}, P)$ with respect to the σ-subalgebra $\mathfrak{B}_h$.

---

[1] Douglas [3] obtained the analogous result for the case of L^1-spaces with a finite measure.

Let us form one more class of sets

$$\mathfrak{B}_k = \{A: A \in \mathfrak{A};\ k\chi_A \in \operatorname{clos} Y\}$$

and prove that $\mathfrak{B}_h = \mathfrak{B}_k$. To prove this it suffices to observe (similarly to [3]) that for any set $T \in \mathfrak{B}_k$ ($T \in \mathfrak{B}_h$), the sequence $\{(h - nk\chi_T)^+\}_{n\geqslant 0}$ ($\{(k - nh\chi_T)^+\}_{n\geqslant 0}$) is such that

$$(h - nk\chi_T)^+ \to h\chi_{\Omega\setminus T} \qquad \left((k - nh\chi_T)^+ \to k\chi_{\Omega\setminus T}\right)$$

in the L^1-norm, and, consequently, $h\chi_{\Omega\setminus T} \in \operatorname{clos} Y$ ($k\chi_{\Omega\setminus T} \in \operatorname{clos} Y$). Hence, $T \in \mathfrak{B}_h$ ($T \in \mathfrak{B}_k$).

We denote the σ-subalgebra $\mathfrak{B}_h = \mathfrak{B}_k$ by $\mathfrak{B}$ and prove now that

$$Y = kX(\mathfrak{B}) \cap X. \tag{1}$$

Let $g \in kX(\mathfrak{B}) \cap X$ be an arbitrary nonnegative function. This means, first, that $g \in X$, and, second, that $g = kf$ for some $f \in X(\mathfrak{B})$, $f \geqslant \mathbf{0}$. Since f is $\mathfrak{B}$-measurable, there exists a sequence $\{f_n\}_{n\geqslant 0}$ of nonnegative $\mathfrak{B}$-measurable simple functions such that $f_n \uparrow f$. Then $\mathbf{0} \leqslant kf_n \leqslant kf = g \in X$ for any n; the subspace X is normal* by assumption, and, consequently, all the functions kf_n are in X. But then the convergence $kf_n \uparrow kf$ implies that $kf_n \to kf$ in the norm $\|\cdot\|_X$.

We remark that $\{kf_n\}_{n\geqslant 0} \subset \operatorname{clos} Y \cap X$, and, hence, $\{kf_n\}_{n\geqslant 0} \subset Y$. Indeed, if kf_n is an element of $\operatorname{clos} Y$, then there exists a sequence $\{g_N^{(n)}\}_{N\geqslant 0} \subset Y$ such that $g_N^{(n)} \to kf_n$ in the L^1-norm. It can always be assumed that this is a monotonically increasing sequence. Then $g_N^{(n)} \to kf_n$ in the norm $\|\cdot\|$ and, hence, $\{kf_n\}_{n\geqslant 0} \subset Y$ (Y is closed in X). We have, in essence, proved that $Y = (\operatorname{clos} Y) \cap X$. Consequently, $kf \in Y$. The inclusion $kX(\mathfrak{B}) \cap X \subset Y$ is proved.

We prove the reverse inclusion. Let us first show that for any $f \in Y$, $f \geqslant \mathbf{0}$, a sequence $\{g_N\}_{N\geqslant 0}$ of nonnegative $\mathfrak{B}$-measurable step functions such that $kg_N \uparrow f$ can be constructed as follows:

$$g_N = \sum_{n=0}^{N} \lambda_n^{(N)} \chi_{S[(f - \lambda_n^{(N)} k)^+]}$$

for any N, where $\lambda_i^{(j)} > 0$. For this it suffices to show that $S[(f - \lambda k)^+] \in \mathfrak{B}$ for any $\lambda > 0$. We form the sequence $\{h_n\}_{n\geqslant 0}$: $h_n = [n(f - \lambda k)^+] \wedge k \in \operatorname{clos} Y$. The sequence of h_n is increasing and bounded in L^1; therefore, it converges in the L^1-norm. The limit function has the form $k\chi_{S[(f-\lambda k)^+]}$ and is in $\operatorname{clos} Y$. Hence, $S[(f - \lambda k)^+] \in \mathfrak{B}$ for any $\lambda > 0$. Then the convergence $kg_N \uparrow f$ implies that $kg_N \to f$ in the norm $\|\cdot\|_X$, and $S(g_N) \subset S(k)$ for each N.

Let us now consider the CE operator $\mathrm{E}^{\mathfrak{B}}$ with respect to $\mathfrak{B}$. Since symmetric KB-spaces have the interpolation property in [4], X is invariant under $E^{\mathfrak{B}}$ and $\|E^{\mathfrak{B}}\|_{X\to X} = 1$. Consequently, for any n and m

$$\begin{aligned}\|kg_n - kg_m\|_X &\geqslant \left\|E^{\mathfrak{B}}(kg_n - kg_m)\right\|_X = \left\|(g_n - g_m)E^{\mathfrak{B}}k\right\|_X \\ &= \left\|(g_n - g_m)\chi_{S(h)}\right\|_X = \|g_n - g_m\|_X\end{aligned}$$

* *Editor's note.* In the Russian literature *normal subspace* means *solid subspace*.

(since $S(g_n - g_m) \subset S(k) = S(h)$ by construction). Thus, the sequence $\{g_N\}_{N \geqslant 0}$ converges in $\|\cdot\|_X$ to some function $g \in X(\mathfrak{B})$. Hence $f = kg$, i.e., $Y \subset kX(\mathfrak{B})$. The first assertion of Theorem 1 is proved.

We prove the second assertion. By (1) and the fact that Y is nontrivial, there certainly exists, for any set $e \subset S(k)$, $e \in \mathfrak{B}$, $P(e) > 0$, a positive element $x_0 \in X(\mathfrak{B})$, such that $P(S(x_0) \cap e) > 0$ and $kx_0 \in Y \subset X$. Then there exist a subset $e' \subset e$, $e' \in \mathfrak{B}$, $P(e') > 0$, and a number $c > 0$ such that $\mathbf{0} < \bar{x} = c\chi_{e'} \leqslant x_0$.

Therefore, since the subspace X is normal and since $kc\chi_{e'} \leqslant kx_0$, the function $kc\chi_{e'}$ is contained in X; hence $k\chi_{e'} \in X$. But $\chi_{e'} \in X(\mathfrak{B})$, because X contains all the bounded functions. Thus, $k\chi_{e'} \in X \cap kX(\mathfrak{B}) = Y$, i.e., the trace of k on the component corresponding to e' is in Y. Any set e ($e \in \mathfrak{B}$, $P(e) > 0$) contains an e' with this property. Hence, $S(k)$ can be partitioned (the principle of exhaustion) into sets Ω_n ($n = 1, 2, \ldots$) on which the traces of k belong to Y. The function k is equal to zero on $\Omega_0 = \Omega \setminus S(k)$; therefore, $k\chi_{\Omega_0} \in Y$. We have obtained the required $\mathfrak{B}$-measurable partition of the measure space. Finally, to prove that $Y_n = k_n X_n(\mathfrak{B})$ it suffices to verify that $k\chi_{e_n} \in X_n$ for any $e_n \in \mathfrak{B}$, $e_n \subset \Omega_n$. But this holds because X is normal and $k_n = k\chi_{\Omega_n} \in X$. It is easy to show also that the pair $(k, \mathfrak{B})$ is unique.

§2. The general form of positive projections

In what follows we shall be interested in sublattices of the form $Y = T(X) \subset X$, where T is a strictly positive projection in X (Theorem 2).

DEFINITION 3. In the case just described we say that a function $k = k_P$ such that $T(X) = kX(\mathfrak{B}) \cap X$ is a *weight function of the operator* T (with respect to P).

DEFINITION 4. Let $X = X(\Omega, \mathfrak{A})$ be a Banach space of $\mathfrak{A}$-measurable functions on Ω with the usual identification. We say that A is an *operator of CE type* in X if $A = V_r B V_s$, where V_r is the operator of multiplication by r, r and s are strictly positive([2]) $\mathfrak{A}$-measurable functions, and B is the restriction to X of the CE operator with respect to some σ-subalgebra $\mathfrak{B}$ of $\mathfrak{A}$ and to some measure μ on $\mathfrak{A}$. If the measure μ on $\mathfrak{A}$ is fixed, then we say that A is an operator of CE type on $X(\Omega, \mathfrak{A})$ with respect to μ.

It can be shown that the concept of an operator of CE type is invariant with respect to the class of absolutely continuous measures on $\mathfrak{A}$, i.e., if A is an operator of CE type with respect to a measure P, then it is the same with respect to any measure μ absolutely continuous with respect to P. It is also useful to note that if A is an operator of CE type in $X(\Omega, \mathfrak{A}, \mu)$, then the function r (the function s is determined by r) can be chosen in such a way that $r = k_\mu$, where k_μ is a weight function for the σ-subalgebra $\mathfrak{B}$ with respect to μ.

THEOREM 2. *Let $(\Omega, \mathfrak{A}, P)$ be the underlying probability space, and X a symmetric KB-space of measurable functions on Ω. Then the general form of a positive projection in X is given by*

$$T = T_0 U_R + T_1 U_{\tilde{X}} + T_2 U_{R \cap \tilde{X}^d}. \tag{2}$$

Here R is the component of essential positivity of T, i.e., $R = \{f \colon f \in X^+ ; Tf = \mathbf{0}\}^d$; $\tilde{X} = \{k\}^{dd}$ is the component generated by the weight function k of the operator $T - T_0 (\tilde{X} \neq \{\mathbf{0}\}, \tilde{X} \subset R)$; T_1 is an operator of CE type in X; T_2 is a strictly positive

---

([2]) That is, $r(\omega) > 0$ a.e.

operator from $(R \cap \tilde{X}^d)$ *to* $\tilde{X}$ *such that* $T_2(R \cap \tilde{X}^d) \subset T_1(\tilde{X})$; T_0 *is any operator with the properties*

$$T_0 \geqslant \mathbf{0}, \qquad U_R T_0 = T_0^2 = \mathbf{0}, \qquad T_0 U_R = T_0(T_1 U_{\tilde{X}} + T_2 U_{R \cap \tilde{X}^d}) = T_0;$$

and U *is the canonical projection onto a component.*

We give *some examples* of how Theorem 2 can be applied in the most typical situations for mathematical analysis. Let T be a strictly positive projection. Then $T(X)$ is a linear sublattice of X, and (2) takes the form

$$T = T_1 U_{\tilde{X}} + T_2 U_{R \cap \tilde{X}^d}.$$

This formula is valid whenever $T(X)$ is a sublattice. If, in addition, the linear sublattice $T(X)$ has disjoint complement zero in X, then one can asssert that T is an operator of CE type. For example, every positive idempotent contraction that preserves the unit is an operator of CE type.

If P is a σ-finite measure, then Theorem 2 can be applied by decomposing Ω into countably many sets of finite measure, and the space X into corresponding components $X_1, X_2, \dots$ to which the theorem is now immediately applicable. The result is the same formula (2), which is thus true also in the σ-finite case.

Finally, we mention an important fact: The formula (2) is preserved when P is replaced by another measure absolutely continuous with respect to it. This follows from the invariance of the concept of an operator of CE type under such a substitution.

Proof of Theorem 2. A positive linear indempotent operator T on a KB-space is order-continuous and has component of essential positivity

$$R = \{f : f \in X^+ ;\ Tf = \mathbf{0}\}^d,$$

and there exists a set $\Omega_R \subset \Omega$ such that $R = \{f\chi_{\Omega_R} \colon f \in X\}$. Then T can be represented as the sum of a projection T_R that is strictly positive on R ($T_R x > \mathbf{0}$ if $x \in R$ and $x > \mathbf{0}$) and a certain supplementary operator T_0. Moreover, $T_R = U_R T$, where U_R is the canonical projection onto R, and T_0 can be any operator with the properties $T_0 \geqslant \mathbf{0}$, $T_0 U_R = T_0 T_R = T_0$ and $U_R T_0 = T_0^2 = \mathbf{0}$. The properties of T_0 can be checked by the direct substitution $T_0 = T - T_R$. The converse is also true: If R is a component in X, T_R is a linear idempotent operator that is strictly positive on R, and T_0 has the properties listed above, then the sum $T = T_0 + T_R$ is a positive projection in X.

Let us consider a strictly positive projection, i.e., the restriction of T_R to R. We denote this operator by the same letter T and assume that it acts in the independent space $X = X(\Omega_R, \mathfrak{A}\Omega_R, P \mid \mathfrak{A}\Omega_R)$. The set $T(X) = \{f\colon\ Tf = f\}$ is a closed linear sublattice of X (see, for example, [5]). By Theorem 1, there exist a unique weight function ik and a unique σ-subalgebra $\mathfrak{B}$ such that $T(X) = kX(\mathfrak{B}) \cap X$. Moreover, there exists a partition $\Omega_R = \bigcup_0^\infty \Omega_n$, $\Omega_n \cap \Omega_m = \varnothing$ $(n \neq m)$, $\Omega_0 = \Omega_R \setminus S(k)$, $\Omega_n \in \mathfrak{B}$ $(n \geqslant 0)$, such that $k_n = k\chi_{\Omega_n} \in T(X)$ on each Ω_n.

Let X_n be the component of X generated by k_n. Denote by T_n the restriction of T to X_n. Since $T = T^2$ and $k_n \in T(X)$, each component X_n is invariant under T, i.e., $T_n(X_n) \subset X_n$. But then, by Theorem 1, $T_n(X_n) = k_n X_n(\mathfrak{B})$.

For the support we write $S(k) = \tilde{\Omega} = \bigcup_1^\infty \Omega_n$; $\tilde{X} = X(\tilde{\Omega}, \mathfrak{A}\tilde{\Omega}) = \{k\}^{dd}$ is the component generated by k. We observe that $\tilde{x} \subset R$. Fix a positive integer n. Henceforth, we assume that we are working with the space $(\Omega_n, \mathfrak{A}\Omega_n, P \mid \mathfrak{A}\Omega_n)$ and regard the component x_n as the independent space $X_n = X(\Omega_n, \mathfrak{A}\Omega_n, P \mid \mathfrak{A}\Omega_n)$, and T_n as an operator acting in this space. Let $\tilde{X}_n$ be the space $k_n^{-1}X_n$ and consider in it the new operator $\tilde{T}_n = V_{k_n^{-1}} T_n V_{k_n}$; this is a strictly positive projection having the properties

$$\tilde{T}_n(\tilde{X}_n) = X_n(\mathfrak{B}), \qquad \tilde{T}_n \mathbf{1} = \mathbf{1}\ (T_n k_n = k_n).$$

Let $\tilde{T}_n^*$ be the adjoint operator with respect to the measure P, i.e.,

$$\int_{\Omega_n} \tilde{T}_n x \cdot y\, dP = \int_{\Omega_n} x \cdot \tilde{T}_n^* y\, dP$$

for any $x \in \tilde{X}_n$ and $y \in \tilde{X}_n^*$ ($\tilde{X}_n^*$ is the space dual to $\tilde{X}_n$). The idempotence of $\tilde{T}_n$ implies that

$$\int_{\Omega_n} x \cdot \tilde{T}_n^* y\, dP = \int_{\Omega_n} \tilde{T}_n x \cdot y\, dP = \int_{\Omega_n} \tilde{T}_n^2 x \cdot y\, dP = \int_{\Omega_n} x \cdot \tilde{T}_n^{*2} y\, dP$$

for any $x \in \tilde{X}_n$ and $y \in \tilde{X}_n^*$. Consequently, $\tilde{T}_n^* = \tilde{T}_n^{*2}$. It is clear that $\tilde{T}_n^*$ is strictly positive. Let $\rho = \tilde{T}_n^* \mathbf{1}$. From the equality

$$\int_{\Omega_n} x \cdot \tilde{T}_n^* \mathbf{1}\, dP = \int_{\Omega_n} \tilde{T}_n x\, dP = \int_{\Omega_n} \tilde{T}_n^2 x\, dP = \int_{\Omega_n} \tilde{T}_n x \cdot \tilde{T}_n^* \mathbf{1}\, dP$$

we have that

$$\int_{\Omega_n} \tilde{T}_n x \cdot \rho\, dP = \int_{\Omega_n} x\rho\, dP \tag{3}$$

for all $x \in \tilde{X}_n$.

We prove that

$$\int_B \tilde{T}_n x \cdot \rho\, dP = \int_B x\rho\, dP \tag{4}$$

for any $B \in \mathfrak{B}$, $B \subset \Omega_n$, and $x \in \tilde{X}_n$. The elements of $\tilde{X}_n$ can be approximated by characteristic functions in the norm of this space. Therefore, by the norm continuity of $\tilde{T}_n$ it suffices to prove (4) only for characteristic functions.

First of all we remark that $\tilde{T}_n g = g$ for all $g \in X_n(\mathfrak{B})$, since $\tilde{T}_n(\tilde{X}_n) = X_n(\mathfrak{B})$ and $\tilde{T}_n = \tilde{T}_n^2$. Further,

$$\tilde{T}_n \chi_{S \cap CB} \leqslant \tilde{T}_n \chi_{CB} = \chi_{CB}$$

for any $B \in \mathfrak{B}$ and $S \in \mathfrak{A}\,(B, S \subset \Omega_n)$; here $CB = \Omega_n \setminus B$ and $CS = \Omega_n \setminus S$. Then

$$\begin{aligned}
\int_B \tilde{T}_n \chi_S \cdot \rho\, dP &= \int_B \tilde{T}_n(\chi_{S\cap B}) \cdot \rho\, dP + \int_B \tilde{T}_n(\chi_{S \cap CB}) \cdot \rho\, dP \\
&= \int_B \tilde{T}_n(\chi_{S\cap B}) \cdot \rho\, dP \leqslant \int_{\Omega_n} \tilde{T}_n(\chi_{S\cap B}) \cdot \rho\, dP \\
&= \int_{\Omega_n} \chi_{S\cap B} \cdot \rho\, dP = \int_B \chi_S \cdot \rho\, dP.
\end{aligned} \tag{5}$$

Similarly, replacement of S by CS leads to

$$\int_B \tilde{T}_n\chi_{CB}\cdot\rho\,dP \leqslant \int_B \chi_{CB}\cdot\rho\,dP \tag{5'}$$

for all $B \in \mathfrak{B}$ and $S \in \mathfrak{A}$. Using the equality $\tilde{T}_n\mathbf{1} = \mathbf{1}$, we get that

$$\int_B \rho\,dP = \int_B \chi_S\rho\,dP + \int_S \chi_{CS}\rho\,dP \geqslant \int_B \tilde{T}_n\chi_S\cdot\rho\,dP + \int_B \tilde{T}_n\chi_{CS}\cdot\rho\,dP$$
$$= \int_B \tilde{T}_n\mathbf{1}\cdot\rho\,dP = \int_B \rho\,dP.$$

From this, by (5) and (5′), it follows that

$$\int_B \tilde{T}_n\chi_S\cdot\rho\,dP = \int_B \chi_S\rho\,dP$$

for any $B \in \mathfrak{B}$ and $S \in \mathfrak{A}\,(B, S \subset \Omega_n)$. But, as we remarked, this proves (4).

The function $\tilde{T}_nx\cdot\chi_B$ is $\mathfrak{B}$-measurable for any $B \in \mathfrak{B}$; therefore, $\tilde{T}_n(\tilde{T}_nx\cdot\chi_B) = \tilde{T}_nx\cdot\chi_B$. Consequently,

$$\int_B x\rho\,dP = \int_B \tilde{T}_nx\cdot\rho\,dP = \int_{\Omega_n}(\tilde{T}_nx\cdot\chi_B)\tilde{T}_n^*\mathbf{1}\,dP = \int_{\Omega_n}\tilde{T}_nx\cdot\chi_B\,dP = \int_B \tilde{T}_nx\,dP$$

for all $B \in \mathfrak{B}$ and $x \in \tilde{X}_n$. Hence, $\tilde{T}_nx = E^{\mathfrak{B}}(x\rho)$ for any $x \in \tilde{X}_n$, where $E^{\mathfrak{B}}$ is the CE operator with respect to $\mathfrak{B}$ and P. Then for all $x \in k_n^{-1}X(\Omega_n, \mathfrak{A}\Omega_n, P \mid \mathfrak{A}\Omega_n)$, we have $T_n(xk_n) = k_nE^{\mathfrak{B}}(x\rho)$, but

$$k_nx \in X_n = X(\Omega_n, \mathfrak{A}\Omega_n, P \mid \mathfrak{A}\Omega_n).$$

Thus, $T_n = V_{k_n}E^{\mathfrak{B}}V_\rho V_{k_n^{-1}}$ on X_n. We note also that $\rho = k_nT_n^*(k_n^{-1})$, i.e., $T_n = V_{k_n}E^{\mathfrak{B}}V_{S_n}$ on X_n, where $S_n = T_n^*(k_n^{-1})$. We carry out analogous constructions for each Ω_n. But since $\tilde{\Omega} = \bigcup_1^\infty \Omega_n$, $\Omega_n \cap \Omega_m = \varnothing$ $(n \neq m)$ and $\Omega_n \in \mathfrak{B}$ for any n, it is possible to construct a simultaneous extension of the operators T_n. Then we get an operator T_1 defined on the component $\tilde{X} = X(\tilde{\Omega}, \mathfrak{A}\tilde{\Omega}, P \mid \mathfrak{A}\tilde{\Omega})$ and having the form $T_1 = V_kE^{\mathfrak{B}}V_S$, where $S = \sum_n S_n$.

Recall that k is a weight function of a projection that is strictly positive on the component R, i.e., $T_R = T - T_0$.

Thus, we have constructed a decomposition of the original operator into the components T_0 and T_1. As for the term $T - T_0U_R - T_1U_{\tilde{X}}$, it is easy to show that it can be any operator of the form $T_2U_{R\cap\tilde{X}^d}$, where T_2: $R \cap \tilde{X}^d \to \tilde{X}$ is strictly positive and $T_2(R \cap \tilde{X}^d) \subset T_1(\tilde{X})$. Indeed, the fact that $T_R - T_1U_{\tilde{X}}$ has the properties listed is obvious. The converse is also true: If $\tilde{X}$ is some component of R, T_1 is an operator of CE type on $\tilde{X}$, and T_2: $R \cap \tilde{X}^d \to \tilde{X}$ is strictly positive and satisfies $T_2(R \cap \tilde{X}^d) \subset T_1(\tilde{X})$, then the sum $T_R = T_1U_{\tilde{X}} + T_2U_{R\cap\tilde{X}^d}$ is a strictly positive projection in R. To prove this it suffices to note that

$$T_R^2 = T_1U_{\tilde{X}}(T_1U_{\tilde{X}} + T_2U_{R\cap\tilde{X}^d}) + T_2U_{R\cap\tilde{X}^d}(T_1U_{\tilde{X}} + T_2U_{R\cap\tilde{X}^d}),$$
$$(T_1U_{\tilde{X}} + T_2U_{R\cap\tilde{X}^d})(R) \subset T_1(\tilde{X}) \subset \tilde{X}, \quad \text{and} \quad T_1^2 = T.$$

Thus, we have obtained the required representation of T.

In conclusion, I regard it as a pleasant duty to thank D. A. Vladimirov for posing the problem and for valuable discussions.

BIBLIOGRAPHY

1. B. Z. Vulikh, *Introduction to the theory of partially ordered spaces*, Fizmatgiz, Moscow, 1961; English transl., Noordhoff, 1967.

2. J. Kampé de Fériet, *Problèmes mathématiques posés par la méchanique statistique de la turbulence*, Proc. Internat. Congr. Math. (Amsterdam, 1954), Vol. III, Noordhoff and North-Holland, 1956, pp. 237–242.

3. R. G. Douglas, *Contractive projections on an* $\mathfrak{L}_1$ *space*, Pacific J. Math. **15** (1965), 443–462.

4. B. S. Mitjagin, *An interpolation theorem for modular spaces*, Mat. Sb. **66(108)** (1965), 473–482. (Russian)

5. V. G. Kulakova, *Positive projection operators and conditional expectation*, Zap. Nauchn. Sem. Leningrad. Otdel. Mat. Inst. Steklov. (LOMI) **47** (1974), 172–174; English transl. in J. Soviet Math. **9** (1978), no. 2.

Translated by H. H. McFADEN

Труды Матем. Инст. Стеклов.
155(1980)

Proceedings of the Steklov
Institute of Mathematics
1983, Issue 1

AN ANALOGUE OF AN INEQUALITY OF J. VON NEUMANN, ISOMETRIC DILATION OF CONTRACTIONS, AND APPROXIMATION BY ISOMETRIES IN SPACES OF MEASURABLE FUNCTIONS

UDC 517.948: 513.8

V. V. PELLER

ABSTRACT. The principal results of this paper are a characterization of the operators in an L^p-space that have a unitary dilation in L^p-space, and a characterization of the operators in the space $L^p[0,1]$ that can be approximated in the pw-topology by unitary operators ($T_\alpha \overset{pw}{\to} T \overset{\text{def}}{\Leftrightarrow} \lim_\alpha (T_\alpha^n f, g) = (T^n f, g)$, $\forall n \geq 0$, $\forall f \in L^p$, $\forall g \in L^{p'}$). It is proved that both of these sets of operators coincide with the set $\mathfrak{M}$ of operators T that admit a contractive majorant (i.e., operators T for which there exists a positive contraction $\tilde{T}$ such that $|Tf| \leq \tilde{T}|f|$, $\forall f \in L^p$). For operators of class $\mathfrak{M}$, an analogue of an inequality of von Neumann is obtained:

$$\|\varphi(T)\|_{L^p} \leq |\varphi|_p,$$

where $T \in \mathfrak{M}$, φ is a complex polynomial, and $|\varphi|_p$ is the norm of φ as a multiplier for the space l^p, i.e., $|\varphi|_p = \|\varphi(S)\|_{l^p}$, where S is the shift operator in l^p. Analogues of these results for symmetric spaces of measurable functions are also obtained. The principal results of this paper were announced in a note by the author: C. R. Acad. Sci. Paris Sér. A-B **287** (1978), A311-A314. The present paper contains a review of all known results on estimates of operator polynomials and questions associated with these estimates in the L^p-spaces.

Bibliography: 46 titles.

Contents

1980 *Mathematics Subject Classification.* Primary 47A60, 47A20, 47B38; Secondary 47B37, 43A22, 47A35.

©1983 American Mathematical Society
0077-1554/83 $1.00 + $.25 per page

0. Introduction

The principal results of the classical spectral theory of normal operators in a Hilbert space can be obtained by various means, one of which consists in the construction of a rich functional calculus. For larger classes of operators in a Hilbert (and moreover, in a general Banach) space, this becomes one of the most fruitful methods. In this connection, the main difficulty in the construction of a maximal possible calculus $\varphi \to \varphi(T)$ consists in obtaining sharp estimates of the operator polynomial norms $\|\varphi(T)\|$. A classical example of such an estimate is Murray's lemma

$$\|\varphi(T)\| \leqslant \sup\{|\varphi(\zeta)| : \zeta \in \sigma(T)\}$$

for any normal operator T and any complex polynomial φ (we will denote the set of all such polynomials by $\mathcal{P}_A$; $\sigma(T)$ is the spectrum of T).

In 1951, von Neumann [1] proved an inequality which has found various applications in contemporary spectral theory in Hilbert space:

$$\|\varphi(T)\| \leqslant \max\{|\varphi(\zeta)| : |\zeta| \leqslant 1\}, \qquad \varphi \in \mathcal{P}_A,$$

for any contraction T (i.e., $\|T\| \leqslant 1$). This inequality became the basis of numerous investigations, from the theory of spectral sets (in von Neumann's sense) to the study of unconditionally convergent spectral decompositions. A maximal functional calculus for contractions in Hilbert space is built upon it. There exist many different proofs of von Neumann's inequality (see, for example, [1]–[3]). Here (see §1), we give yet one more such proof, based on the approximation of contractions by unitary operators.

Attempts to find sharp estimates of $\|\varphi(T)\|$ for other classes of Hilbert space operators T (for example, operators with a prescribed rate of growth for the powers $\|T^n\|$, $n \geqslant 0$, or rate of growth for the resolvent near the spectrum), or even for contractions, but in spaces with non-Hilbert norm, run into substantial difficulties. An analogue of von Neumann's inequality in a Banach space X can be seen in the explicit computation of the norm

$$|||\varphi|||_X \stackrel{\text{def}}{=} \sup\{\|\varphi(T)\|_X : \|T\|_X \leqslant 1\},$$

where T is a linear operator in X, and $\|T\|_X$ is its norm.

A trivial upper bound in terms of the Taylor coefficients $\hat{\varphi}(n)$ of the polynomial φ, and the obvious inequality $|||\varphi|||_X \geqslant \|\varphi(\lambda I)\|_X$ (where $|\lambda| = 1$, and I is the identity operator) show that

$$\max\{|\varphi(\zeta)| : |\zeta| \leqslant 1\} \leqslant |||\varphi|||_X \leqslant \sum_{n \geqslant 0} |\hat{\varphi}(n)|, \qquad \varphi \in \mathcal{P}_A.$$

It is known that in certain spaces X, the right-hand inequality turns into equality ($X = l^1, X = C[a, b]$), while this can happen for the left-hand inequality only when X is a Hilbert space [4]. Nontrivial estimates of $|||\cdot|||_X$ in the non-Hilbert situation are known only in the case $\dim X < +\infty$ [5].

This paper is a survey of all the known results on estimates of operator polynomials, and on related questions in L^p-spaces and in symmetric spaces of measurable functions. The principal new results are the criterion for the existence of a unitary dilation in L^p-space of a contraction (§3), a characterization of operators that can be approximated by unitary operators (§4), and the deduction, with their aid, of an estimate of the norm $|\varphi(T)|$ in terms of the multiplier norm of the polynomial φ (§7). These results were announced in the note [6], and set forth in the preprint [7]. We also give analogues of these results for symmetric spaces, and several corollaries of the multiplier estimates. In order to make the exposition complete and self-contained, we include certain results from [8]–[11].

The role of multipliers. Let l^p_A be the space of all functions $f = \sum_{n\geqslant 0}\hat{f}(n)z^n$ that are analytic in the disc $\mathbb{D} = \{\zeta: |\zeta| < 1\}$ and have finite norm $\|f\|^p_{l^p_A} = \sum_{n\geqslant 0}|\hat{f}(n)|^p < +\infty$. A function g, analytic in $\mathbb{D}$, is called a *p-multiplier* if

$$f \in l^p_A \Rightarrow fg \in l^p_A.$$

We will denote the space of all p-multipliers by M_p, and as the norm $|g|_p$ of an element $g \in M_p$ we will take the norm of the corresponding multiplication operator $f \to fg$ in the space l^p_A. We note that $|\varphi|_p = |\varphi(S)|_{l^p}$ for every polynomial φ, where S is the shift operator in the sequence space l^p:

$$S(x_0, x_1, x_2, \ldots) = (0, x_0, x_1, x_2, \ldots).$$

In 1966, V. I. Matsaev formulated a conjecture (see [12]) according to which, for infinite-dimensional L^p-spaces, $|||\cdot|||_{L^p}$ is the p-multiplier norm, i.e.,

$$|||\varphi|||_{L^p} = |\varphi|_p, \qquad \varphi \in \mathcal{P}_A.$$

In other words, Matsaev's conjecture asserts that the operator S is the "worst" operator among all contractions in L^p-spaces.

Let us recall, in addition, that p-multipliers are also remarkable in many of their other properties, and are an important object of investigation in harmonic analysis and in the theory of integrodifferential equations, while for classical mathematical analysis or the theory of functions of a complex variable they frequently serve as an object on which new methods for these disciplines are forged.

It is well known that

$$\max\{|\varphi(\zeta)|: |\zeta| \leqslant 1\} = |\varphi|_2 \leqslant |\varphi|_p = |\varphi|_{p'} \leqslant |\varphi|_1 = \sum_{n\geqslant 0}|\hat{\varphi}(n)|,$$

$$\varphi \in \mathcal{P}_A,\ 1 \leqslant p \leqslant +\infty,\ 1/p + 1/p' = 1,$$

where the left-hand inequality is a commonplace: its analogue is valid in any functional space (see [13], Problem 53). Consequently, for $p = 2$, Matsaev's conjecture turns into von Neumann's inequality. For $p = 1$ or $+\infty$, it is also clearly valid. Therefore, if not otherwise stated, it is assumed everywhere below that $1 < p < +\infty$.

Coifman and Weiss [14] have proved that the inequality

$$|\varphi(T)|_{L^p} \leqslant |\varphi|_p, \qquad \varphi \in \mathcal{P}_A, \tag{1}$$

is valid for invertible isometries T (unitary operators) in an L^p space. A. K. Kitover (see [9]) showed that this inequality is valid for arbitrary isometries T. The author [7] proved inequality (1) for operators T *that admit a contractive majorant* (i.e., for T such that there exists a *positive contraction* $\tilde{T}$ ($f \geqslant \mathbf{0} \Rightarrow \tilde{T}f \geqslant \mathbf{0}$) for which $|Tf| \leqslant \tilde{T}|f|$ for $f \in L^p$); we denote by $\mathfrak{M}$ the set of all operators that admit a contractive majorant. The class $\mathfrak{M}$ contains operators of the following types for which inequality (1) was proved previously in the author's papers [8]–[10]: *disjoint contractions* T (i.e., such that $f \cdot g = \mathbf{0} \Rightarrow Tf \cdot Tg = \mathbf{0}$), *absolute contractions* T (i.e., $|T|_{L^1} \leqslant 1$ and $|T|_{L^\infty} \leqslant 1$), and positive contractions. We note that, independently of the author,

Coifman and Weiss [15] proved that (1) is valid for positive contractions. Recently, Coifman, Rochberg, and Weiss reported [16] that they had also proved (1) for operators of class $\mathfrak{M}$.

Dilation and the *pw*-topology. We will now describe two methods (of the many known ones) for proving von Neumann's inequality that are suitable for use in more general situations. The first method consists in the construction of a unitary dilation for a contraction. Allowing for some looseness of speech, a *dilation* of an operator T: $X \to X$ can be defined as an operator U: $Y \to Y$, where X is isometrically imbedded in Y as an orthogonally complemented subspace, and $PU^n \mid X = T^n$, $n \geqslant 0$ (where P is a projection with unit norm of Y onto the subspace X). The second method consists in the approximation of contractions in the *pw-topology* ($T_\alpha \overset{pw}{\to} T \overset{\text{def}}{\Leftrightarrow}$ the operator powers T_α^n converge weakly to T^n for all natural numbers n) by unitary operators (see §1). Both of these methods can also be used for the proof of (1) in an L^p space, since this inequality holds for isometries, as can be proved directly (see §10).

Dilations of operators have been used for a comparatively long time in the spectral theory of operators. The functional model of Sz.-Nagy and Foiaş [2], [17], which lets us reduce the study of a contraction to the study of its characteristic function, is constructed upon the basis of unitary dilation. Normal dilations in Hilbert space are investigated in connection with the study of spectral sets (see, for example, [18]). The pointwise ergodic theorem of M. A. Akcoglu for positive contractions in L^p was obtained as a corollary of a theorem on dilation [19]. It can be expected that the construction and study of dilations with suitable properties will also be useful in other problems connected with linear operators. We should remark right now, however, that the ambient space in which the dilation of an operator is constructed must be of the same "metric type" as the space in which the operator acts, since otherwise one cannot hope to recover properties of operators from corresponding properties of their dilations, and in fact the existence of a unitary dilation in some Banach space becomes almost trivially commonplace [20].

The maximal functional calculus. Let T be an operator in a space $L^p(\mathfrak{X}, \mathfrak{F}, \mu)$ for which inequality (1) holds. Of great importance is the question of the characterization of the operators T for which this inequality is valid within its natural limits, i.e., for which the calculus $\varphi \to \varphi(T)$, $\varphi \in \mathfrak{P}_A$, can be extended to a contractive representation of the algebra M_p ($\mathfrak{P}_A$ is not dense in M_p). If $p = 2$ (i.e., T is a contraction in a Hilbert space), then all completely nonunitary contractions admit such a calculus (the calculus of Sz.-Nagy and Foiaş; see [2]). Analogues of the complete nonunitariness condition for L^p-spaces are found in §8.

Description of the results. In §1, we illustrate the second of the main ideas on which the estimates proved in this paper are based; namely, we give a new proof of the classical inequality of von Neumann, which follows immediately from the following theorem.

THEOREM 1.([1]) *The set of unitary operators is pw-dense in the unit ball of the space of operators acting in an infinite-dimensional Hilbert space.*

([1]) In this paper, we use straight indexing of theorems, lemmas, and selected displayed formulas. This indexing is preserved in the Introduction.

§2 is devoted to the problem of constructing an isometric dilation for disjoint contractions in an L^p space. It turns out that these, and only these, admit a "trivial" isometric dilation (in L^p spaces). More precisely, in §2 we characterize the operators in a space $L^p(\mathcal{X}, \mathcal{F}, \mu)$ that have an isometric dilation in a space $L^p(\mathcal{Y}, \mathcal{G}, \nu)$ where $\mathcal{X} \subset \mathcal{Y}, \mathcal{F} = \{F \in \mathcal{G}\colon F \subset \mathcal{X}\}$ and $\mu = \nu \mid \mathcal{F}$ (in this case $L^p(\mathcal{X}, \mathcal{F}, \mu)$ is imbedded in a natural way in $L^p(\mathcal{Y}, \mathcal{G}, \nu)$, and an operator T in $L^p(\mathcal{X}, \mathcal{F}, \mu)$ is related to its dilation U in $L^p(\mathcal{Y}, \mathcal{G}, \nu)$ by the very simple relation $(T^n f)(x) = (U^n f)(x)$, $x \in \mathcal{X}$, $f \in L^p(\mathcal{X}, \mathcal{F}, \mu)$, $n \geqslant 0$).

THEOREM 2. *For $p \neq 2$, the class of operators described above coincides with the class of all disjoint contractions of the space $L^p(\mathcal{X}, \mathcal{F}, \mu)$.*

§3 is devoted to the characterization of the operators in $L^p(\mathcal{X}, \mathcal{F}, \mu)$, $p \neq 2$, that have a unitary dilation in L^p-space.(²) The main result of this section is the following theorem.

THEOREM 3. *The operator T in the space $L^p(\mathcal{X}, \mathcal{F}, \mu)$ has a unitary dilation in L^p-space if and only if $T \in \mathfrak{M}$.*

This theorem was proved by the author in [7]. Previously, Akcoglu proved [19] that positive contractions in the space l^p_n that satisfy certain additional restrictions have a unitary dilation in L^p-space. After [7] was published, the papers [21] and [22], in which it was proved that every positive contraction in an L^p-space has a unitary dilation, became available to the author.

In the case where T is a positive contraction, the proof we give implies the existence of a positive unitary dilation. Thus, Theorem 3 strengthens a theorem of Akcoglu and Sucheston [19], [22].

The proof is carried out in two stages. First, the assertion is proved for finite-dimensional spaces, and then the general case is obtained with the aid of the construction of an ultraproduct of Banach spaces (see [23]). Independently of the author, W. B. Johnson noted the possibility of using the techniques of ultraproducts of Banach spaces in the construction of dilations of positive contractions (see [22]).

In §4 we describe the operators in $L^p[0, 1]$ that can be approximated by unitary operators in the pw-topology. The main result of the section is the following theorem.

THEOREM 4. *The closure of the set of unitary operators in $L^p[0, 1]$ in the pw-topology coincides with the set $\mathfrak{M}$.*

This theorem is contained in [7]. In the earlier papers [8] and [9] it was proved that every absolute contraction in $L^p[0, 1]$ can be pw-approximated by isometric operators.

The concept of the pw-topology was introduced in [8]. Theorems on pw-approximation can also be useful in other situations. For example, with the aid of Theorem 4, one can obtain a proof of the pointwise ergodic theorem for contractions of an L^p space that admit a contractive majorant (this proof is given in §4).

Thus, the two methods of proof described above for inequality (1) lead to a single result, which is the principal assertion of this paper (see §7).

(²) By l^p-space we mean a Banach space that is isometric to some space $L^p(\mathcal{X}, \mathcal{F}, \mu)$.

THEOREM 9. *Inequality* (1) *is true for operators on an L^p space that admit a contractive majorant.*

This theorem proves the truth of a conjecture of T. Ando (private communication).

§5 is devoted to questions of approximation and dilation in an L^1-space. It is proved that, in this case, every contraction of a space $L^1(\mathcal{X}, \mathcal{F}, \mu)$ has a unitary dilation in L^1-space, and the *pw*-closure of the set of unitary operators on $L^1[0,1]$ coincides with the set of all contractions. We note that for positive contractions of an L^1 space, the existence of a unitary dilation was proved by Akcoglu [24].

In §6, we investigate the questions of approximation by unitary operators and the construction of a unitary dilation in symmetric spaces of measurable functions on the interval [0, 1] (see [25]). Here we prove Theorems 7 and 8, which assert that *every absolute contraction* (i.e., operator that is a contraction in the L^1- and L^∞-norms) *of a separable symmetric space X has a unitary dilation in a space isometric to X, and can be approximated in the pw-topology by unitary operators.*

The proof of the existence of a dilation also is first given for the finite-dimensional case, but in the transition to the general case, in place of the construction of an ultraproduct of Banach spaces, we make use of the construction of an ultraproduct of measure spaces [26].

In §7, we obtain an analogue of inequality (1) for absolute contractions of a separable symmetric space. Let $M(X)$ be the space of multipliers on Fourier coefficients of the space X, i.e.

$$\{c_n\}_{n\in\mathbf{Z}} \in M(X) \overset{\text{def}}{\Leftrightarrow} \sum_{n\in\mathbf{Z}} c_n \hat{f}(n) e^{2\pi i n t} \in X \quad \text{if} \quad \sum_{n\in\mathbf{Z}} \hat{f}(n) e^{2\pi i n t} \in X.$$

Let θ be an irrational number, and $\alpha = e^{2\pi i\theta}$. We have the following assertion.

THEOREM 10. *Let T be an absolute contraction in X. Then*

$$|\varphi(T)|_X \leqslant |\varphi|_X \overset{\text{def}}{=} \|\{\varphi(\alpha^n)\}_{n\in\mathbf{Z}}\|_{M(X)}, \qquad \varphi \in \mathcal{P}_A;$$

the definition of the norm $|\cdot|_X$ does not depend on the choice of the number θ.

The proof is carried out by means of the approximation of absolute contractions by operators that are unitarily equivalent to a rotation operator. If $X = L^p[0,1]$, then $|\varphi|_X = |\varphi|_p$ for $\varphi \in \mathcal{P}_A$.

In §8, we find "complete nonunitariness" conditions for an operator $T \in \mathfrak{M}$ that are sufficient for the construction of an M_p-calculus. We give analogues of these assertions for symmetric spaces of measurable functions. From the results obtained, it follows that if the sequence of powers $\{T^n\}_{n\geqslant 0}$ converges strongly to the zero operator, then the operator T admits an M_p-calculus. In the case where T is an absolute contraction in an L^p-space, it is possible to assert that T admits an M_p-calculus under a weaker condition: if $f \in L^p$, $f \neq \mathbf{0}$, then there exists an $n \in \mathbb{N}$ such that $\|T^n f\|_p < \|f\|_p$.

§9 is devoted to estimates of compositions in spaces of multipliers. Since p-multipliers are operators in l^p_A, inequality (1) can be employed in the study of p-multipliers, also. Here we prove the following theorem.

THEOREM 15. *If* $f \in l^1_A$, $\|f\|_{l^1_A} \leq 1$, *and* $|\hat{f}(0)| < 1$, *then*

$$\varphi \in M_p \Rightarrow \varphi \circ f \in M_p \text{ and } |\varphi \circ f|_p \leq |\varphi|_p.$$

We obtain analogues of this theorem for multipliers of L^p-spaces on a locally compact abelian group G, and for multipliers of symmetric spaces.

In §10, we investigate shifts on the weighted spaces $l^p(w)$ and $l^p(\mathbb{Z}, w)$. In the case where the sequence of weights satisfies the conditions

$$w_n \downarrow, \quad \varlimsup_{n \to \infty} \sqrt[n]{w_n^{-1}} \leq 1,$$

we prove that for the shift operator in such a space, inequality (1) reduces to an equality. This circumstance makes it possible to prove that the space of multipliers of $l^p_A(w)$ coincides with M_p. A device of A. K. Kitover is used in the proof of these assertions (previously, similar arguments in the estimation of operators in spaces have been used by Cotlar [27], Calderón [28], and Coifman and Weiss [14]; see also the survey [15]), which leads, in particular, to a proof of inequality (1) for isometries that is independent of the previous results.

I take this opportunity to express deep gratitude to N. K. Nikol'skiĭ for helpful discussions. I am also grateful to S. V. Kislyakov for several useful remarks.

Principal notation

$|T|_X$ is the norm of the operator T in the space X.

If $(\mathcal{X}, \mathcal{F}, \mu)$ is a measure space, then $L^p = L^p(\mathcal{X}, \mathcal{F}, \mu)$, $\|\cdot\|_p \stackrel{\text{def}}{=} \|\cdot\|_{L^p}$, and $|\cdot|_p \stackrel{\text{def}}{=} |\cdot|_{L^p}$.

$\mathfrak{M}$ is the class of operators in a space L^p that admit a contractive majorant.

$L^p[0, 1]$ is the L^p-space (with respect to Lebesgue measure) of functions on the interval $[0, 1]$.

$$l^p(w) \stackrel{\text{def}}{=} \left\{x = \{x_k\}_{k \geq 0}\colon \|x\|^p_{l^p(w)} = \sum_{k \geq 0} |x_k|^p w_k < +\infty\right\},$$

$$l^p_n(w) \stackrel{\text{def}}{=} \left\{x = \{x_k\}_{1 \leq k \leq n}\colon \|x\|^p_{l^p_n(w)} = \sum_{1 \leq k \leq n} |x_k|^p w_k\right\},$$

$$l^p_n \stackrel{\text{def}}{=} l^p_n(w) \quad \text{for } w_k \equiv 1,$$

$$l^p(\mathbb{Z}, W) \stackrel{\text{def}}{=} \left\{x = \{x_k\}_{k \in \mathbb{Z}}\colon \|x\|^p_{l^p(\mathbb{Z}, w)} = \sum_{k \in \mathbb{Z}} |x_k|^p w_k < +\infty\right\},$$

$$l^p(w, X) \stackrel{\text{def}}{=} \left\{f = \{f_k\}_{k \geq 0}, f_k \in X, \|f\|^p_{l^p(w, X)} = \sum_{k \geq 0} \|f_k\|^p_X w_k < +\infty\right\}$$

(here the w_k are positive numbers, and X is a Banach space).

z is the function defined in the unit disc by $z(\zeta) = \zeta$, $\zeta \in \mathbb{D}$.

$\mathcal{P}_A$ is the set of all complex polynomials.

If $\overline{\lim}_n \sqrt[n]{w_n^{-1}} \leqslant 1$, then $l_A^p(w)$ is the space of all functions $f = \sum_{k\geqslant 0} \hat{f}(k)z^k$ analytic in the disc $\mathbb{D}$ with finite norm

$$\|f\|_{l_A^p(w)}^p \stackrel{\text{def}}{=} \sum_{k\geqslant 0} |\hat{f}(k)|_{w_k}^p < +\infty; \qquad l_A^p \stackrel{\text{def}}{=} l_A^p(w) \quad \text{for } w_k \equiv 1.$$

M_p is the space of all p-multipliers.

$|f|_p$ is the norm of the p-multiplier f.

clos $\mathcal{Q}$ is the closure of the set $\mathcal{Q}$.

χ_Δ is the characteristic function of the set Δ.

All measures, unless otherwise stated, are assumed to be σ-finite.

§1. An approximation proof of von Neumann's inequality

In this section we introduce the concept of the pw-topology, and obtain a proof of von Neumann's inequality based on the approximation, in this topology, of contractions by unitary operators.

DEFINITION. Let X be a Banach space. We will say that a generalized sequence $\{T_\alpha\}$ of operators *converges to the operator T in the pw-topology* if

$$\lim_\alpha (T_\alpha^n x, y) = (T^n x, y)$$

for all $x \in X$, $y \in X^*$, and $n \geqslant 0$.

We note that strong convergence for contractions implies convergence in the pw-topology.

The following lemma demonstrates the significance of the pw-topology for estimates of norms of operator polynomials.

LEMMA 1. *Let $\{T_\alpha\}$ be a generalized sequence of operators, in a Banach space X, that converges to an operator T in the pw-topology. Then*

$$|\varphi(T)|_X \leqslant \underline{\lim}_\alpha |\varphi(T_\alpha)|_X \quad \textit{for } \varphi \in \mathcal{P}_A.$$

PROOF. Let $x \in X$, $y \in X^*$. Then

$$|(\varphi(T)x, y)| = \lim_\alpha |(\varphi(T_\alpha)x, y)| \leqslant \underline{\lim}_\alpha |\varphi(T_\alpha)|_X \|x\|_X \|y\|_{X^*}.$$

Consequently, $|\varphi(T)|_X \leqslant \underline{\lim}_\alpha |\varphi(T_\alpha)|_X$. ●

The following approximation theorem permits us to prove von Neumann's inequality.

THEOREM 1. *The set of unitary operators in an infinite-dimensional Hilbert space is pw-dense in the unit ball of operators.*

PROOF. (I). Let $\mathcal{K}$ be a Hilbert space, and R a contraction in $\mathcal{K}$. Let us define an operator $\tilde{R}$ in the space $l^2(\mathcal{K})$ by setting $\tilde{R}(x_0, x_1, \ldots) = (Rx_0, \mathbf{0}, \mathbf{0}, \ldots)$. Let I be the identity operator in $\mathcal{K}$. Set $\mathcal{D}_R \stackrel{\text{def}}{=} (I - R^*R)^{1/2}$. Then

$$\|\mathcal{D}_R x\|_{\mathcal{K}}^2 + \|Rx\|_{\mathcal{K}}^2 = (x - R^*Rx, x) + (Rx, Rx) = \|x\|_{\mathcal{K}}^2.$$

Define an operator A_n: $l^2(\mathcal{K}) \to l^2(\mathcal{K})$ by

$$A_n(x_0, x_1, x_2, \ldots) = \Big(Rx_0, \underbrace{\mathbf{0}, \mathbf{0}, \ldots, \mathbf{0}}_{n}, \mathcal{D}_R x_0, x_1, x_2, \ldots\Big).$$

A_n is an isometric operator, since

$$\|A_n(x_0, x_1, x_2, \dots)\|^2_{l^2(\mathcal{K})} = \|Rx_0\|^2_{\mathcal{K}} + \|\mathfrak{D}_R x_0\|^2_{\mathcal{K}} + \sum_{i\geq 1} \|x_i\|^2_{\mathcal{K}} = \sum_{i \geq 0} \|x_i\|^2_{\mathcal{K}}.$$

We have

$$|(A_n^k f, g) - (\tilde{R}^k f, g)| \leq \|f\|_K \Big(\sum_{i \geq n} \|g_i\|^2_{\mathcal{K}} \Big)^{1/2} \to 0$$

for $f = \{f_i\}_{i \geq 0}$, $g = \{g_i\}_{i\geq 0} \in l^2(\mathcal{K})$, and consequently the operator $\tilde{R}$ in $l^2(\mathcal{K})$ is the *pw*-limit of isometric operators.

(II). Let T be a contraction in an infinite-dimensional Hilbert space $\mathcal{K}$. We will show that T can be approximated by isometric operators.

Let $\{P_n\}_{n\geq 0}$ be a sequence of orthoprojections in $\mathcal{K}$ such that $\mathcal{K}_n = P_n\mathcal{K} \subset \mathcal{K}_{n+1}$, $\mathcal{K}_n$ is isomorphic to $\mathcal{K}_n^{\perp}$, and the sequence $\{P_n\}_{n \geq 0}$ converges strongly to the identity operator. Then $\mathcal{K} = l^2(\mathcal{K}_n) = \mathcal{K}_n \oplus \mathcal{K}_n^{(1)} \oplus \mathcal{K}_n^{(2)} \oplus \cdots$, where $\mathcal{K}_n^{(i)}$ is a subspace of $\mathcal{K}$ that is isomorphic to $\mathcal{K}_n$; let $T_n \stackrel{\text{def}}{=} P_nT \,|\, \mathcal{K}_n$. Also let $\tilde{T}_n$: $\mathcal{K} \to \mathcal{K}$, where $\tilde{T}_n(x_0, x_1, \dots) = (T_n x_0, \mathbf{0}, \mathbf{0}, \dots)$.

Let $f \in \mathcal{K}_m$. Then

$$\|\tilde{T}_n f - Tf\|_{\mathcal{K}} = \|P_n Tf - Tf\|_{\mathcal{K}} \to 0 \quad \text{for } n \geq m.$$

Since $\operatorname{clos} \bigcup_m \mathcal{K}_m = \mathcal{K}$, it follows that the sequence $\{\tilde{T}_n\}_{n\geq 0}$ converges to T strongly, and, consequently, also in the *pw*-topology. From (I) it follows that every operator $\tilde{T}_n$ can be *pw*-approximated by isometries, and thus T also can be approximated by isometric operators.

(III). It remains to prove that for any isometric operator V in a Hilbert space $\mathcal{K}$, there exists a sequence of unitary operators that is strongly convergent to V.

By a theorem of Wold [15],* $\mathcal{K} = \mathcal{K}_1 \oplus \mathcal{K}_2$, where $\mathcal{K}_1$ and $\mathcal{K}_2$ are invariant subspaces for the operator V, $V \,|\, \mathcal{K}_1 = U$, a unitary operator, and $V \,|\, \mathcal{K}_2 = V_0$, a unilateral shift in the space $\mathcal{K}_2$, i.e., $\mathcal{K}_2 = l^2(\mathcal{K})$ and $V_0(k_0, k_1 \dots) = (\mathbf{0}, k_0, k_1 \dots)$. Let us define a sequence $\{V_n\}_{n \geq 1}$ of unitary operators in $\mathcal{K}_2$ by

$$V_n(k_0, k_1, k_2, \dots) = (k_n, k_0, k_1, k_2, \dots, k_{n-1}, k_{n+1}, k_{n+2}, \dots).$$

The sequence $\{V_n\}_{n\geq 1}$ converges strongly to the operator V_0:

$$\|(V_n - V_0)(k_0, k_1, k_2, \dots)\|^2_{\mathcal{K}_2} \leq 4 \sum_{m \geq n} \|k_m\|^2_{\mathcal{K}} \to 0.$$

Let $U_n = U \oplus V_n$. Then the sequence $\{U_n\}_{n\geq 1}$ converges strongly to the operator $U \oplus V_0 = V$. ●

We note that Halmos [29] proved that the set of unitary operators is dense in the unit ball in the weak topology.

COROLLARY (von Neumann's inequality). *Let T be a contraction in a Hilbert space $\mathcal{K}$. Then*

$$|\varphi(T)|_{\mathcal{K}} \leq \max\{|\varphi(\zeta)|: |\zeta| \leq 1\}$$

for every polynomial φ.

---

* *Editor's note.* The result cited here is Theorem 1.1 in Chapter I of [2].

PROOF. If the space $\mathcal{H}$ is infinite dimensional, then the assertion follows from Theorem 1 and Lemma 1, since von Neumann's inequality is true for unitary operators, in view of the spectral theorem.

Now let $\mathcal{H}$ be a finite-dimensional space. Imbed it in an infinite-dimensional Hilbert space $\tilde{\mathcal{H}}$. Let P be the orthoprojection of $\tilde{\mathcal{H}}$ onto $\mathcal{H}$, and $\tilde{T} = TP$. Then $\tilde{T}$ is a contraction in $\tilde{\mathcal{H}}$, and

$$\|\varphi(T)f\|_{\mathcal{H}} = \|\varphi(\tilde{T})f\|_{\tilde{\mathcal{H}}} \leqslant \max\{|\varphi(\zeta)|: |\zeta| \leqslant 1\} \cdot \|f\|_{\mathcal{H}}$$

for $f \in \mathcal{H}$. ●

§2. Disjoint contractions

DEFINITION. Let J be an isometric imbedding of a Banach space X into a Banach space Y, onto an orthogonally complemented subspace, i.e., such that there exists an orthoprojection P (a projection with unit norm) of Y onto the image JX; and let T be an operator in X. An operator U in the space Y is called a *dilation of the operator* T if $PU^nJ = JT^n$ for every natural number n ($PU^n \mid X = T^n$, if X is a subspace of Y).

The following lemma allows us to reduce the proof of inequality (1) to the construction of an isometric dilation.

LEMMA 2. *Let the operator U in the space Y be a dilation of the operator T in the space X. Then $|\varphi(T)|_X \leqslant |\varphi(U)|_Y$ for every polynomial φ.*

PROOF. For $f \in X$ we have

$$\|\varphi(T)f\|_X = \|J\varphi(T)f\|_Y = \|P\varphi(U)Jf\|_Y \leqslant \|\varphi(U)Jf\|_Y \leqslant |\varphi(U)|_Y \|f\|_X. \quad ●$$

Let $(\mathcal{X}, \mathcal{F}, \mu)$ and $(\mathcal{Y}, \mathcal{G}, \nu)$ be measure spaces, and $(\mathcal{X}, \mathcal{F}, \mu) \subset (\mathcal{Y}, \mathcal{G}, \nu)$ (i.e., $\mathcal{X} \subset \mathcal{Y}$, $\mathcal{F} = \{F \in \mathcal{G}: F \subset \mathcal{X}\}$, and $\mu = \nu \mid \mathcal{F}$). If every function in $L^p(\mathcal{X}, \mathcal{F}, \mu)$ is extended to zero on $\mathcal{Y} \setminus \mathcal{X}$, then $L^p(\mathcal{X}, \mathcal{F}, \mu)$ will be isometrically imbedded in $L^p(\mathcal{Y}, \mathcal{G}, \nu)$, and the corresponding projection P: $L^p(\mathcal{Y}, \mathcal{G}, \nu) \to L^p(\mathcal{X}, \mathcal{F}, \mu)$ ($Pf = \chi_{\mathcal{X}} f$) will be orthogonal. In this section, we consider isometric dilations of operators in $L^p(\mathcal{X}, \mathcal{F}, \mu)$ with respect to such canonical imbeddings. In this case, an operator T in $L^p(\mathcal{X}, \mathcal{F}, \mu)$ and its dilation U in $L^p(\mathcal{Y}, \mathcal{G}, \nu)$ are linked together by the elementary relation

$$(T^n f)(x) = (U^n \tilde{f})(x), \qquad f \in L^p(\mathcal{X}, \mathcal{F}, \mu),\ x \in \mathcal{X},\ n \geqslant 0.$$

Here, $\tilde{f}$ is the extension of f by zero on the set $\mathcal{Y} \setminus \mathcal{X}$.

We note that every contraction in $L^2(\mathcal{X}, \mathcal{F}, \mu)$ has a dilation with respect to the canonical imbedding, and this circumstance makes it possible to prove the existence of a unitary operator in $L^2(\mathcal{X}, \mathcal{F}, \mu)$ for which the pointwise ergodic theorem does not hold [30].

Let us recall that an operator T in an L^p-space is called *disjoint* if

$$f \cdot g = \mathbf{0} \Rightarrow (Tf) \cdot (Tg) = \mathbf{0}.$$

THEOREM 2. *Let $1 \leqslant p < +\infty$, $p \neq 2$. A contraction T of a space $L^p(\mathcal{X}, \mathcal{F}, \mu)$ has an isometric dilation in the space $L^p(\mathcal{Y}, \mathcal{G}, \nu)$ over some measure space $(\mathcal{Y}, \mathcal{G}, \nu)$ with $(\mathcal{Y}, \mathcal{G}, \nu) \supset (\mathcal{X}, \mathcal{F}, \mu)$, with respect to the canonical imbedding, if and only if T is a disjoint operator.*

For the proof of the theorem, we need the following characterization of disjoint operators.

LEMMA 3. *Let T be a disjoint operator in a space $L^p(\mathfrak{X}, \mathfrak{F}, \mu)$ with finite measure μ. Then there exist a measurable function h and a mapping S: $\mathfrak{F} \to \mathfrak{F}$ (mod 0) that satisfy the conditions $S(\mathfrak{X}\setminus\mathfrak{A}) = S\mathfrak{X}\setminus S\mathfrak{A}$, $S(\bigcup_1^\infty \mathfrak{A}_i) = \bigcup_1^\infty S\mathfrak{A}_i$ for disjoint sets $\mathfrak{A}_i$, and $Tf = h\mathbf{S}f$, where $\mathbf{S}$ is the operator that is continuous in the topology of convergence in measure and defined on characteristic functions by $\mathbf{S}\chi_{\mathfrak{A}} = \chi_{S\mathfrak{A}}$.*

PROOF OF LEMMA 3. Put $S\mathfrak{A} \overset{\text{def}}{=} \{x \in \mathfrak{X}: (T\chi_{\mathfrak{A}})(x) \neq 0\}$. Let $\mathfrak{A}$ and $\mathfrak{B}$ be disjoint sets. Then $\mu(S\mathfrak{A} \cap S\mathfrak{B}) = 0$, and consequently, $S(\mathfrak{A} \cup \mathfrak{B}) = S\mathfrak{A} \cup S\mathfrak{B}$ to within a set of measure zero. From the continuity of the operator T we obtain that $S(\bigcup_1^\infty \mathfrak{A}_i) = \bigcup_1^\infty S\mathfrak{A}_i$ (mod 0) for disjoint sets $\mathfrak{A}_i$. In particular, $S\mathfrak{X} = S\mathfrak{A} \cup S(\mathfrak{X}\setminus\mathfrak{A})$. Since $S\mathfrak{A}$ and $S(\mathfrak{X}\setminus\mathfrak{A})$ are disjoint, we have $S(\mathfrak{X}\setminus\mathfrak{A}) = S\mathfrak{X}\setminus S\mathfrak{A}$.

Let us prove that the operator $\mathbf{S}$ defined on characteristic functions by the equality $\mathbf{S}\chi_{\mathfrak{A}} = \chi_{S\mathfrak{A}}$ and extended linearly to the set of all simple functions is continuous on this set in the topology of convergence in measure. Let $f_n = \sum_{k=1}^{m_n} c_k^{(n)}\chi_{\mathfrak{A}_k^{(n)}}$, where the $\mathfrak{A}_k^{(n)}$ are disjoint sets, and the sequence $\{f_n\}_{n\geq 0}$ converges in measure to the zero function. Then $\mathbf{S}f_n = \sum_{k=1}^{m_n} c_k^{(n)}\chi_{S\mathfrak{A}_k^{(n)}}$, and

$$\{x \in \mathfrak{X}: |(\mathbf{S}f_n)(x)| \geq \varepsilon\} = S\{x \in \mathfrak{X}: |f_n(x)| \geq \varepsilon\}.$$

Since $f_n \to \mathbf{0}$ (in measure) and the measure σ defined by $\sigma(\mathfrak{A}) = \mu(S\mathfrak{A})$ is absolutely continuous with respect to the measure μ, it follows that $\mathbf{S}f_n \to \mathbf{0}$ (in measure). Now let us extend the operator $\mathbf{S}$, by continuity, to the set of all measurable functions.

Put $h = T\chi_{\mathfrak{X}}$. Then for every $E \in \mathfrak{F}$ we have $h = T\chi_E + T\chi_{\mathfrak{X}\setminus E}$. Consequently, $T\chi_E = h\chi_{SE} = h\mathbf{S}\chi_E$ (disjointness). Since T is continuous in $L^p(\mathfrak{X}, \mathfrak{F}, \mu)$ and $\mathbf{S}$ is continuous in the topology of convergence in measure, we have $Tf = h \cdot \mathbf{S}f$ for every function f in $L^p(\mathfrak{X}, \mathfrak{F}, \mu)$. ●

PROOF OF THEOREM 2. Let T be a disjoint contraction. Let us first assume that $\mu\mathfrak{X} < +\infty$. Then, by Lemma 3, $Tf = h\mathbf{S}f$. We will show that the operator T has an isometric dilation in $l^p(L^p(\mathfrak{X}, \mathfrak{F}, \mu))$(3) with respect to the imbedding $f \to (f, \mathbf{0}, \mathbf{0}, \ldots)$. Let P: $l^p(L^p(\mathfrak{X}, \mathfrak{F}, \mu)) \to L^p(\mathfrak{X}, \mathfrak{F}, \mu)$, where $P(f_0, f_1, f_2, \ldots) = f_0$. Then P is an orthoprojection from $l^p(L^p(\mathfrak{X}, \mathfrak{F}, \mu))$ onto the subspace $L^p(\mathfrak{X}, \mathfrak{F}, \mu)$. Let us put $d\sigma \overset{\text{def}}{=} |h|^p\, d\mu$ and $\lambda(\mathfrak{A}) \overset{\text{def}}{=} \sigma(S\mathfrak{A})$. Clearly, λ is a measure, and

$$\|Tf\|_p^p = \int_{\mathfrak{X}} |h|^p |\mathbf{S}f|^p\, d\mu = \int_{\mathfrak{X}} |\mathbf{S}f|^p\, d\sigma = \int_{\mathfrak{X}} |f|^p\, d\lambda.$$

Let $\mathfrak{A} \in \mathfrak{F}$. Since T is a contraction, we have $\lambda(\mathfrak{A}) = \|T\chi_{\mathfrak{A}}\|_p^p \leq \|\chi_{\mathfrak{A}}\|_p^p = \mu(\mathfrak{A})$. Consequently, $\lambda \leq \mu$, and the measure λ is absolutely continuous with respect to μ. So, by the Radon-Nikodým theorem, there exists an $\mathfrak{F}$-measurable function g such that $d\lambda = g\, d\mu$ and $0 \leq g \leq 1$. Hence,

$$\|f\|_p^p - \|Tf\|_p^p = \int_{\mathfrak{X}} |f|^p\, d\mu - \int_{\mathfrak{X}} |f|^p g\, d\mu = \int_{\mathfrak{X}} (1 - g)|f|^p\, d\mu.$$

(3) $l^p(L^p(\mathfrak{X}, \mathfrak{F}, \mu))$ is the L^p-space over the measure space $\sum_{k\geq 0} \oplus (\mathfrak{X}, \mathfrak{F}, \mu)$.

Define the operator R: $L^p(\mathfrak{X}, \mathfrak{F}, \mu) \to L^p(\mathfrak{X}, \mathfrak{F}, \mu)$ by $Rf = (1-g)^{1/p}f$. Then $\|Tf\|_p^p + \|Rf\|_p^p = \|f\|_p^p$. We now construct an isometric dilation of the operator T in $l^p(L^p(\mathfrak{X}, \mathfrak{F}, \mu))$ by putting

$$U: l^p(L^p(\mathfrak{X}, \mathfrak{F}, \mu)) \to l^p(L^p(\mathfrak{X}, \mathfrak{F}, \mu)),$$

where

$$U(f_0, f_1, f_2, \dots) = (Tf_0, Rf_0, f_1, f_2, \dots).$$

U is an isometric operator, since

$$\|U(f_0, f_1, f_2, \dots)\|_{l^p(L^p)}^p = \|Tf_0\|_p^p + \|Rf_0\|_p^p + \sum_{i \geqslant 1} \|f_i\|_p^p = \sum_{i \geqslant 0} \|f_i\|_p^p.$$

It is also clear that $PU^n \mid L^p(\mathfrak{X}, \mathfrak{F}, \mu) = T^n$ for all natural numbers n.

Now let μ be an arbitrary σ-finite measure. There exists an $\omega \in L^1(\mathfrak{X}, \mathfrak{F}, \mu)$ such that $\omega(x) > 0$ for all $x \in \mathfrak{X}$. Put $d\sigma = \omega\, d\mu$. Clearly, σ is a finite measure. Let $Vf = f\omega^{-1/p}$; then V is an isometry from $L^p(\mathfrak{X}, \mathfrak{F}, \mu)$ onto $L^p(\mathfrak{X}, \mathfrak{F}, \sigma)$. Consider the operator $\tilde{T}$: $L^p(\mathfrak{X}, \mathfrak{F}, \sigma) \to L^p(\mathfrak{X}, \mathfrak{F}, \sigma)$, where $\tilde{T} = VTV^{-1}$. Then $\tilde{T}$ is a disjoint contraction in $L^p(\mathfrak{X}, \mathfrak{F}, \sigma)$, and the operators T and $\tilde{T}$ are unitarily equivalent. Since $\tilde{T}$ has an isometric dilation in $l^p(L^p(\mathfrak{X}, \mathfrak{F}, \sigma))$, it follows that T also has an isometric dilation in $l^p(L^p(\mathfrak{X}, \mathfrak{F}, \mu))$.

The converse assertion is obvious, since for $p \neq 2$ isometric operators are disjoint [31], [32]. ●

From Theorem 2, Lemma 2, and a theorem of A. K. Kitover (see §10) it now follows that inequality (1) is valid for disjoint contractions in an L^p-space.

§3. A characterization of the operators in an L^p-space that have a unitary dilation

The principal result of this section (Theorem 3), a characterization of the operators that have a unitary dilation in L^p-space, allows us to prove inequality (1) for such operators in §7. Here, we assume that $1 < p < +\infty, p \neq 2$.

Let us recall the definitions of certain classes of operators.

DEFINITION. Let T be an operator in a space $L^p(\mathfrak{X}, \mathfrak{F}, \mu)$. It is called an *absolute contraction* if $|T|_1 \leqslant 1$ and $|T|_\infty \leqslant 1$, i.e., $\|Tf\|_1 \leqslant \|f\|_1$ and $\|Tf\|_\infty \leqslant \|f\|_\infty$ for $f \in L^1 \cap L^\infty$, and a *positive operator* if $f \geqslant \mathbf{0} \Rightarrow Tf \geqslant \mathbf{0}$. We will say that T *admits a contractive majorant* if there exists a positive contraction $\tilde{T}$ in $L^p(\mathfrak{X}, \mathfrak{F}, \mu)$ such that $|(Tf)(x)| \leqslant (\tilde{T}|f|)(x)$ a.e. for $f \in L^p(\mathfrak{X}, \mathfrak{F}, \mu)$. In this case, $\tilde{T}$ is called a *contractive majorant* of T. We denote by $\mathfrak{M}$ the set of all operators that admit a contractive majorant.

THEOREM 3. *An operator T in a space $L^p(\mathfrak{X}, \mathfrak{F}, \mu)$ has a unitary dilation in L^p-space if and only if $T \in \mathfrak{M}$. If the space $L^p(\mathfrak{X}, \mathfrak{F}, \mu)$ is separable, then it is possible to construct a unitary dilation of T in $L^p[0, 1]$.*

The following assertions, which are used both in the proof of Theorem 3 and in the succeeding sections, give some idea of the operators of class $\mathfrak{M}$.

LEMMA 4. *The class $\mathfrak{M}$ contains the sets of isometries, disjoint contractions, positive contractions, and absolute contractions.*

LEMMA 5. *If $T, R \in \mathfrak{M}$, then $TR \in \mathfrak{M}$.*

LEMMA 6. *If T is an operator of class $\mathfrak{M}$ in a space $L^p(\mathfrak{X}, \mathfrak{F}, \mu)$, where $p \neq 2$, and U is an isometry from $L^p(\mathfrak{X}, \mathfrak{F}, \mu)$ onto $L^p(\mathfrak{Y}, \mathfrak{S}, \nu)$, then the operator UTU^{-1} in $L^p(\mathfrak{Y}, \mathfrak{S}, \nu)$ belongs to the class $\mathfrak{M}$.*

We remark that the set of absolute contractions does not have a similar property of unitary invariance.

PROOF OF PART OF LEMMA 4 (for isometries, and disjoint and positive contractions). That isometries and disjoint contractions are included in the class $\mathfrak{M}$ can be seen immediately from the known parametric representation of operators of these classes in an L^p-space (see [31], [32], and Lemma 3 of this paper). If T is a positive contraction, then $|(Tf)(x)| \leq (T|f|)(x)$ a.e. for $f \in L^p$, and we can put $\tilde{T} = T$. ●

The proof of the last inclusion of Lemma 4 (for absolute contractions) will be carried out after the proof of Theorem 4 (see §4).

PROOF OF LEMMA 5. Let $\tilde{T}$ and $\tilde{R}$ be contractive majorants for T and R, respectively. Let $f \in L^p$. Then $|TRf| \leq \tilde{T}|Rf| \leq \tilde{T}\tilde{R}|f|$. Thus, the product $\tilde{T}\tilde{R}$ is a contractive majorant of the operator TR. ●

PROOF OF LEMMA 6. From the characterization of isometries of L^p-spaces [32], it follows that there exist isometries

$$V: L^p(\mathfrak{X}, \mathfrak{F}, \mu) \to L^p(\mathfrak{Y}, \mathfrak{S}, \nu)$$

and

$$W: L^p(\mathfrak{Y}, \mathfrak{S}, \nu) \to L^p(\mathfrak{X}, \mathfrak{F}, \mu)$$

such that $|Uf| \leq V|f|$ a.e. ($\forall f \in L^p(\mathfrak{X}, \mathfrak{F}, \mu)$) and $|U^{-1}g| \leq W|g|$ a.e. ($\forall g \in L^p(\mathfrak{Y}, \mathfrak{S}, \nu)$). So, $|UTU^{-1}g| \leq V|TU^{-1}g| \leq V\tilde{T}|U^{-1}g| \leq V\tilde{T}W|g|$, where $\tilde{T}$ is a contractive majorant of T. Thus, $V\tilde{T}W$ is a contractive majorant of UTU^{-1}, and consequently, $UTU^{-1} \in \mathfrak{M}$. ●

Before passing to the proof of Theorem 3, let us recall the definitions of certain concepts which we will need later on.

DEFINITION (see [23] and [33]). Let Γ be a set on which there is given an ultrafilter λ (see [34]), and let $\{X_\gamma\}_{\gamma \in \Gamma}$ be a collection of Banach spaces. Let $\Pi_0(\{X_\gamma\}_{\gamma\in\Gamma})$ be the linear space of all collections $f = \{f_\gamma\}_{\gamma\in\Gamma}$ such that $f_\gamma \in X_\gamma$ and $\sup_{\gamma\in\Gamma} \|f_\gamma\|_{X_\gamma} < +\infty$. Let us define a seminorm $\|f\| = \lim_\lambda \|f_\gamma\|_{X_\gamma}$ on the set $\Pi_0(\{X_\gamma\}_{\gamma\in\Gamma})$ (this limit over the ultrafilter exists in view of the compactness of the interval $[0, \sup_{\gamma\in\Gamma} \|f_\gamma\|_{X_\gamma}]$). By the *ultraproduct* of the spaces X_γ over the ultrafilter λ, we mean the space $\Pi_{\gamma\in\Gamma} X_\gamma/\lambda$ obtained as the result of completion of the normed (quotient) space $\Pi_0(\{X_\gamma\}_{\gamma\in\Gamma})/\{f: \|f\| = 0\}$.

If the spaces X_γ are Banach lattices, then a Banach lattice structure can be introduced on the ultraproduct $\Pi_{\gamma\in\Gamma} X_\gamma/\lambda$ in a natural way. If X_γ is an L^p-space for every $\gamma \in \Gamma$, then the ultraproduct $\Pi_{\gamma\in\Gamma} X_\gamma/\lambda$ is isometric to some L^p-space (see [33]; this follows from the lattice characterization of L^p-spaces: it is enough to show that $\|f + g\|^p = \|f\|^p + \|g\|^p$ for disjoint elements $f, g \in \Pi_{\gamma\in\Gamma} X_\gamma/\lambda$). However, in all nontrivial cases, it is impossible to realize this space as a space $L^p(\mathfrak{X}, \mathfrak{F}, \mu)$ where μ is a σ-finite measure.

DEFINITION. Let $(\mathfrak{X}, \mathfrak{F}, \mu)$ be a measure space, $\mathfrak{Y} \in \mathfrak{F}$, $\mathfrak{S}$ a σ-algebra of subsets of $\mathfrak{Y}$, $\mathfrak{S} \subset \mathfrak{F}$, and $\nu = \mu|\mathfrak{S}$. Then the space $L^p(\mathfrak{Y}, \mathfrak{S}, \nu)$ is isometrically imbedded in $L^p(\mathfrak{X}, \mathfrak{F}, \mu)$ in a natural way. Let $f \in L^p(\mathfrak{X}, \mathfrak{F}, \mu)$. Consider the set function

$\mathcal{Q} \to \int_{\mathcal{Q}} f\,d\mu$, $\mathcal{Q} \in \mathcal{E}$. This set function is absolutely continuous with respect to ν. Let $E_{\mathcal{E}} f$ be its Radon-Nikodým derivative. The operator $E_{\mathcal{E}}$: $L^p(\mathcal{X}, \mathcal{F}, \mu) \to L^p(\mathcal{Y}, \mathcal{E}, \nu)$ is called a *conditional expectation operator*.

We note that $E_{\mathcal{E}}$ is a positive orthoprojection of $L^p(\mathcal{X}, \mathcal{F}, \mu)$ onto the subspace $L^p(\mathcal{Y}, \mathcal{E}, \nu)$ for all p, $1 \leqslant p \leqslant +\infty$.

Let us now proceed to the proof of Theorem 3. The proof will be carried out in two stages. First, the theorem will be proved in the finite-dimensional case, and then, using the construction of an ultraproduct of Banach spaces in a suitable way, we will obtain the general case. The proof of the following lemma is carried out similarly to the way the proof was done for Theorem 2.13 of [19], where the case of positive contractions was under consideration.

LEMMA 7. *Let T be the operator in the space l_n^p that is defined by the matrix $\{t_{ij}\}_{1\leqslant i,j\leqslant n}$ in the standard basis. Suppose that $\|\{|t_{ij}|\}_{1\leqslant i,j\leqslant n}\|_p = 1$ and $t_{ij} \neq 0$, $1 \leqslant i, j \leqslant n$. Then T has a unitary dilation in L^p-space.*

PROOF. Let $\tilde{T}$ be the operator in l_n^p given by the matrix $\{|t_{ij}|\}_{1\leqslant i,j\leqslant n}$, and let $\tilde{T}^*$ be the adjoint operator in $l_n^{p'}$. For every vector $c = \{c_i\}_{1\leqslant i\leqslant n} \in l_n^p$ with nonnegative coordinates, there exists a unique vector $c^* = \{c_i^*\}_{1\leqslant i\leqslant n} \in l_n^{p'}$ such that $(c, c^*) = \|c\|_p^p = \|c^*\|_{p'}^{p'}$ (namely, $c_i^* = c_i^{p-1}$).

Since the operator $\tilde{T}$ is positive, there exists a vector $u = \{u_i\}_{1\leqslant i\leqslant n} \in l_n^p$ such that $u_i \geqslant 0$ for $1 \leqslant i \leqslant n$ and $\|u\|_p = \|\tilde{T}u\|_p = 1$. Let $v = \tilde{T}u$. Then $1 = (v, v^*) = (T^*u, v^*) = (u, \tilde{T}^*v^*)$, and consequently, $u^* = \tilde{T}^*v^*$. In view of the fact that $t_{ij} \neq 0$ for $1 \leqslant i, j \leqslant n$, we have $u_i, v_i > 0$ for $1 \leqslant i \leqslant n$. Thus, we have constructed vectors $u = \{u_i\}_{1\leqslant i\leqslant n}$ and $v = \{v_i\}_{1\leqslant i\leqslant n}$ such that $\|u\|_p = \|v\|_{p'} = 1$, $v_i, u_i > 0$, and

$$v_i = \sum_{1\leqslant j\leqslant n} |t_{ij}| u_j, \quad u_j^{p-1} = \sum_{1\leqslant i\leqslant n} |t_{ij}| v_i^{p-1} \quad \text{for } 1 \leqslant i, j \leqslant n.$$

Put $\xi_{ij} = |t_{ij}|(u_j/v_i)$ and $\eta_{ij} = |t_{ij}|(v_i/u_j)^{p-1}$. Then, $\sum_{1\leqslant j\leqslant n} \xi_{ij} = 1$ and $\sum_{1\leqslant i\leqslant n} \eta_{ij} = 1$. Let the $F_i = [i-1, i] \times [i-1, i]$ and $F = \bigcup_{1\leqslant i\leqslant n} F_i$ be sets in the plane. Let us divide every interval $[i-1, i]$ into disjoint intervals $I_{i1}, \ldots, I_{in}$ of lengths $\xi_{i1}, \ldots, \xi_{in}$ and into disjoint intervals $J_{1i}, \ldots, J_{ni}$ of lengths $\eta_{1i}, \ldots, \eta_{ni}$. Put

$$R_{ij} = [i-1, i] \times J_{ij}, \qquad S_{ij} = I_{ij} \times [i-1, i].$$

For every pair (i, j), there exist real numbers a_{ij}, b_{ij}, c_{ij} and d_{ij} such that the transformation

$$\tau_{ij}(x, y) = (a_{ij}x + b_{ij}, c_{ij}y + d_{ij})$$

maps R_{ij} onto S_{ij} (see the figure).

Define the mapping τ: $F \to F$ by $\tau | R_{ij} = \tau_{ij}$. Consider $L^p(F)$, the L^p-space on the set F with respect to the plane Lebesgue measure m_2. We define an operator U in this space by the equality

$$(Uf)(x, y) = \frac{v_i}{u_j} \frac{t_{ij}}{|t_{ij}|} f(\tau^{-1}(x, y)) \quad \text{for } (x, y) \in S_{ij}.$$

It follows from the definitions of τ, ξ_{ij}, and η_{ij} that U is a unitary operator.

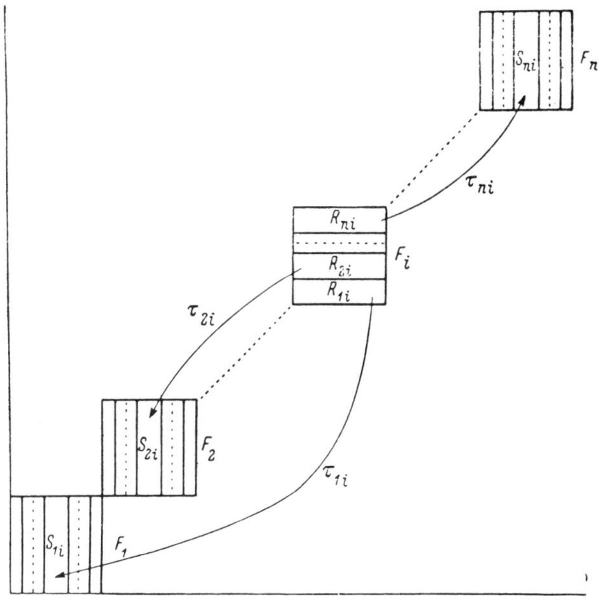

Let $c = \{c_i\}_{1 \leqslant i \leqslant n} \in l_n^p$. Put $Jc = \sum_{1 \leqslant i \leqslant n} c_i \chi_{F_i}$. Then J is an isometric imbedding of l_n^p into $L^p(F)$. We define an orthoprojection E: $L^p(F) \to J l_n^p$ by

$$Ef = \sum_{1 \leqslant i \leqslant n} \chi_{F_i} \int_{F_i} f \, dm_2, \qquad f \in L^p(F).$$

Let $f \in L^p(F)$, where the function f depends only on the coordinate x, i.e., $f(x, y) = \varphi(x)$ for $(x, y) \in F$. Then

$$\begin{aligned} \int_{F_i} Uf \, dm_2 &= \sum_{1 \leqslant j \leqslant n} \frac{v_i}{u_j} \frac{t_{ij}}{|t_{ij}|} \int_{S_{ij}} f\left(\tau_{ij}^{-1}(x, y)\right) dx\, dy \\ &= \sum_{1 \leqslant j \leqslant n} \frac{v_i}{u_j} \frac{t_{ij}}{|t_{ij}|} \frac{m_2(S_{ij})}{m_2(R_{ij})} \int_{R_{ij}} f(x, y)\, dx\, dy \\ &= \sum_{1 \leqslant j \leqslant n} \left(\frac{v_i}{u_j}\right)^{1-p} \frac{t_{ij}}{|t_{ij}|} \eta_{ij} \int_{[j-1, j]} \varphi(x)\, dx = \sum_{1 \leqslant j \leqslant n} t_{ij} \int_{F_j} f \, dm_2. \end{aligned}$$

Consequently,

$$EUf = \sum_{1 \leqslant i \leqslant n} \chi_{F_i} \sum_{1 \leqslant j \leqslant n} t_{ij} \int_{F_j} f \, dm_2. \tag{2}$$

Thus, if $f \in L^p(F)$ and f depends only on the coordinate x, then $EUf = EUEf$. We note that in this case Uf also depends only on the coordinate x, and consequently $EU^k f = (EU)^k f$ for $k \geqslant 0$.

If $c = \{c_i\}_{1 \leqslant i \leqslant n} \in l_n^p$, it follows from (2) that

$$EU\left(\sum_{1 \leqslant i \leqslant n} c_i \chi_{F_i}\right) = \sum_{1 \leqslant i \leqslant n} (Tc)_i \chi_{F_i}. \tag{3}$$

We have

$$EU^k\Big(\sum_{1\leqslant i\leqslant n} c_i\chi_{F_i}\Big) = (EU)^k\Big(\sum_{1\leqslant i\leqslant n} c_i\chi_{F_i}\Big) = \sum_{1\leqslant i\leqslant n} (T^kc)_i\chi_{F_i}, \qquad k \geqslant 0,$$

applying (3) k times.

Thus, the operator U is a unitary dilation of T with respect to the imbedding J and the projection P. ●

LEMMA 8. *Let T be an operator of class $\mathfrak{M}$ in l^p_n. Then T has a unitary dilation in L^p-space.*

PROOF. Let $\{t_{ij}\}_{1\leqslant i,j\leqslant n}$ be the matrix of the operator T in the standard basis. We may assume that $|\{|t_{ij}|\}_{1\leqslant i,j\leqslant n}|_p = 1$; otherwise we could extend the operator T to l^p_{n+1} by putting $t_{n+1,n+1} = 1$ and $t_{i,n+1} = t_{n+1,i} = 0$ for $i \leqslant n$, and make use of the following simple observation: if B is a dilation of A and C is a dilation of B, then C is a dilation of A. Now let $|\{|t_{ij}|\}_{1\leqslant i,j\leqslant n}|_p = 1$. In this case, there exists a sequence of operators

$$\{T_m\}_{m\geqslant 1} = \Big\{\{t^{(m)}_{ij}\}_{1\leqslant i,j\leqslant n}\Big\}_{m\geqslant 1}$$

converging to T such that $|\{|t^{(m)}_{ij}|\}_{1\leqslant i,j\leqslant n}|_p = 1$ and $t^{(m)}_{ij} \neq 0$ for all i, j and m.

By Lemma 7, each operator T_m has a unitary dilation in L^p-space. Consequently, there exist a sequence of L^p-spaces $\{X_m\}_{m\geqslant 1}$, a sequence of isometric imbeddings J_m: $l^p_n \to X_m$, and a sequence of operators P_m: $X_m \to l^p_n$ of unit norm such that

$$P_m U^k_m J_m = T^k_m, \qquad k \geqslant 0. \tag{4}$$

Consider the filter on the set $\mathbb{N}$ of natural numbers generated by the sets $\{i \in \mathbb{N}: i \geqslant m\}$, $m \geqslant 1$ (see [34]), and let λ be an ultrafilter that majorizes it. Let $X = \prod_{m\in\mathbb{N}} X_m/\lambda$ be the ultraproduct of the spaces X_m over the ultrafilter λ; then X is an L^p-space. Let us define an imbedding J: $l^p_n \to X$ by

$$Jc = \{J_m c\}_{m\geqslant 1}, \qquad c \in l^p_n.$$

Then $\|Jc\|_X = \lim_\lambda \|J_m c\|_{X_m} = \|c\|_p$, i.e., J is an isometry.

We define an operator P: $X \to l^p_n$ on a dense subset by

$$P(\{x_m\}_{m\geqslant 1}) = \lim_\lambda P_m x_m, \qquad \{x_m\}_{m\geqslant 1} \in \Pi_0(\{X_m\}_{m\geqslant 1})\text{(4)}$$

(the limit exists in view of the compactness of the unit ball in l^p_n). Then

$$\|P(\{x_m\}_{m\geqslant 1})\|_p = \Big\|\lim_\lambda P_m x_m\Big\|_p \leqslant \lim_\lambda \|P_m x_m\|_p \leqslant \lim_\lambda \|x_m\|_{X_m} = \|\{x_m\}_{m\geqslant 1}\|_X.$$

Consequently, the operator P can be extended to X, with $|P| \leqslant 1$.

We define an operator U on a dense subset of X by

$$U(\{x_m\}_{m\geqslant 1}) = \{U_m x_m\}_{m\geqslant 1}, \qquad \{x_m\}_{m\geqslant 1} \in \Pi_0(\{X_m\}_{m\geqslant 1}).$$

It is clear that U can be extended to a unitary operator, since U is isometric and invertible, where the inverse operator is defined by

$$U^{-1}(\{x_m\}_{m\geqslant 1}) = \{U^{-1}_m x_m\}_{m\geqslant 1}, \qquad \{x_m\}_{m\geqslant 1} \in \Pi_0(\{X_m\}_{m\geqslant 1}).$$

---

(4) It is easily seen that if the element $\{x_m\}_{m\geqslant 1}$ of the space $\Pi_0(\{X_m\}_{m\geqslant 1})$ passes into the zero element when the quotient space with respect to the subspace $\{f \in \Pi_0: \|f\| = 0\}$ is taken, then $P(\{x_m\}_{m\geqslant 1}) = 0$, and consequently, this formula gives a well-defined operator P: $X \to l^p_n$.

We will show that

$$PU^kJ = T^k, \qquad k \geqslant 0.$$

Let $c \in l_n^p$. Then, in view of (4),

$$PU^kJc = \lim_\lambda P_m U_m^k J_m c = \lim_\lambda T_m^k c = T^k c, \qquad k \geqslant 0,$$

i.e., U is a unitary dilation of the operator T in X. ●

REMARK. The space X in which the unitary dilation has been constructed cannot be realized as an L^p-space over a space with σ-finite measure. Concerning how it is possible to pass to an L^p-space with a finite measure, see the proof of Theorem 3.

PROOF OF THEOREM 3. *Sufficiency.* Since an L^p-space on a measure space with σ-finite measure is isometric to an L^p-space on a measure space with finite measure and the class $\mathfrak{M}$ is unitarily invariant, we will assume measures to be finite.

Let T be an operator of class $\mathfrak{M}$ in $L^p(\mathcal{X}, \mathcal{F}, \mu)$. Consider the directed set Γ of finite partitions of $\mathcal{X}$ into measurable sets (if $\gamma_1, \gamma_2 \in \Gamma$ then $(\gamma_1 \leqslant \gamma_2) \overset{\text{def}}{\Leftrightarrow} (\gamma_2$ is a refinement of $\gamma_1)$). Let $\gamma = \{\mathcal{X}_1^\gamma, \dots, \mathcal{X}_{n_\gamma}^\gamma\} \in \Gamma$. Define an operator Q_γ in $L^p(\mathcal{X}, \mathcal{F}, \mu)$ by

$$(Q_\gamma f)(x) = \frac{1}{\mu(\mathcal{X}_i^\gamma)} \int_{\mathcal{X}_i^\gamma} f \, d\mu, \qquad x \in \mathcal{X}_i^\gamma.$$

Then Q_γ is a positive orthoprojection.

Put $w_i^\gamma = \mu(\mathcal{X}_i^\gamma)$, $1 \leqslant i \leqslant n_\gamma$, and consider the space $l_{n_\gamma}^p(w^\gamma)$. Let us identify it with the subspace of $L^p(\mathcal{X}, \mathcal{F}, \mu)$ that consists of the functions that are constant on the elements of the partition γ:

$$c = \{c_i\}_{1 \leqslant i \leqslant n_\gamma} \to \sum_{1 \leqslant i \leqslant n_\gamma} c_i \chi_{\mathcal{X}_i^\gamma}.$$

The operator $Q_\gamma T Q_\gamma$ leaves this subspace invariant, and, consequently, induces an operator $T_\gamma = \{t_{ij}^{(\gamma)}\}_{1 \leqslant i,j \leqslant n_\gamma}$ in $l_{n_\gamma}^p(w^\gamma)$.

In view of Lemma 5, we have $Q_\gamma T Q_\gamma \in \mathfrak{M}$, and consequently $T_\gamma \in \mathfrak{M}$. We note that the generalized sequence $\{Q_\gamma T Q_\gamma\}_{\gamma \in \Gamma}$ converges to T in the strong operator topology.

Since the space $l_{n_\gamma}^p(w^\gamma)$ is isometric to $l_{n_\gamma}^p$, we have, by Lemma 8, that T_γ has a unitary dilation in the L^p-space X_γ, i.e., there exist an isometric imbedding J_γ: $l_{n_\gamma}^p(w^\gamma) \to X_\gamma$, an operator P_γ: $X_\gamma \to l_{n_\gamma}^p(w^\gamma)$ of unit norm, and a unitary operator U_γ in X_γ such that

$$P_\gamma U_\gamma^k J_\gamma = T_\gamma^k, \qquad k \geqslant 0. \tag{5}$$

Consider the filter on the set Γ generated by the sets $\{\beta \in \Gamma : \beta \geqslant \gamma\}$, $\gamma \in \Gamma$, and let λ be an ultrafilter that majorizes it. Consider the space $X = \prod_{\gamma \in \Gamma} X_\gamma / \lambda$. Then X is an L^p-space.

Define the operator J: $L^p(\mathcal{X}, \mathcal{F}, \mu) \to X$ by setting

$$Jf = \{J_\gamma Q_\gamma f\}_{\gamma \in \Gamma}, \qquad f \in L^p(\mathcal{X}, \mathcal{F}, \mu).$$

Then, if f is constant on the elements of the partition $\gamma_0 \in \Gamma$, it follows that $Q_\gamma f = f$ for $\gamma \geqslant \gamma_0$. Consequently,

$$\|\gamma f\|_X = \lim_\lambda \|J_\gamma Q_\gamma f\|_{X_\gamma} = \lim_\lambda \|J_\gamma f\|_{X_\gamma} = \|f\|_p.$$

In view of the fact that the simple functions form a dense set, we conclude that J is an isometric imbedding of $L^p(\mathfrak{X}, \mathfrak{F}, \mu)$ into X. Define the operator U in X by

$$U(\{x_\gamma\}_{\gamma\in\Gamma}) = \{U_\gamma x_\gamma\}_{\gamma\in\Gamma}, \qquad \{x_\gamma\}_{\gamma\in\Gamma} \in \Pi_0(\{X_\gamma\}_{\gamma\in\Gamma}).$$

Clearly, U is a unitary operator in X.

Finally, define the operator P: $X \to L^p(\mathfrak{X}, \mathfrak{F}, \mu)$ by

$$P(\{x_\gamma\}_{\gamma\in\Gamma}) = \lim_\lambda P_\gamma x_\gamma, \qquad \{x_\gamma\}_{\gamma\in\Gamma} \in \Pi_0(\{X_\gamma\}_{\gamma\in\Gamma})$$

(the limit is computed in the weak topology, and exists in view of the weak compactness of the unit ball in an L^p-space). Clearly, $|P| \leqslant 1$.

We will show that

$$PU^kJ = T^k, \qquad k \geqslant 0.$$

Let $f \in L^p(\mathfrak{X}, \mathfrak{F}, \mu)$, and let f be constant on the elements of the partition $\gamma_0 \in \Gamma$. Then equality (5) yields

$$\begin{aligned} PU^kJf &= \lim_\lambda P_\gamma U_\gamma^k J_\gamma Q_\gamma f = \lim_\lambda P_\gamma U_\gamma^k J_\gamma f = \lim_\lambda T_\gamma^k f \\ &= \lim_\lambda (Q_\gamma T Q_\gamma)^k f = T^k f, \qquad k \geqslant 0, \end{aligned}$$

since the generalized sequence $\{Q_\gamma T Q_\gamma\}_{\gamma\in\Gamma}$ of operators converges strongly to T, and the multiplication operation is continuous in the strong operator topology on bounded sets. Thus, in view of the density of the set of simple functions, we obtain that T has a unitary dilation in L^p-space.

We will now show that a dilation of T can be constructed in an L^p-space over a finite measure space. Let $Y_0 = JL^p(\mathfrak{X}, \mathfrak{F}, \mu)$, and let Y_{n+1} be the closed vector lattice generated by the set

$$\{U^m Y_n : m \in \mathbb{Z}\}.$$

Put $Z = \operatorname{clos} \bigcup_{n\geqslant 1} Y_n$. Then Z is a closed vector sublattice of X that is invariant with respect to U and U^{-1}, and isometric to a space $L^p(\mathfrak{Z}, \mathfrak{S}, \nu)$, where ν is a finite measure (see [33]). Let $\tilde{U} = U \mid Z$ and $\tilde{P} = P \mid Z$. Then $\tilde{U}$ is a unitary dilation of T with respect to $\tilde{P}$ and J.

If the space X is separable, so is Z. We will verify that in this case a dilation of T can be constructed in $L^p[0, 1]$ (and with this, the proof of sufficiency will be complete).

Since every separable L^p-space is isometric to one of the spaces l_n^p, $1 \leqslant n \leqslant +\infty$, $L^p[0, 1]$ or $L^p[0, 1] \oplus_p l_n^p$, $1 \leqslant n \leqslant +\infty$ (the symbol $\oplus_p$ denotes a sum of l^p type: $\| f \oplus_p g \|^p = \| f \|^p + \| g \|^p$, $f \in L^p[0, 1]$, $g \in l_n^p$), and every unitary operator U in the space $L^p[0, 1] \oplus_p l_n^p$ decomposes into a direct sum $U_1 \oplus U_2$, where U_1 is a unitary operator in $L^p[0, 1]$ and U_2 is a unitary operator in l_n^p (see [33]), it remains to show that any unitary operator acting in l_n^p, $1 \leqslant n \leqslant +\infty$, has a unitary dilation in $L^p[0, 1]$. Clearly, we may assume that $n = +\infty$.

Let U be a unitary operator in l^p defined by a matrix $\{u_{ij}\}_{0\leqslant i,\, j<+\infty}$ with respect to the standard basis. Let us imbed l^p isometrically into $L^p[0, +\infty)$ as follows:

$$\{c_i\}_{i\geqslant 0} \to \sum_{i\geqslant 0} c_i \chi_{[i,i+1]}, \qquad \{c_i\}_{i\geqslant 0} \in l^p.$$

Clearly, this is an imbedding onto an orthogonally complemented subspace. We define an operator $\tilde{U}$ in $L^p[0, +\infty)$ by

$$(\tilde{U}f)(x) = \sum_{j \geqslant 0} u_{ij} f(x + j - i), \qquad x \in [i, i+1].$$

Then it is easy to verify that $\tilde{U}$ is a unitary operator in $L^p[0, +\infty)$ that is a dilation of U. Since $L^p[0, +\infty)$ is isometric to $L^p[0, 1]$, the assertion has been proved.

Necessity. Let T be an operator in a space $L^p(\mathfrak{X}, \mathfrak{F}, \mu)$, and U a unitary dilation of T in a space $L^p(\tilde{\mathfrak{X}}, \tilde{\mathfrak{F}}, \tilde{\mu})$ with respect to an imbedding J and an orthoprojection P. Then

$$PUJ = JT. \tag{6}$$

By a theorem of Ando [35], the operator P is unitarily equivalent to a conditional expectation operator, i.e., there exist a measure space $(\tilde{\mathfrak{Y}}, \tilde{\mathfrak{S}}, \tilde{\nu})$, $\mathfrak{Y} \subset \tilde{\mathfrak{Y}}$, a σ-algebra $\mathfrak{S}$ of subsets of $\mathfrak{Y}$ with $\mathfrak{S} \subset \tilde{\mathfrak{S}}$, $\nu = \tilde{\nu} \mid \mathfrak{S}$, and an isometry V of $L^p(\tilde{\mathfrak{X}}, \tilde{\mathfrak{F}}, \tilde{\mu})$ onto $L^p(\tilde{\mathfrak{Y}}, \tilde{\mathfrak{S}}, \tilde{\nu})$ such that $VPV^{-1} = E_{\mathfrak{S}}$, where $E_{\mathfrak{S}}$ is the conditional expectation operator with respect to $\mathfrak{S}$. Let $\tilde{U} = VUV^{-1}$ and $\tilde{J} = VJ$; then it follows from (6) that

$$E_{\mathfrak{S}} \tilde{U} \tilde{J} = \tilde{J} T.$$

$\tilde{J}$ is an isometry of $L^p(\mathfrak{X}, \mathfrak{F}, \mu)$ onto $L^p(\mathfrak{Y}, \mathfrak{S}, \nu)$. Let $\tilde{T} = \tilde{J} T \tilde{J}^{-1}$, which is an operator in $L^p(\mathfrak{Y}, \mathfrak{S}, \nu)$. Then $E_{\mathfrak{S}} \tilde{U} \mid L^p(\mathfrak{Y}, \mathfrak{S}, \nu) = \tilde{T}$, and by virtue of Lemmas 4 and 5 and the positiveness of the operator $E_{\mathfrak{S}}$ we have $\tilde{T} \in \mathfrak{M}$ and consequently, by Lemma 6, $T \in \mathfrak{M}$. ●

REMARK 1. The constructions presented in Theorem 3 lead to a dilation with respect to a positive orthoprojection. Using the theorem of Ando [35] on the unitary equivalence of an orthoprojection in L^p to a conditional expectation operator (if the orthoprojection is positive, then the operators that realize the unitary equivalence can also be chosen to be positive), it is not hard to show that if T is an operator of class $\mathfrak{M}$ in $L^p(\mathfrak{X}, \mathfrak{F}, \mu)$, then there exist a measure space $(\tilde{\mathfrak{Y}}, \tilde{\mathfrak{S}}, \tilde{\nu})$ (in the case that $L^p(\mathfrak{X}, \mathfrak{F}, \mu)$ is separable, one can choose for $L^p(\tilde{\mathfrak{Y}}, \tilde{\mathfrak{S}}, \tilde{\nu})$ the space $L^p[0, 1]$), $\mathfrak{Y} \subset \tilde{\mathfrak{Y}}$, a σ-algebra $\mathfrak{S}$ of subsets of $\mathfrak{Y}$ with $\mathfrak{S} \subset \tilde{\mathfrak{S}}$, $\nu = \tilde{\nu} \mid \mathfrak{S}$, a positive isometry J of $L^p(\mathfrak{X}, \mathfrak{F}, \mu)$ onto $L^p(\mathfrak{Y}, \mathfrak{S}, \nu)$, and a unitary operator U in $L^p(\tilde{\mathfrak{Y}}, \tilde{\mathfrak{S}}, \tilde{\nu})$ such that

$$E_{\mathfrak{S}} U^k J = J T^k, \qquad k \geqslant 0,$$

where $E_{\mathfrak{S}}$ is the conditional expectation operator.

REMARK 2. If T is a positive contraction in $L^p(\mathfrak{X}, \mathfrak{F}, \mu)$, then the above construction leads to the existence of a positive unitary dilation in L^p-space. Thus, Theorem 3 strengthens the theorem of Akcoglu and Sucheston on dilations of positive contractions [19], [22].

REMARK 3. Theorem 3 is valid for $p = 2$ if, in the formulation, we replace unitary dilation by disjoint unitary dilation.

§4. Approximations by unitary operators in the space $L^p[0, 1]$

In this section, we will describe the pw-closure of the set of all unitary operators in $L^p[0, 1]$, $1 < p < +\infty$, $p \neq 2$ (Theorem 4), which will lead in §7 to yet one more proof of inequality (1) for operators that admit a contractive majorant. This theorem also makes it possible to give an approximation proof of the pointwise ergodic theorem for contractions of class $\mathfrak{M}$ in L^p that was proved by Akcoglu [19] for positive contractions with the aid of a theorem on dilation.

THEOREM 4. *The closure in the pw-topology (and also in the weak operator topology) of the set of unitary operators of the space $L^p[0, 1]$ coincides with the set* $\mathfrak{M}$.

REMARK 1. The theorem holds also for $p = 2$, if one considers the closure of the set of disjoint unitary operators.

PROOF. Let T be an operator of class $\mathfrak{M}$ in $L^p[0, 1]$. Consider the orthoprojection P_n defined by

$$(P_n f)(x) = \begin{cases} k\int_{(i-1)/k}^{i/k} f(x)\,dx, & x \in \left[\frac{i-1}{k}, \frac{i}{k}\right], 1 \leqslant i < k, \\ 0, & x \in \left[\frac{k-1}{k}, 1\right], \end{cases}$$

where $k = 2^n$. Let X_n be the subspace of $L^p[0, 1]$ consisting of the functions that are constant on the intervals $[(i-1)/k, i/k]$, $1 \leqslant i < k$, and vanish on $[(k-1)/k, 1]$. Then X_n is isometric to l^p_{k-1}, and the operator $P_n T P_n$ leaves X_n invariant. Put $T_n = P_n T P_n \mid X_n$. Then T_n is an operator of class $\mathfrak{M}$ in X_n (see Lemmas 4 and 5). By Remark 1 after Theorem 3, there exists a unitary dilation of T_n in $L^p[0, 1]$ with respect to the orthoprojection E defined by

$$(Ef)(x) = \begin{cases} \frac{1}{|F_i|}\int_{F_i} f(x)\,dx, & x \in F_i, \\ 0, & x \in [0, 1] \setminus \bigcup_{1 \leqslant i < k} F_i, \end{cases}$$

and the imbedding $\chi_{[(i-1)/k, i/k]} \to (k|F_i|)^{1/p}\chi_{F_i}$, where $\{F_i\}_{1 \leqslant i \leqslant k-1}$ is a disjoint collection of subsets of $[0, 1]$ and $|F_i|$ is the Lebesgue measure of F_i. Since there exists a one-to-one (mod 0) measure-preserving mapping of the set F_i onto the interval $[0, |F_i|]$ (see [36]), it follows that the operator $f \to (1/|F_i|)\int_{F_i} f(x)\,dx$ in $L^p(F_i)$ is unitarily equivalent to the operator $f \to (1/|F_i|)\int_0^{|F_i|} f(x)\,dx$ in $L^p[0, |F_i|]$, which in turn is unitarily equivalent to the operator $f \to k\int_{(i-1)/k}^{i/k} f(x)\,dx$ in $L^p[(i-1)/k, i/k]$ (in the latter case, the unitary equivalence is induced by a linear mapping of $[0, |F_i|]$ onto $[(i-1)/k, i/k]$). Consequently, we can assert that there exists a unitary dilation U_n of T_n with respect to the inclusion of X_n in $L^p[0, 1]$ and the orthoprojection P_n, i.e.

$$P_n U_n^m \mid X_n = T_n^m, \qquad m \geqslant 0. \tag{7}$$

We will show that the sequence of operators $\{U_n\}_{n \geqslant 1}$ converges to T in the *pw*-topology. Let $f, g \in X_j$ for some j. Then for $n \geqslant j$ we have

$$\int_0^1 (U_n^m f)g = \sum_{1 \leqslant i < k} \int_{(i-1)/k}^{i/k} (U_n^m f)g = \sum_{1 \leqslant i < k} \left(g \left| \left[\frac{i-1}{k}, \frac{i}{k}\right]\right.\right) \int_{(i-1)/k}^{i/k} U_n^m f$$
$$= \sum_{1 \leqslant i < k} \left(g \left| \left[\frac{i-1}{k}, \frac{i}{k}\right]\right.\right) \int_{(i-1)/k}^{i/k} T_n^m f$$

in view of (7). But since $T_n^m f = (P_n T P_n)^m f$, we have

$$\int_0^1 (U_n^m f)g = \sum_{1 \leqslant i < k} \int_{(i-1)/k}^{i/k} (T_n^m f)g = \int_0^1 (P_n T P_n)^m f \cdot g,$$

i.e., $(U_n^m f, g) = ((P_n T P_n)^m f, g)$, and since $\bigcup_{j\geqslant 1} X_j$ is dense in $L^p[0,1]$ and in $L^{p'}[0,1]$, it is sufficient to show that

$$pw\text{-}\lim_n P_n T P_n = T.$$

But this is clear, in view of the strong convergence of the sequence $\{P_n\}_{n\geqslant 1}$ to the identity operator and the continuity of the multiplication operation in the strong operator topology on bounded sets.

Let us prove that the set $\mathfrak{M}$ is weakly closed in the space of operators acting in $L^p(\mathfrak{X}, \mathfrak{F}, \mu)$. Let the generalized sequence of operators $\{T_\alpha\}$ converge weakly to the operator T, and let $\tilde{T}_\alpha$ be a contractive majorant of the operator T_α. In view of the compactness of the unit ball of the space of operators in the weak operator topology, there exists a generalized subsequence $\{\tilde{T}_{\alpha_\beta}\}$ that converges weakly to an operator $\tilde{T}$.

We show that $\tilde{T}$ is a contractive majorant of T. It is clear that $\tilde{T}$ is a contraction. Suppose that there exist a function $f \in L^p(\mathfrak{X}, \mathfrak{F}, \mu)$ and a set e of positive measure on which $|Tf| \geqslant \tilde{T}|f| + \varepsilon$, $\varepsilon > 0$. Then (it is easy to see) there exists a set of positive measure $e_1 \subset e$ such that $|\int_{e_1} Tf| < \int_{e_1} \tilde{T}|f|$. Let $g = \chi_{e_1}$. Then $g \in L^{p'}(\mathfrak{X}, \mathfrak{F}, \mu)$, and

$$\lim_\beta \left(T_{\alpha_\beta} f, g\right) = (Tf, g); \quad \lim_\beta \left(\tilde{T}_{\alpha_\beta}|f|, g\right) = \left(\tilde{T}|f|, g\right),$$

i.e.,

$$\lim_\beta \left|\int_{e_1} T_{\alpha_\beta} f\right| = \left|\int_{e_1} Tf\right|; \quad \lim_\beta \int_{e_1} \tilde{T}_{\alpha_\beta}|f| = \int_{e_1} \tilde{T}|f|.$$

But since $\tilde{T}_{\alpha_\beta}$ is a contractive majorant of T_{α_β}, we have $|T_{\alpha_\beta} f| \leqslant \tilde{T}_{\alpha_\beta}|f|$ a.e., and consequently,

$$\left|\int_{e_1} T_{\alpha_\beta} f\right| \leqslant \int_{e_1} |T_{\alpha_\beta} f| \leqslant \int_{e_1} \tilde{T}_{\alpha_\beta}|f|.$$

Passing to the limit in this inequality, we obtain $|\int_{e_1} Tf| \leqslant \int_{e_1} \tilde{T}|f|$. Contradiction. ●

REMARK 2. Note that we have proved the following assertion: *the set $\mathfrak{M}$ is closed in the weak topology in the space of operators in $L^p(\mathfrak{X}, \mathfrak{F}, \mu)$.*

REMARK 3. In this paper, Theorem 4 is deduced from Theorem 3. It is possible to prove these theorems in the reverse order. Theorem 4 can be proved, for example, in the same manner as in [10] (Theorem 5), where it is proved that every absolute contraction can be pw-approximated by isometries. The difference is that it is necessary to consider not only "interval shifting" operators, but also stretching operators. (This proof is close to the proof of Theorem 3 in the present paper in conceptual slant.) Theorem 3 can now be derived from Theorem 4.

Let T be an operator of class $\mathfrak{M}$ in $L^p[0,1]$ (the case of an arbitrary measure space $(\mathfrak{X}, \mathfrak{F}, \mu)$ is easily reduced to this case), and let the generalized sequence of unitary operators $\{U_\alpha\}_{\alpha\in\Gamma}$ converge to T in the pw-topology. Let λ be an ultrafilter that majorizes the filter generated by the sets $\{\alpha \in \Gamma\colon \alpha \geqslant \beta\}$, $\beta \in \Gamma$. Define an operator U in the ultraproduct of spaces $X = \prod_{\alpha\in\Gamma} X_\alpha/\lambda$ (where X_α is a copy of $L^p[0,1]$) by

$$U(\{f_\alpha\}_{\alpha\in\Gamma}) = \{U_\alpha f_\alpha\}_{\alpha\in\Gamma}.$$

Let J be the natural imbedding of $L^p[0,1]$ into X, and P: $X \to L^p[0,1]$ the operator defined by

$$P(\{f_\alpha\}_{\alpha\in\Gamma}) = \lim_\lambda f_\alpha$$

in the weak topology in $L^p[0,1]$. Then for $f \in L^p[0,1]$ we have

$$PU^nJf = \lim_\lambda U_\alpha^n f = T^n f, \qquad n \geqslant 0,$$

since $pw\text{-}\lim_\alpha U_\alpha = T$; consequently, U is a unitary dilation of T with respect to the imbedding J and the projection P. ●

COROLLARY (Pointwise Ergodic Theorem [19]). *Let T be an operator of class $\mathfrak{M}$ in the space $L^p[0,1]$, $1 < p < +\infty$. Then the sequence of functions*

$$\left\{(1/m)\sum_{0\leqslant k\leqslant m-1} T^k f\right\}_{m\geqslant 1}$$

converges a.e. when $f \in L^p[0,1]$.

PROOF. It is well known (see, for example, [37]) that it is sufficient to prove the following maximal ergodic theorem:

$$\tilde{f}(x) \overset{\text{def}}{=} \sup_m \frac{1}{m}\left|\sum_{0\leqslant k\leqslant m-1}(T^k f)(x)\right| \Rightarrow \|\tilde{f}\|_p \leqslant c_p\|f\|_p.$$

This estimate is valid for unitary operators in L^p (disjoint unitary operators, for $p = 2$) with $c_p = p/(p-1)$ [37]. (Recently, a simple proof of this assertion was obtained in [38].)

Let $f \in L^p$ and

$$\tilde{f}_N(x) \overset{\text{def}}{=} \sup_{1\leqslant m\leqslant N} \frac{1}{m}\left|\sum_{0\leqslant k\leqslant m-1}(T^k f)(x)\right|.$$

It is enough to prove that $\|\tilde{f}_N\|_p \leqslant (p/(p-1))\|f\|_p$ for all natural numbers N.

Let us divide the interval $[0,1]$ into disjoint sets $\{e_m\}_{1\leqslant m\leqslant N}$ such that

$$\tilde{f}_N(x) = \frac{1}{m}\left|\sum_{0\leqslant k\leqslant m-1}(T^k f)(x)\right| \quad \text{for } x \in e_m.$$

Let $\{U_n\}_{n\geqslant 1}$ be a sequence of unitary operators in L^p that converges to T in the pw-topology. Then, it follows from the maximal ergodic theorem for unitary operators that

$$\left(\sum_{1\leqslant m\leqslant N}\int_{e_m}\frac{1}{m}\left|\sum_{0\leqslant k\leqslant m-1}(U_n^k f)(x)\right|^p dx\right)^{1/p} \leqslant \frac{p}{p-1}\|f\|_p.$$

Let $g \in L^{p'}[0,1]$, $\|g\|_{p'} = 1$, and $\int_0^1 \tilde{f}_N g = \|\tilde{f}_N\|_p$. Then

$$\lim_n \sum_{1\leqslant m\leqslant N}\int_{e_m}\frac{1}{m}\left(\sum_{0\leqslant k\leqslant m-1}(U_n^k f)(x)\right)g(x)\,dx$$

$$= \sum_{1\leqslant m\leqslant N}\int_{e_m}\frac{1}{m}\left(\sum_{0\leqslant k\leqslant m-1}(T^k f)(x)\right)g(x)\,dx = \int_0^1 \tilde{f}_N g = \|\tilde{f}_N\|_p.$$

Consequently,

$$\|\tilde{f}_N\|_p \leqslant \varliminf_n \left\| \sum_{1 \leqslant m \leqslant N} \frac{1}{m} \chi_{e_m} \left| \sum_{0 \leqslant k \leqslant m-1} U_n^k f \right| \right\|_p \leqslant \frac{p}{p-1} \|f\|_p . \quad \bullet$$

To conclude this section, we will finish the proof of Lemma 4 begun in §3.

CONCLUSION OF THE PROOF OF LEMMA 4 (absolute contractions). First let T be an operator in the finite-dimensional space $l_n^p(w)$, and let $\{t_{ij}\}_{1 \leqslant i,j \leqslant n}$ be the matrix of T with respect to the standard basis of $l_n^p(w)$. Then T is an absolute contraction if and only if

$$\sum_{1 \leqslant j \leqslant n} |t_{ij}| \leqslant 1, \qquad \sum_{1 \leqslant i \leqslant n} |t_{ij}| w_i \leqslant w_j, \qquad 1 \leqslant i, j \leqslant n.$$

Let $\tilde{T}$ be the operator determined by the matrix $\{|t_{ij}|\}_{1 \leqslant i,j \leqslant n}$. Then $\tilde{T}$ is a positive contraction (even a positive absolute contraction) of $l_n^p(w)$, and serves as a contractive majorant for T.

Now we consider the case of an absolute contraction in an arbitrary space $L^p(\mathfrak{X}, \mathfrak{F}, \mu)$. Let us denote by Γ the directed set of finite partitions of $\mathfrak{X}$ into measurable subsets. Let $\gamma = \{\mathfrak{X}_1, \dots, \mathfrak{X}_n\} \in \Gamma$. We define an operator P_γ by

$$(P_\gamma f)(x) = \begin{cases} \dfrac{1}{\mu \mathfrak{X}_i} \displaystyle\int_{\mathfrak{X}_i} f \, d\mu, & x \in \mathfrak{X}_i, \ \mu \mathfrak{X}_i < +\infty, \\ 0, & x \in \mathfrak{X}_i, \ \mu \mathfrak{X}_i = +\infty. \end{cases}$$

Then the operator $P_\gamma T P_\gamma$ induces an absolute contraction T_γ in the finite-dimensional space $P_\gamma L^p(\mathfrak{X}, \mathfrak{F}, \mu)$, which is isometric to a finite-dimensional L^p-space. From the part of Lemma 4 already proved, it follows that T_γ admits a contractive majorant $\tilde{T}_\gamma$ in the space $P_\gamma L^p(\mathfrak{X}, \mathfrak{F}, \mu)$. Then, $|P_\gamma T P_\gamma f| \leqslant \tilde{T}_\gamma |P_\gamma f| \leqslant \tilde{T}_\gamma P_\gamma |f|$. Thus, $\tilde{T}_\gamma P_\gamma$ is a contractive majorant of $P_\gamma T P_\gamma$, i.e., $P_\gamma T P_\gamma \in \mathfrak{M}$. Since the generalized sequence of operators $\{P_\gamma T P_\gamma\}_{\gamma \in \Gamma}$ converges to T in the strong operator topology, it follows from Remark 2 after Theorem 4 that $T \in \mathfrak{M}$. ●

§5. The case of the space L^1

In this section, it is proved that Theorems 3 and 4 are also valid for L^1. Let us recall that every contraction in L^1 admits a contractive majorant (see, for example, [39]; the real case is examined there, but it is easy to see that the assertion is also valid for complex L^1-spaces).

THEOREM 5. *Every contraction in a space $L^1(\mathfrak{X}, \mathfrak{F}, \mu)$ admits a unitary dilation in L^1-space; if $L^1(\mathfrak{X}, \mathfrak{F}, \mu)$ is separable, then the dilation can be constructed in* $L^1[0, 1]$.

This theorem strengthens the result of Akcoglu [24] establishing the existence of a unitary dilation for positive contractions of L^1.

THEOREM 6. *The set of unitary transformations of* $L^1[0, 1]$ *is pw-dense in the unit sphere of the space of operators in* $L^1[0, 1]$.

The proofs of these assertions are carried out using the same scheme as for $p > 1$, but with certain alterations.

LEMMA 9. *Let T be an operator in the finite-dimensional space l_n^1, with matrix $\{t_{ij}\}_{1\leqslant i,j\leqslant n}$ with respect to the standard basis. Suppose that $t_{ij}\neq 0$, $1\leqslant i,j\leqslant n$, and $\Sigma_{1\leqslant i\leqslant n}|t_{ij}|=1$, $1\leqslant j\leqslant n$. Then T has a unitary dilation in an L^1-space.*

PROOF. Let $u=\{u_i\}_{1\leqslant i\leqslant n}$ be an arbitrary vector with strictly positive coordinates. Put $v=\tilde{T}u$, where the operator $\tilde{T}$ is determined by the matrix $\{|t_{ij}|\}_{1\leqslant i,j\leqslant n}$. Now put $\xi_{ij}=|t_{ij}|(u_j/v_i)$ and $\eta_{ij}=|t_{ij}|$. Then $\Sigma_{1\leqslant j\leqslant n}\xi_{ij}=1$ and $\Sigma_{1\leqslant i\leqslant n}\eta_{ij}=1$. The subsequent construction is realized just as in the proof of Lemma 7. ●

REMARK. Using the isomorphism (mod 0) of an interval with linear Lebesgue measure and a rectangle with plane Lebesgue measure as measure spaces, one can assert the existence of a unitary dilation if $L^1[0,1]$ with respect to a positive imbedding and a conditional expectation operator.

LEMMA 10. *Let T be a contraction in the space l_n^1. Then T has a unitary dilation in L^1-space.*

PROOF. If T does not satisfy the condition $\Sigma_{1\leqslant i\leqslant n}|t_{ij}|=1$, $1\leqslant j\leqslant n$, then it can be extended to an operator in the space l_{n+1}^1 by putting $t_{n+1,n+1}=1$, $t_{n+1,j}=1-\Sigma_{1\leqslant i\leqslant n}|t_{ij}|$, $1\leqslant j\leqslant n$, and $t_{i,n+1}=0$, $1\leqslant i\leqslant n$; the extension operator will be a contraction in l_{n+1}^1. After this, the proof is carried out just as in the proof of Lemma 8. ●

PROOF OF THEOREM 5. The proof follows the scheme of the proof of Theorem 3. The single change that needs to be made is in connection with the definition of the operator $P\colon \Pi_{\gamma\in\Gamma}X_\gamma/\lambda\to L^1(\mathfrak{X},\mathfrak{F},\mu)$.([5]) In the definition of this operator in the case $p>1$, the reflexivity of an L^p-space is used. For $p=1$, this property is absent, but we can assert the existence of a limit $\lim_\lambda P_\gamma x_\gamma$ in the second conjugate space $(L^1(\mathfrak{X},\mathfrak{F},\mu))^{**}$ (in the topology $\sigma((L^1)^{**},(L^1)^*)$). The space L^1 is orthogonally complemented in $(L^1)^{**}$ (see [33]). Let $\mathbf{P}$ be an orthoprojection of $(L^1)^{**}$ onto L^1. We define the operator $P\colon X\to L^1(\mathfrak{X},\mathfrak{F},\mu)$ by

$$P(\{x_\gamma\}_{\gamma\in\Gamma})=\mathbf{P}\lim_\lambda P_\gamma x_\gamma,\qquad \{x_\gamma\}_{\gamma\in\Gamma}\in\Pi_0(\{X_\gamma\}_{\gamma\in\Gamma}),$$

where the limit is taken in the topology $\sigma((L^1)^{**},(L^1)^*)$.

We will show that the operator U is a dilation of T with respect to J and P. Let f be a simple function. Then

$$\begin{aligned}PU^kJf&=\mathbf{P}\lim_\lambda P_\gamma U_\gamma^k J_\gamma Q_\gamma f=\mathbf{P}\lim_\lambda P_\gamma U_\gamma^k J_\gamma f=\mathbf{P}\lim_\lambda T_\gamma^k f\\&=\mathbf{P}\lim_\lambda (Q_\gamma T Q_\gamma)^k f=\mathbf{P}T^k f=T^k f.\end{aligned}$$

The rest of the proof is carried out just as in Theorem 3. ●

PROOF OF THEOREM 6. Let T be a contraction in $L^1[0,1]$. Consider the directed set Γ of finite partitions of the interval $[0,1]$ into measurable subsets of positive

([5]) The notation is borrowed from the proof of Theorem 3.

measure. Let us choose an arbitrary element $\mathcal{X}^\gamma_{n_\gamma}$ in each partition $\gamma = (\mathcal{X}^\gamma_1, \dots, \mathcal{X}^\gamma_{n_\gamma}) \in \Gamma$. Put

$$w_i^\gamma \stackrel{\text{def}}{=} |\mathcal{X}_i^\gamma|; \quad \varepsilon_\gamma \stackrel{\text{def}}{=} \max_{1 \leqslant i \leqslant n_\gamma} w_i^\gamma; \quad (P_\gamma f)(x) = \frac{1}{w_i^\gamma} \int_{\mathcal{X}_i^\gamma} f, \qquad x \in \mathcal{X}_i^\gamma;$$

$$(\tilde{P}_\gamma f)(x) = \begin{cases} (P_\gamma f)(x), & x \notin \mathcal{X}^\gamma_{n_\gamma}, \\ 0, & x \in \mathcal{X}^\gamma_{n_\gamma}; \end{cases} \qquad \tilde{T}_\gamma = \tilde{P}_\gamma T \tilde{P}_\gamma.$$

Then $\lim_\gamma \varepsilon_\gamma = 0$, and P_γ and $\tilde{P}_\gamma$ are conditional expectation operators. Put $X_\gamma = P_\gamma L^1[0,1]$, $\tilde{X}_\gamma = \tilde{P}_\gamma L^1[0,1]$, $w_\gamma = \{w_i^\gamma\}_{1 \leqslant i \leqslant n_\gamma}$, and $\tilde{w}_\gamma = \{w_i^\gamma\}_{1 \leqslant i \leqslant n_\gamma - 1}$. Then the spaces X_γ and $\tilde{X}_\gamma$ are naturally isometric to $l^1_{n_\gamma}(w_\gamma)$ and $l^1_{n_\gamma - 1}(\tilde{w}_\gamma)$ respectively. The space $\tilde{X}_\gamma$ is invariant under $\tilde{T}_\gamma$; let $\tilde{R}_\gamma = \tilde{T}_\gamma \mid \tilde{X}_\gamma$, a contraction in the space $\tilde{X}_\gamma$, and let $\{\tilde{t}^\gamma_{ij}\}_{1 \leqslant i,j \leqslant n_\gamma - 1}$ be its matrix with respect to the natural basis. Let the contraction $\hat{R}_\gamma$ in the space X_γ be determined by the matrix $\{\hat{t}^\gamma_{ij}\}_{1 \leqslant i,j \leqslant n_\gamma}$ with respect to the natural basis, where $\hat{t}^\gamma_{ij} = \tilde{t}^\gamma_{ij}$ for $1 \leqslant i, j \leqslant n_\gamma - 1$, $\hat{t}^\gamma_{i,n_\gamma} = 0$ for $i < n_\gamma$, $\hat{t}^\gamma_{n_\gamma j} = w_i^\gamma - \Sigma_{1 \leqslant i \leqslant n_\gamma - 1} |\tilde{t}^\gamma_{ij}| w_i^\gamma$ for $1 \leqslant j \leqslant n_\gamma - 1$, and $\hat{t}_{n_\gamma n_\gamma} = w^\gamma_{n_\gamma}$. Choose an operator $R_\gamma = \{t^\gamma_{ij}\}_{1 \leqslant i,j \leqslant n_\gamma}$ in X_γ such that $t^\gamma_{ij} \neq 0$ and $\Sigma_{1 \leqslant i \leqslant n_\gamma} |t^\gamma_{ij}| w_i^\gamma = w_j^\gamma$ for $1 \leqslant i, j \leqslant n_\gamma$, and $|R_\gamma - \hat{R}_\gamma| \leqslant \varepsilon_\gamma$.

Since $l^1_{n_\gamma}(w_\gamma)$ is isometric to $l^1_{n_\gamma}$, it follows from the remark after Lemma 9 that R_γ has a unitary dilation U_γ in $L^1[0,1]$ with respect to the inclusion of X_γ in $L^1[0,1]$ and the projection P_γ, i.e.,

$$P_\gamma U_\gamma^k f = R_\gamma^k f, \qquad f \in X_\gamma. \tag{8}$$

Define the operator T_γ in $L^1[0,1]$ by $T_\gamma = R_\gamma P_\gamma$. We show that the generalized sequence of operators $\{T_\gamma\}_{\gamma \in \Gamma}$ converges strongly to T. Let f be a simple function, i.e., $f \in X_{\gamma_0}$, $\gamma_0 \in \Gamma$. Then for $\gamma \geqslant \gamma_0$

$$\|T_\gamma f - Tf\|_1 = \|R_\gamma P_\gamma f - Tf\|_1 = \|R_\gamma f - Tf\|_1 \leqslant \|\tilde{R}_\gamma f - R_\gamma f\|_1 + \|\tilde{R}_\gamma f - Tf\|_1;$$

$$\begin{aligned} \|\tilde{R}_\gamma f - R_\gamma f\|_1 &\leqslant \|\tilde{R}_\gamma f - \hat{R}_\gamma f\|_1 + \|\hat{R}_\gamma f - R_\gamma f\|_1 \\ &\leqslant \max_{1 \leqslant i \leqslant n_\gamma} w_i^\gamma \|f\|_1 + \varepsilon_\gamma \|f\|_1 = 2\varepsilon_\gamma \|f\|_1 \to 0; \end{aligned}$$

$$\|\tilde{R}_\gamma f - Tf\|_1 = \|\tilde{P}_\gamma T \tilde{P}_\gamma f - Tf\|_1 \to 0$$

in view of the strong convergence of the generalized sequence $\{\tilde{P}_\gamma\}_{\gamma \in \Gamma}$ to the identity operator.

We now show that the generalized sequence $\{U_\gamma\}_{\gamma \in \Gamma}$ converges to T in the *pw*-topology. Let $f, g \in X_{\gamma_0}$. For $\gamma \geqslant \gamma_0$ we have

$$(U_\gamma^k f, g) = \sum_{1 \leqslant i \leqslant n_\gamma} \int_{\mathcal{X}_i^\gamma} U_\gamma^k f \cdot g = \sum_{1 \leqslant i \leqslant n_\gamma} (g \mid \mathcal{X}_i^\gamma) \int_{\mathcal{X}_i^\gamma} U_\gamma^k f = \sum_{1 \leqslant i \leqslant n_\gamma} (g \mid \mathcal{X}_i^\gamma) \int_{\mathcal{X}_i^\gamma} R_\gamma^k f$$

in view of (8). Consequently,

$$(U_\gamma^k f, g) = \sum_{1 \leqslant i \leqslant n_\gamma} (g \mid \mathcal{X}_i^\gamma) \int_{\mathcal{X}_i^\gamma} R_\gamma^k f = \sum_{1 \leqslant i \leqslant n_\gamma} \int_{\mathcal{X}_i^\gamma} R_\gamma^k f \cdot g = (R_\gamma^k f, g) = (T_\gamma^k f, g).$$

The weak convergence of the generalized sequence $\{U_\gamma^k\}_{\gamma\in\Gamma}$ of operators to T^k now follows from the already proved strong convergence of the generalized sequence $\{T_\gamma\}_{\gamma\in\Gamma}$ to T. ●

§6. Symmetric spaces of measurable functions

In this section, we obtain results on approximation by unitary operators and unitary dilation of contractions in symmetric spaces of measurable functions, based upon which we will prove, in §7, an analogue of inequality (1) for these spaces.

DEFINITION [25], [40]. A Banach space X of measurable functions on the interval $[0, 1]$ is called *symmetric* if it has the following properties:

1) If $f \in X$ and g is a measurable function such that $|g(x)| \leqslant |f(x)|$ for almost all $x \in [0, 1]$, then $g \in X$ and $\|g\|_X \leqslant \|f\|_X$.

2) If $|f|$ and $|g|$ are equimeasurable functions and $f \in X$, then $g \in X$, and $\|f\|_X = \|g\|_X$.

If X is a symmetric space, then $L^\infty[0, 1] \subset X \subset L^1[0, 1]$ (see [25]). Examples of symmetric spaces are Orlicz, Lorentz, and Marcinkiewicz spaces.

In this section, X always denotes a separable symmetric space. We will need the following properties of separable symmetric spaces (see [40]).

(α) *The conjugate space X^* can be identified with the space of functions g measurable on* $[0, 1]$ *and such that*

$$\|g\|_{X^*} = \sup\left\{\int_0^1 |fg| : f \in X, \|f\|_X \leqslant 1\right\} < +\infty;$$

the duality between X and X^* is realized by the form $(f, g) = \int_0^1 fg$, $f \in X$, $g \in X^*$; and the space X^*, equipped with such a norm, becomes a symmetric space.

(β) *The set* $C[0, 1]$ *of continuous functions is dense in X.*

(γ) *If $f \in X$, then $\lim_{|e|\to 0} \|f\chi_e\|_X = 0$, where $|e|$ denotes the Lebesgue measure of a measurable set e.*

We recall that the operator T is called an *absolute contraction* if $|T|_{L^1} \leqslant 1$ and $|T|_{L^\infty} \leqslant 1$ (in this case, by an interpolation theorem of Mitjagin [41], T is also a contraction in X).

Let $\mathfrak{A}$ be the set of absolute contractions of X, $\mathfrak{U}$ the set of unitary operators in X, and $\operatorname{clos}_{pw}\mathfrak{U}$ the closure of $\mathfrak{U}$ in the *pw*-topology.

We now formulate the principal results of this section.

THEOREM 7. *Let $T \in \mathfrak{A}$. Then T has a unitary dilation in a space isometric to X, i.e., there exist an isometric mapping $J\colon X \to X$, an orthoprojection P of X onto JX, and a unitary operator U in X such that $PU^nJ = JT^n$, $n \geqslant 0$.*

THEOREM 8. $\mathfrak{A} \subset \operatorname{clos}_{pw}\mathfrak{U}$. *If the space X does not coincide elementwise with any space L^p, $1 \leqslant p \leqslant +\infty$, then $\mathfrak{A} = \operatorname{clos}_{pw}\mathfrak{U}$.*

For the construction of dilations in symmetric spaces, we will consider scales of L^p-spaces, $1 \leqslant p \leqslant +\infty$, similar to ultraproducts. The ultraproducts

$$\prod_{\gamma\in\Gamma} L^p(\mathcal{X}_\gamma, \mathcal{F}_\gamma, \mu_\gamma)/\lambda, \qquad 1 \leqslant p \leqslant +\infty,$$

themselves do not suit our indicated purpose, since $\prod_{\gamma\in\Gamma} L^\infty(\mathcal{X}_\gamma, \mathcal{F}_\gamma, \mu_\gamma)/\lambda$ is not dense in $\prod_{\gamma\in\Gamma} L^1(\mathcal{X}_\gamma, \mathcal{F}_\gamma, \mu_\gamma)/\lambda$ (and such a property is essentially needed). We will need the concept of an ultraproduct of measure spaces, and we will consider scales of L^p spaces, $1 \leqslant p \leqslant +\infty$, over this ultraproduct.

DEFINITION (see [26]). Let Γ be a set on which an ultrafilter λ is given, and let $(\mathcal{X}_\gamma, \mathcal{F}_\gamma, \mu_\gamma)$ be a collection of spaces with probability measures. The set of elements of the ultraproduct

$$\Lambda = \prod_{\gamma \in \Gamma} L^1(\mathcal{X}_\gamma, \mathcal{F}_\gamma, \mu_\gamma)/\lambda$$

of the form $\{\chi_{A_\gamma}\}_{\gamma \in \Gamma}$, where $A_\gamma \in \mathcal{F}_\gamma$, forms a Boolean σ-algebra $\mathcal{F}$ with probability measure μ, where

$$\mu\left(\{\chi_{A_\gamma}\}_{\gamma \in \Gamma}\right) \overset{\text{def}}{=} \lim_{\lambda} \mu_\gamma(A_\gamma) = \left\| \{\chi_{A_\gamma}\}_{\gamma \in \Gamma} \right\|_\Lambda.$$

The subspace of Λ generated by $\mathcal{F}$ can be realized as a space $L^1(\mathcal{X}, \mathcal{F}, \mu)$, where $(\mathcal{X}, \mathcal{F}, \mu)$ is a space with probability measure, called the *ultraproduct* of the measure spaces over the ultrafilter λ.

We have the equality $\Lambda = L^1(\mathcal{X}, \mathcal{F}, \mu) \oplus \Lambda'$, where Λ' is the set of elements of Λ that are disjoint from $\chi_{\mathcal{X}} = \{\chi_{\mathcal{X}_\gamma}\}_{\gamma \in \Gamma}$. The set of elements of the form $\{f_\gamma\}_{\gamma \in \Gamma}$ with $\| f_\gamma \|_{L^\infty} \leqslant \text{const}$ generates the space $L^\infty(\mathcal{X}, \mathcal{F}, \mu)$, and $\|\{f_\gamma\}_{\gamma \in \Gamma}\|_{L^\infty} = \lim_\lambda \| f_\gamma \|_{L^\infty}$ (see [26]).

Before proving Theorems 7 and 8, let us introduce the following notation. Let Φ be a one-to-one (mod 0) mapping of [0, 1] onto itself such that for every measurable set e the sets $\Phi^{-1}(e)$ and $\Phi(e)$ are measurable, and $|\Phi^{-1}(e)| = |\Phi(e)|$ (such a mapping Φ is called a *measure space automorphism*). We will denote by U_Φ the operator in the space X defined by $U_\Phi f = f \circ \Phi$; U_Φ is a unitary operator in X.

Let X_n be the subspace of X consisting of all functions that are constant on the intervals $[(i-1)/2^n, i/2^n]$, $1 \leqslant i \leqslant 2^n$. A projection P_n onto X_n is defined by

$$(P_n f)(x) = 2^n \int_{(i-1)/2^n}^{i/2^n} f, \qquad x \in \left[\frac{i-1}{2^n}, \frac{i}{2^n}\right].$$

Then P_n is a contraction in $L^1[0, 1]$ and in $L^\infty[0, 1]$.

We will call an operator T in the space X_n an *absolute contraction* if it is a contraction with respect to the norms on X_n induced by $L^1[0, 1]$ and $L^\infty[0, 1]$.

LEMMA 11. *Let T be an absolute contraction in the space X_n. Then T has a dilation in X, with respect to the inclusion of X_n in X and the projection P_n, that is a unitary operator in L^1 and L^∞.*(6)

PROOF. Let $\{t_{ij}\}_{1 \leqslant i,j \leqslant 2^n}$ be the matrix of T in the basis $\{\chi_{[(i-1)/2^n, i/2^n]}\}_{1 \leqslant i \leqslant 2^n}$. Then $\sum_{1 \leqslant i \leqslant 2^n} |t_{ij}| \leqslant 1$, $1 \leqslant j \leqslant 2^n$, and $\sum_{1 \leqslant j \leqslant 2^n} |t_{ij}| \leqslant 1$, $1 \leqslant i \leqslant 2^n$.

First let $\sum_{1 \leqslant i \leqslant 2^n} |t_{ij}| = \sum_{1 \leqslant j \leqslant 2^n} |t_{ij}| = 1$ for all i and j. Then the assertion is proved exactly as in Lemma 7; it is enough to set $u_i = v_i = 1$, $1 \leqslant i \leqslant 2^n$; in this case the unitary operator U defined in Lemma 7 is induced by a measure-preserving permutation, and consequently is a unitary operator in L^1 and in L^∞.

---

(6) Note that it follows from the characterization of unitary operators in L^p (see [32]) that operators that are unitary in $L^1[0, 1]$ and in $L^\infty[0, 1]$ have the form $Uf = hU_\Phi f$, where Φ is a measure space automorphism and h is a measurable function such that $|h(x)| = 1$ for almost all $x \in [0, 1]$, and, consequently, are also unitary operators in X.

Now let us consider the general case. It is easy to show that there exists a matrix $\{\mu_{ij}\}_{1\le i,j\le 2^n}$ such that $|t_{ij}|\le \mu_{ij}$, and $\sum_{1\le i\le 2^n}\mu_{ij}=\sum_{1\le j\le 2^n}\mu_{ij}=1$, $1\le i,j\le 2^n$. For every pair (i,j), clearly, there exist numbers α_{ij} and β_{ij} such that

$$|\alpha_{ij}|=|\beta_{ij}|=\mu_{ij}/2, \qquad \alpha_{ij}+\beta_{ij}=t_{ij}. \tag{9}$$

We have $X_n\subset X_{n+1}$. We will define operators in X_{n+1} by means of block matrices

$$\left\{\begin{pmatrix} a_{ij} & b_{ij}\\ c_{ij} & d_{ij}\end{pmatrix}\right\}_{1\le i,j\le 2^n},$$

where the matrix

$$\begin{pmatrix} a_{ij} & b_{ij}\\ c_{ij} & d_{ij}\end{pmatrix}$$

defines an operator from the subspace of X_{n+1} generated by the functions

$$\chi_{[(j-1)/2^n,(2j-1)/2^{n+1}]}, \qquad \chi_{[(2j-1)/2^{n+1},j/2^n]}$$

to the subspace generated by the functions

$$\chi_{[(i-1)/2^n,(2i-1)/2^{n+1}]}, \qquad \chi_{[(2i-1)/2^{n+1},i/2^n]}.$$

Let us define the operator $\tilde T$ in X_{n+1} by

$$\tilde T=\left\{\begin{pmatrix} \alpha_{ij} & \beta_{ij}\\ \beta_{ij} & \alpha_{ij}\end{pmatrix}\right\}_{1\le i,j\le 2^n},$$

where α_{ij} and β_{ij} satisfy equalities (9). Then $\tilde T$ is an absolute contraction in X_{n+1}, $\tilde T X_n\subset X_n$, $\tilde T\,|\,X_n=T$, and, consequently, $\tilde T$ is a dilation of T with respect to the inclusion of X_n in X_{n+1} and the projection $P_n\,|\,X_{n+1}$.

From the part of the lemma already proved it follows that $\tilde T$ has a unitary dilation in X with respect to the projection P_{n+1}. Since $P_nP_{n+1}=P_n$, it follows that T also has a dilation in X that is a unitary operator in L^1 and in L^∞. ●

PROOF OF THEOREM 7. Put $T_n\overset{\text{def}}{=}P_nTP_n$. Then $TX_n\subset X_n$, and $T_n\,|\,X_n$ is an absolute contraction. By Lemma 11, there exists an operator U_n that is unitary in $L^1[0,1]$ and in $L^\infty[0,1]$ and is a dilation of $T_n\,|\,X_n$.

Let $\mathcal{X}_n=[0,1]$, let $\mathcal{F}_n$ be the σ-algebra of Lebesgue subsets of $[0,1]$, let μ_n be Lebesgue measure, and let λ be an ultrafilter on $\mathbb{N}$ that majorizes the filter generated by the sets $\{m\in\mathbb{N}: m\ge n\}$. We will denote by $(\mathcal{X},\mathcal{F},\mu)$ the ultraproduct over λ of the measure spaces $(\mathcal{X}_n,\mathcal{F}_n,\mu_n)$.

We define the following operators:(7)

$$\tilde J: L^1[0,1]\to L^1(\mathcal{X},\mathcal{F},\mu),\quad \tilde J f\overset{\text{def}}{=}\{f_n\}_{n\in\mathbf{N}},\qquad f_n\equiv f;$$

$$\tilde U: L^1(\mathcal{X},\mathcal{F},\mu)\to L^1(\mathcal{X},\mathcal{F},\mu),\quad \tilde U f\overset{\text{def}}{=}\{U_nf_n\}_{n\in\mathbf{N}};$$

(7) The element $\{U_nf_n\}_{n\in\mathbb{N}}$ of the ultraproduct $\prod_{n\in\mathbb{N}}L^1(\mathcal{X}_n,\mathcal{F}_n,\mu_n)/\lambda$ belongs to the space $L^1(\mathcal{X},\mathcal{F},\mu)$, since the functions $|f_n|$ and $|U_nf_n|$ are equimeasurable.

$$\tilde{P}\colon L^p(\mathfrak{X},\mathfrak{F},\mu) \to L^p[0,1],\quad \tilde{P}\{f_n\}_{n\in\mathbf{N}} \overset{\text{def}}{=} \lim_{\lambda} P_n f_n,\qquad 1<p<+\infty$$

(the limit is taken in the weak topology of L^p).

Then $\tilde{J}$ is an isometric imbedding of $L^p[0,1]$ into $L^p(\mathfrak{X},\mathfrak{F},\mu)$ for $p\in[1,+\infty]$; $\tilde{U}$ is a unitary operator in $L^p(\mathfrak{X},\mathfrak{F},\mu)$, $1\leqslant p\leqslant+\infty$; and $\tilde{P}$ is a positive orthoprojection for $p\in(1,+\infty)$, and consequently $\tilde{P}$ is a conditional expectation operator (see [35]), and therefore is continuous from $L^1(\mathfrak{X},\mathfrak{F},\mu)$ to $L^1[0,1]$. Exactly as in Theorem 3, it can be proved that $\tilde{U}$ is a dilation of T with respect to $\tilde{J}$ and $\tilde{P}$.

It is easy to show (using the characterization of unitary operators in L^p [32]) that there exists a σ-subalgebra $\tilde{\mathfrak{S}}\subset\mathfrak{F}$, generated by a countable collection of sets in $\mathfrak{F}$, such that

$$\tilde{J}L^1[0,1]\subset L^1(\mathfrak{X},\tilde{\mathfrak{S}},\nu),\quad \tilde{U}L^1(\mathfrak{X},\tilde{\mathfrak{S}},\nu)\subset L^1(\mathfrak{X},\tilde{\mathfrak{S}},\nu),$$

and

$$\tilde{U}^{-1}L^1(\mathfrak{X},\tilde{\mathfrak{S}},\nu)\subset L^1(\mathfrak{X},\tilde{\mathfrak{S}},\nu),$$

where $\nu=\mu\,|\,\tilde{\mathfrak{S}}$. But in this case, the measure space $(\mathfrak{X},\tilde{\mathfrak{S}},\nu)$ is isomorphic to the interval $[0,1]$ with Lebesgue measure (see [36]).

Thus, using the characterization of L^p-isometries and the positiveness of the operator $\tilde{J}$, we can assert that there exists a mapping of the σ-algebra of all measurable subsets of $[0,1]$ onto some σ-algebra $\mathfrak{S}$ of measurable subsets of $[0,1]$ such that

$$\Psi([0,1]\backslash A)=[0,1]\backslash\Psi A,\qquad \Psi\left(\bigcup_{i=1}^{\infty}A_i\right)=\bigcup_{i=1}^{\infty}\Psi A_i$$

for disjoint sets A_i, and $|\Psi A|=|A|$, there exists a measure space automorphism Φ: $[0,1]\to[0,1]$, and there exists a measurable function h such that $|h(x)|=1$ a.e., for which we have the relation

$$JT^n=E_{\mathfrak{S}}U^nJ,\qquad n\geqslant 0,$$

where the operator J is defined on characteristic functions by $J\chi_A=\chi_{\Psi A}$, the operator U is defined by $Uf=hU_\Phi f$ for $f\in L^1[0,1]$, and $E_{\mathfrak{S}}$ is the conditional expectation operator with respect to the σ-algebra $\mathfrak{S}$. Then J is an isometric imbedding of X into X, $P=E_{\mathfrak{S}}$ is an orthoprojection of X onto JX (by a theorem of Mitjagin [41]), and the unitary operator U is a dilation of the operator T with respect to J and P. ●

PROOF OF THEOREM 8. If T is an absolute contraction in the space X, then, using properties (α) and (β) of separable symmetric spaces and Theorem 7, we can prove, exactly as in Theorem 4, the existence of a sequence $\{U_n\}_{n\geqslant1}$ of operators, unitary in L^1 and in L^∞, that is pw-convergent to T.

If X does not coincide elementwise with any space $L^p[0,1]$, then every unitary operator in X is a unitary operator in $L^1[0,1]$ and in $L^\infty[0,1]$ (see [42]). The assertion of the theorem now follows from the fact that the set of absolute contractions is closed in the weak operator topology. ●

REMARK. Note that, in fact, we have proved that every absolute contraction of X can be pw-approximated by operators that are unitary in $L^1[0,1]$ and in $L^\infty[0,1]$.

In the next section, we will prove a more precise assertion about approximations of absolute contractions by unitary operators (Theorem 11).

§7. Estimates of operator polynomials

In this section we obtain the principal result of this paper, namely, inequality (1) for operators in an L^p space that admit a contractive majorant. Further, we establish an analogue of this inequality for absolute contractions in separable symmetric spaces.

Let us recall (a theorem of A. K. Kitover; see §10 for the proof) that inequality (1) is valid for isometric operators T, i.e.,

$$|\varphi(T)|_{L^p} \leqslant |\varphi|_p \stackrel{\text{def}}{=} |\varphi(S)|_{l^p}, \qquad \varphi \in \mathcal{P}_A,$$

where S is the shift operator in l^p (see §10).

Consequently, Theorem 3 and Lemma 2 imply the following assertion.

THEOREM 9. *Let T be an operator that admits a contractive majorant in the space $L^p(\mathfrak{X}, \mathfrak{F}, \mu)$. Then, for every polynomial φ,*

$$|\varphi(T)|_{L^p} \leqslant |\varphi|_p.$$

This theorem also follows from Theorem 4 and Lemma 1; in fact, from Theorem 4 and Lemma 1 the validity of inequality (1) for operators of class $\mathfrak{M}$ in $L^p[0,1]$ is seen immediately. But from this it is easy to pass to the case of an arbitrary measure space. First, using operators of a special kind, inequality (1) can be proved for operators of class $\mathfrak{M}$ in the finite-dimensional space l^p_n, and then, with the aid of approximation by "finite-dimensional operators" (in analogy to the way it was done in the proof of Lemma 4 in the case of absolute contractions; see §4), we obtain the general case.

Thus, both methods of estimation of operator polynomials described in the Introduction lead, in the case of an L^p-space, to one result, namely, inequality (1) for operators that admit a contractive majorant.

We note, in addition, that for $p = 2$ the following classes of operators in an L^p-space coincide: the set $\mathfrak{D}$ of operators that have a unitary dilation; the set $\mathfrak{I}$ of operators that satisfy (1); and the set $\mathcal{C}$ of contractions. Clearly, for any p, $\mathfrak{D} \subset \mathfrak{I} \subset \mathcal{C}$. We will show that $\mathfrak{D} \neq \mathfrak{I}$ for $p \neq 2$, $1 < p < +\infty$. Let T be a contraction in $L^p[0,1]$ such that $T \notin \mathfrak{M}$.(8) Define an operator R in $L^p[0,2]$ by

$$(Rf)(x) = \begin{cases} 0, & x \in [0,1], \\ (T(f\,|\,[0,1]))(x-1), & x \in [1,2]. \end{cases}$$

Then $R \notin \mathfrak{D}$, since $\mathfrak{D} = \mathfrak{M}$ by virtue of Theorem 3. Since $R^2 = \mathbf{0}$, it is easily shown that $R \in \mathfrak{I}$. A similar example can be constructed in the same way in any L^p-space of dimension greater than three. It is not known whether the sets $\mathfrak{I}$ and $\mathcal{C}$ coincide for $p \neq 2$, $1 < p < +\infty$.

---

(8) It is easy to show that for $1 < p < +\infty$ the set $\mathcal{C} \setminus \mathfrak{M}$ is already nonempty in the two-dimensional space l^p_2. Let A be the operator in l^p_2 that is defined in the natural basis by the matrix $\left(\begin{smallmatrix} 1 & -1 \\ 1 & 1 \end{smallmatrix}\right)$. Then it is easily verified that the operator $B = A/|A|_p$ belongs to the set $\mathcal{C} \setminus \mathfrak{M}$.

We now pass to estimating operator polynomials in symmetric spaces of measurable functions. Let X be a symmetric space. We may assume that the functions in X are given on the circle $\mathbb{T} = \{e^{2\pi i t}: t \in [0, 1]\}$. A numerical sequence $c = \{c_n\}_{n \in \mathbf{Z}}$ is called a *multiplier* for the space X if for every function $f \in X$ the sequence $\{\hat{f}(n)c_n\}_{n \in \mathbf{Z}}$ is the sequence of Fourier coefficients of a function in X. We denote by $M(X)$ the space of all multipliers for X; we define the norm $\|c\|_{M(X)}$ of a multiplier c to be the norm of the corresponding operator Q_c in X, where $(Q_c f)\hat{\ }(n) = c_n \hat{f}(n)$, $n \in \mathbf{Z}$.

In this section, we obtain for $|\varphi(T)|_X$ in a separable symmetric space X an estimate analogous to (1). However, if $X \neq L^p$, it is no longer possible to reduce the case to that of multiplication operators (multipliers) in a suitable space of analytic functions; therefore, we need to use multipliers Q_c of the Fourier coefficients (though they may be less convenient for the description of the calculus).

Let θ be an irrational number, and $\alpha = e^{2\pi i \theta}$. With a symmetric space X of measurable functions, let us associate the following norm, defined on the set of polynomials $\mathcal{P}_A$:

$$|\varphi|_X \overset{\text{def}}{=} \|\{\varphi(\alpha^n)\}_{n \in \mathbf{Z}}\|_{M(X)}, \qquad \varphi \in \mathcal{P}_A.$$

It is easily seen that

$$\sup\{|\varphi(\zeta)|: |\zeta| < 1\} \leqslant |\varphi|_X \leqslant \sum_{n \geqslant 0} |\hat{\varphi}(n)|, \qquad \varphi \in \mathcal{P}_A.$$

THEOREM 10. *Let X be a separable symmetric space. Then the norm $|\cdot|_X$ does not depend on the choice of the number θ, and*

$$|\varphi(T)|_X \leqslant |\varphi|_X, \qquad \varphi \in \mathcal{P}_A, \tag{10}$$

for any absolute contraction T in X.

This theorem will be obtained as a corollary of the approximation theorem (Theorem 11), which sharpens Theorem 8.

Let us recall [43] that a measure space automorphism Ψ is called *aperiodic* if $\Psi^n x \neq x$ for almost all $x \in [0, 1]$, for every natural number n. If Φ is a measure space automorphism, then we denote by U_Φ, as before, the operator in X defined by $(U_\Phi f)(x) = f(\Phi x)$, $f \in X$.

THEOREM 11. *Let T be an absolute contraction of the space X, and let Ψ be an aperiodic measure space automorphism. Then there exists a generalized sequence of operators, each of which is unitarily equivalent*(9) *to the operator U_Ψ, that is pw-convergent to T.*

PROOF OF THEOREM 10. We define an automorphism Ψ_θ by

$$\Psi_\theta(x) = x + \theta \pmod 1, \qquad x \in [0, 1].$$

It is easily seen that $|\varphi|_X = |\varphi(U_{\Psi_\theta})|_X$. If the number θ is irrational, then the automorphism Ψ_θ is aperiodic.

---

(9) Operators T_1 and T_2 are called *unitarily equivalent* if there exists a unitary operator U such that $UT_1U^{-1} = T_2$.

Let T be an absolute contraction in X. By Theorem 11, there exists a generalized sequence of unitary operators $\{U_\alpha\}$ such that $pw\text{-}\lim_\alpha U_\alpha U_{\Psi_\theta} U_\alpha^{-1} = T$. By Lemma 1, we have

$$|\varphi(T)|_X \leq \varliminf_\alpha |\varphi(U_\alpha U_{\Psi_\theta} U_\alpha^{-1})|_X = \varliminf_\alpha |U_\alpha \varphi(U_{\Psi_\theta}) U_\alpha^{-1}|_X$$
$$= |\varphi(U_{\Psi_\theta})|_X = |\varphi|_X, \qquad \varphi \in \mathcal{P}_A.$$

The independence of $|\varphi(U_{\Psi_\theta})|_X$ from the choice of θ follows from the fact that U_{Ψ_θ} can be approximated by operators that are unitarily equivalent to U_{Ψ_τ}, where τ is an irrational number. ●

We note that $|\varphi|_X = |\varphi|_p$ if $X = L^p$, and consequently inequality (10) is a generalization of (1), for absolute contractions.

Before proving Theorem 11, we introduce the concept of the weak topology on the set of measure space automorphisms. A sequence of automorphisms $\{\Phi_n\}_{n\geq 1}$ converges weakly to an automorphism Φ if for every measurable set e the measure of the symmetric difference $|\Phi_n(e)\Delta\Phi(e)|$ converges to zero (see [43]). We note that the weak convergence of the sequence of automorphisms $\{\Phi_n\}_{n\geq 1}$ to the automorphism Φ implies the convergence of the operator sequence $\{U_{\Phi_n}\}_{n\geq 1}$ to the operator U_Φ in the strong operator topology. In fact, if $f = \chi_e$, then

$$\lim_n \|U_{\Phi_n} f - U_\Phi f\|_X = \lim_n \|\chi_{\Phi_n(e)\Delta_\Phi(e)}\|_X = 0,$$

by virtue of property (γ) (see §6) of separable symmetric spaces.

We now pass to the proof of Theorem 11. It will be carried out as follows. We will successively approximate operators of a more complex structure by operators that are unitarily equivalent to simpler operators. To conclude the proof, we will apply a theorem of Rokhlin and Halmos.

Let $\{\delta_r\}_{1\leq r\leq n}$ and $\{\Delta_r\}_{1\leq r\leq n}$ be partitions of $[0, 1]$ into intervals with dyadic rational end points such that $|\Delta_r| = |\delta_r|$, $1 \leq r \leq n$, let $\{\lambda_r\}_{1\leq r\leq n}$ be a collection of complex numbers with moduli equal to one, and let Φ be the mapping of $[0, 1]$ onto itself such that $\Phi | \delta_r$ is the linear, order-preserving mapping of δ_r onto Δ_r. We will call such an automorphism Φ a *permutation*. Define an operator R by

$$(Rf)(x) = \lambda_r f(\Phi(x)), \quad \text{if } x \in \delta_r. \tag{11}$$

We will denote the set of all such operators by $\mathfrak{L}_1$.

LEMMA 12. *If T is an absolute contraction, then* $T \in \operatorname{clos}_{pw} \mathfrak{L}_1$.

PROOF. By virtue of the remark at the end of §6, it is sufficient to prove that operators of the form

$$Tf = e^{ih} U_\Phi f,$$

where Φ is a measure space automorphism and h is a real measurable function, can be pw-approximated by operators from the set $\mathfrak{L}_1$.

It is known (see [43], p. 65) that the set of permutations is weakly everywhere dense in the set of all automorphisms. Therefore, there exist sequences $\{\delta_r^{(m)}\}_{1\leq r\leq n_m}$ and $\{\Delta_r^{(m)}\}_{1\leq r\leq n_m}$ of partitions of $[0, 1]$ into intervals with dyadic rational end points such that $|\delta_r^{(m)}| = |\Delta_r^{(m)}|$, $1 \leqslant r \leqslant n_m$, $m \geqslant 1$, for which the sequence of automorphisms $\{\Phi_m\}_{m\geqslant 1}$ that map each interval $\delta_r^{(m)}$ onto the interval $\Delta_r^{(m)}$ in a linear, order-preserving fashion converges weakly to an automorphism Φ. Consequently, the operator sequence $\{U_{\Phi_m}\}_{m\geqslant 1}$ converges strongly to U_Φ.

Without loss of generality, we may assume that the sequences of partitions $\{\delta_r^{(m)}\}_{1\leqslant r\leqslant n_m}$ and $\{\Delta_r^{(m)}\}_{1\leqslant r\leqslant n_m}$ are successive refinements, and that

$$\lim_m \max_{1\leqslant r\leqslant n_m} |\Delta_r^{(m)}| = 0.$$

In this case, it is easily shown that the operator of multiplication by e^{ih} can be approximated in the strong operator topology by operators of multiplication by e^{ih_m}, where h_m is a real function that is constant on the elements of the partition $\{\delta_r^{(m)}\}_{1\leqslant r\leqslant n_m}$. Define an operator T_m by

$$T_m f = e^{ih_m} U_{\Phi_m} f, \qquad f \in X.$$

Then it is easily seen that $T_m \in \mathfrak{L}_1$ and $\lim_m T_m = T$ in the strong topology (and consequently, also in the *pw*-topology). ●

Let $\{E_r\}_{1\leqslant r\leqslant m}$ be a partition of [0, 1] into intervals, and $\{F_{rj}\}_{1\leqslant j\leqslant n_r}$ a partition of E_r into n_r intervals of equal length, indexed by the subscript j from left to right, and let $\{\Lambda_{rj}\}_{1\leqslant r\leqslant m, 1\leqslant j\leqslant n_r}$ be a collection of complex numbers with moduli equal to one. Let us define an operator Q by

$$(Qf)(x) = \begin{cases} \lambda_{rj} f(x + |F_{rj}|), & x \in F_{rj}, j < n_r, \\ \lambda_{rj} f(x - (n_r - 1)|F_{rj}|), & x \in F_{rj}, j = n_r. \end{cases} \tag{12}$$

We will denote the set of all such operators by $\mathfrak{L}_2$. If $\lambda_{rj_1} = \lambda_{rj_2}$ for all r, j_1, and j_2, $1 \leqslant r \leqslant m$, $1 \leqslant j_1 \leqslant n_r$, $1 \leqslant j_2 \leqslant n_r$, in the definition of the operator Q, then we will say that $Q \in \mathfrak{L}_3$, and if $\lambda_{rj} = 1$ for all r and j, then we will say that $Q \in \mathfrak{L}_4$. The operators in $\mathfrak{L}_4$ are cyclic shifts on each interval E_r with respect to the partition $\{F_{rj}\}_{1\leqslant j\leqslant n_r}$.

LEMMA 13. *If $R \in \mathfrak{L}_1$, then there exists an operator T in the set $\mathfrak{L}_3$ that is unitarily equivalent to R.*

PROOF. Let the operator R be defined by (11). We choose a natural number N such that for all j, $0 \leqslant j < 2^N$, the interval $[j/2^N, (j+1)/2^N]$ is contained in one of the intervals δ_r and in one of the intervals Δ_r from the definition of R. Then

$$\Phi([j/2^N, (j+1)/2^N]) \subset [k/2^N, (k+1)/2^N]$$

for some k, $0 \leqslant k < 2^N$. Consequently, Φ induces a bijection of the partition $\zeta = \{[j/2^N, (j+1)/2^N]\}_{0\leqslant j\leqslant 2^N-1}$ onto itself, and therefore, the elements of this partition can be reindexed so that $\zeta = \{\gamma_{rj}\}_{1\leqslant r\leqslant m, 1\leqslant j\leqslant n_r}$, $\Sigma_{1\leqslant r\leqslant m} n_r = 2^N$, and

$$\Phi(\gamma_{rj}) = \begin{cases} \gamma_{r,j+1}, & j < n_r, \\ \gamma_{r1}, & j = n_r. \end{cases}$$

Let $\lambda_{rj} = \lambda_k$ if $\gamma_{rj} \subset \delta_k$. Define the interval F_{rj} by

$$F_{rj} = \left[\frac{1}{2^N}\left(\sum_{k<r} n_k + j - 1\right), \frac{1}{2^N}\left(\sum_{k<r} n_k + j\right)\right].$$

Let Ψ be the mapping of $[0, 1]$ onto itself such that $\Psi \mid \gamma_{rj}$ is the linear, order-preserving mapping of γ_{rj} onto F_{rj}. Then $U_\Psi^{-1} R U_\Psi = Q$, where Q is the mapping defined by (12).

Put

$$(Tf)(x) = \begin{cases} \mu_r f(x + |F_{rj}|), & x \in F_{rj}, j < n_r, \\ \mu_r f(x - (n_r - 1)|F_{rj}|), & x \in F_{rj}, j = n_r, \end{cases} \tag{13}$$

where the μ_r are complex numbers such that $\mu_r^{n_r} = \prod_{j=1}^{n_r} \lambda_{rj}$, $1 \leqslant r \leqslant m$. We will prove that the operators T and Q are unitarily equivalent. In fact, if $c_{r1} = 1$ and $c_{rj} = (\lambda_{r,j-1}/\mu_r)c_{r,j-1}$, $1 \leqslant r \leqslant m$, $1 \leqslant j \leqslant n_r$, and the unitary operator V is defined by $(Vf)(x) = c_{rj} f(x)$ for $x \in F_{rj}$, then $VQV^{-1} = T$, and $T \in \mathfrak{L}_3$. ●

REMARK. It follows from the proof of the lemma that Q is unitarily eqivalent to an operator in $\mathfrak{L}_4$ if $\prod_{j=1}^{n_r} \lambda_{rj} = 1$, $1 \leqslant r \leqslant m$.

LEMMA 14. *If $T \in \mathfrak{L}_3$, then T can be approximated in the strong topology by operators, each of which is unitarily equivalent to an operator in $\mathfrak{L}_4$.*

PROOF. Let the operator T be defined by (13), and let $\mu_r = e^{2\pi i \theta_r}$. Clearly, we may assume that the θ_r are dyadic rational numbers. Let $E_r = \bigcup_{1 \leqslant j \leqslant n_r} F_{rj}$, $1 \leqslant r \leqslant m$. Let us break up the interval E_r into $2n_r$ equal intervals $\tilde{F}_{rj}$, $1 \leqslant r \leqslant m$, $1 \leqslant j \leqslant \tilde{n}_r \overset{\text{def}}{=} 2n_r$, indexed from left to right by the subscript j. Put

$$(\tilde{T}f)(x) = \begin{cases} \mu_r f(x + |\tilde{F}_{rj}|), & x \in \tilde{F}_{rj}, j < \tilde{n}_r, \\ \mu_r f(x - (\tilde{n}_r - 1)|\tilde{F}_{rj}|), & x \in \tilde{F}_{rj}, j = \tilde{n}_r. \end{cases}$$

It suffices to prove that T can be approximated in the strong topology by operators that are unitarily equivalent to $\tilde{T}$, since applying repeatedly, if necessary, the procedure of doubling the number of intervals, we can achieve that $\mu_r^{\tilde{n}_r} = 1$ for all r, $1 \leqslant r \leqslant m$. Then, by virtue of the remark after Lemma 13, the operator $\tilde{T}$ will be unitarily equivalent to an operator in $\mathfrak{L}_4$.

Let N be a natural number. Let us divide every interval F_{rj}, $1 \leqslant j \leqslant n_r$, into $2N$ intervals $\delta_{rj}^{(k)}$, $1 \leqslant k \leqslant 2N$, of equal length, indexing the latter from left to right by the index k. We define an operator T_N by

$$(T_N f)(x) = \begin{cases} \mu_r f(x + |F_{rj}|), & x \in F_{rj}, j < n_r, \\ \mu_r f\left(x - (n_r - 1)|F_{rj}| + \dfrac{1}{2N}|F_{rj}|\right), & x \in \delta_{rj}^{(k)}, j = n_r, k \text{ odd}, \\ \mu_r f\left(x - (n_r - 1)|F_{rj}| - \dfrac{1}{2N}|F_{rj}|\right), & x \in \delta_{rj}^{(k)}, j = n_r, k \text{ even}. \end{cases}$$

The sequence $\{T_N\}_{N \geqslant 1}$ converges strongly to T. In fact, if $f \in C[0, 1]$ and ω_f is the modulus of continuity of f, then

$$|(T_N f)(x) - (Tf)(x)| \leqslant \omega_f(1/N) \underset{N}{\to} 0 \qquad \text{for } x \in [0, 1].$$

We will prove that T_N is unitarily equivalent to $\tilde{T}$. For this, we divide every interval $\tilde{F}_{rj}$ into N intervals $\Delta_{rj}^{(k)}$, $1 \leqslant k \leqslant N$, of equal length, indexed from left to right by the index k. Let Φ be a mapping of $[0, 1]$ onto itself such that $\Phi \mid \Delta_{rj}^{(k)}$ is the linear, order-preserving mapping of $\Delta_{rj}^{(k)}$ onto $\delta_{rj}^{(2k-1)}$ for $j \leqslant n_r$, and $\Phi \mid \Delta_{rj}^{(k)}$ is the similar superposition of $\Delta_{rj}^{(k)}$ onto $\delta_{r,j-n_r}^{(2k)}$ for $j > n_r$. Then it is easily verified that $U_\Phi^{-1} T_N U_\Phi = \tilde{T}$. ●

PROOF OF THEOREM 11. Let Ψ be an aperiodic automorphism. By virtue of Lemmas 12–14, it is enough to prove that every operator in $\mathfrak{L}_4$ can be strongly approximated by operators that are unitarily equivalent to U_Ψ. Let $T \in \mathfrak{L}_4$. It is clear that in this case $T = U_\Phi$ for some automorphism Φ. By a theorem of Rokhlin and Halmos, there exists a sequence $\{S_n\}_{n\geqslant 1}$ of measure space automorphisms such that the sequence of automorphisms $\{S_n^{-1}\Psi S_n\}_{n\geqslant 1}$ converges weakly to Φ (see [43], p. 77). But then the operator sequence $\{U_{S_n^{-1}\Psi S_n}\}_{n\geqslant 1}$ converges to U_Φ in the strong operator topology, while $U_{S_n^{-1}\Psi S_n}$ is unitarily equivalent to U_Ψ. ●

REMARK. We note that it follows from our results that $|\varphi(U_\Psi)|_X = |\varphi|_X$, $\varphi \in \mathcal{P}_A$, for every aperiodic automorphism Ψ. By means of approximating such an automorphism Ψ by periodic automorphisms (see [43], p. 75), it is easily shown that $|\varphi|_X = \sup_n |\varphi(U_{\Psi_{1/n}})|_X$. (We recall that the automorphism $\Psi_{1/n}$ is defined by $\Psi_{1/n}(x) = x + 1/n \pmod 1$.)

§8. A maximal functional calculus

Let T be a contraction in a Hilbert space $\mathcal{H}$. Then $\mathcal{H} = \mathcal{H}_0 \oplus \mathcal{H}_1$, where $T \mid \mathcal{H}_0$ is a unitary operator and $T \mid \mathcal{H}_1$ is a completely nonunitary contraction (i.e., having no unitary part). If the spectral measure of the operator $T \mid \mathcal{H}_0$ is absolutely continuous with respect to Lebesgue measure, then the polynomial calculus $\varphi \to \varphi(T)$ can be extended to a contractive representation of H^∞, the Hardy algebra of all bounded analytic functions f in the disc $\mathbb{D}$ with norm $\|f\|_{H^\infty} = \sup\{|f(\zeta)| : |\zeta| < 1\}$, i.e., to a correspondence $\varphi \to \varphi(T)$, $\varphi \in H^\infty$, having the properties

1) $(\varphi_1 + \varphi_2)(T) = \varphi_1(T) + \varphi_2(T)$;
2) $(\varphi_1 \cdot \varphi_2)(T) = \varphi_1(T)\varphi_2(T)$;
3) $\varphi(T) = \Sigma_{n\geqslant 0} \hat{\varphi}(n) T^n$, if $\Sigma_{n\geqslant 0} |\hat{\varphi}(n)| < +\infty$;
   $|\varphi(T)|_{\mathcal{H}} \leqslant \|\varphi\|_{H^\infty}$

(the calculus of Sz.-Nagy and Foiaş [2]). In particular, completely nonunitary contractions admit such a calculus.

In this section we obtain analogues of this calculus for contractions of an L^p space and for contractions in symmetric spaces. The problem is to find some sort of conditions like complete nonunitariness under which an operator T of class $\mathfrak{M}$ on an L^p-space admits, in (1), the replacement of the polynomial φ by any p-multiplier, i.e. conditions under which the operator T admits an M_p-calculus. We note that the set $\mathcal{P}_A$ is not dense in M_p, since there exist functions in class M_p which cannot be continuously extended to $\operatorname{clos} \mathbb{D}$ [44]. (Recall that we are assuming that $1 < p < +\infty$.)

LEMMA 15. *Let T be an operator of class $\mathfrak{M}$ in a space $L^p(\mathcal{X}, \mathcal{F}, \mu)$. Suppose that for every sequence $\{\varphi_n\}_{n\geqslant 1}$, $\varphi_n \in l_A^1$, such that $|\varphi_n|_p \leqslant$ const and $\lim_n \varphi_n(\zeta) = 0$ for $\zeta \in \mathbb{D}$, the operator sequence $\{\varphi_n(T)\}_{n\geqslant 1}$ converges weakly to the zero operator. Then T admits an M_p-calculus, i.e., with every function $\varphi \in M_p$, there can be associated an operator $\varphi(T)$ in $L^p(\mathcal{X}, \mathcal{F}, \mu)$ such that conditions 1)–3) hold and*

4) $|\varphi(T)|_{L^p} \leqslant |\varphi|_p$.

PROOF. Let the operator T satisfy the conditions of the lemma, and $\varphi \in M_p$. Then there is a sequence $\{f_n\}_{n\geq 1}$ such that $|f_n|_p \leq |\varphi|_p$, $f_n(\zeta) \to \varphi(\zeta)$, and $f_n \in l^1_A$ (as the f_n we can take the Poisson means φ_{r_n}, where $\varphi_{r_n}(\zeta) = \varphi(r_n\zeta)$ and $r_n \uparrow 1$; see [45]). It is easily shown that the operator sequence $\{f_n(T)\}_{n\geq 1}$ is weakly convergent, and we will denote the limit by $\varphi(T)$. Clearly, the definition of $\varphi(T)$ does not depend on the choice of the approximating sequence $\{f_n\}_{n\geq 1}$. Moreover, if the sequence of functions $\{\varphi_n\}_{n\geq 1}$, $\varphi_n \in M_p$, converges pointwise in $\mathbb{D}$ to a function $\varphi \in M_p$ and $\sup_n |\varphi_n|_p < +\infty$, then $\lim_n \varphi_n(T) = \varphi(T)$ in the weak operator topology. The required properties 1), 3), and 4) are obvious. Let us establish the multiplicativity of the calculus.

Let φ, $\psi \in M_p$, and let the sequences of polynomials $\{\varphi_n\}_{n\geq 1}$ and $\{\psi_n\}_{n\geq 1}$, respectively, converge pointwise in $\mathbb{D}$ to φ and ψ, where $\sup_n |\varphi_n|_p < +\infty$ and $\sup_n |\psi_n|_p < +\infty$. If $g \in L^p$ and $h \in L^{p'}$, then

$$((\varphi_m\psi_n)(T)g, h) = (\varphi_m(T)\psi_n(T)g, h) = (\psi_n(T)g, (\varphi_m(T))^*h).$$

We have

$$\lim_n ((\varphi_m\psi_n)(T)g, h) = ((\varphi_m\psi)(T)g, h),$$

and

$$\lim_n (\psi_n(T)g, (\varphi_m(T))^*h) = (\psi(T)g, (\varphi_m(T))^*h) = (\varphi_m(T)\psi(T)g, h).$$

Hence,

$$((\varphi_m\psi)(T)g, h) = (\varphi_m(T)\psi(T)g, h).$$

The sequence of functions $\{\varphi_m\psi\}_{m\geq 1}$ converges pointwise in $\mathbb{D}$ to the function $\varphi\psi$, and $\sup_m |\varphi_m\psi| < +\infty$, which, again, gives us

$$\lim_m ((\varphi_m\psi)(T)g, h) = ((\varphi\psi)(T)g, h),$$

$$\lim_m (\varphi_m(T)\psi(T)g, h) = (\varphi(T)\psi(T)g, h).$$

In view of the arbitrariness of g and h, we have $(\varphi\psi)(T) = \varphi(T)\psi(T)$. ●

We note that if T has an eigenvalue λ on the unit circle, then it does not satisfy the conditions of Lemma 15, since there exists a function in M_p that does not have a radial limit at λ (see [44]).

THEOREM 12. *Let T be an operator of class $\mathfrak{M}$ in a space $L^p(\mathcal{X}, \mathcal{F}, \mu)$. If one of the sequences $\{T^n\}_{n\geq 1}$ or $\{(T^*)^n\}_{n\geq 1}$ converges strongly to the zero operator, then the conclusion of Lemma* 15 *holds.*

PROOF. It is enough to consider the case where the sequence $\{T^n\}_{n\geq 1}$ converges strongly to the zero operator, since if the operator T admits an M_p-calculus, so does T^*.

We will show that T satisfies the conditions of Lemma 15. Let $\varphi_n \in l^1_A$, $\varphi_n(\zeta) \to 0$ for $\zeta \in \mathbb{D}$, and $|\varphi_n|_p \leq 1$, and let

$$f_n^{(m)} = \sum_{0\leq k\leq m} \hat{\varphi}_n(k)z^k \quad \text{and} \quad g_n^{(m)} = \sum_{k>m} \hat{\varphi}_n(k)z^{k-m}.$$

Then $\varphi_n = f_n^{(m)} + z^m g_n^{(m)}$. We have $\hat{\varphi}_n(k) \underset{n}{\to} 0$ for all $k \in \mathbb{N}$. Consequently, $\| f_n^{(m)} \|_{l_A^1} \leqslant 1/m$ for sufficiently large n, and therefore, $| g_n^{(m)} |_p < 2$. Let $F \in L^p(\mathcal{X}, \mathcal{F}, \mu)$. Then, for $n \geqslant n(m)$,

$$\begin{aligned}\|\varphi_n(T)F\|_p &\leqslant \|f_n^{(m)}(T)F\|_p + \|g_n^{(m)}(T)T^m F\|_p \\ &\leqslant \sum_{0 \leqslant k \leqslant m} |\hat{f}_n(k)| + | g_n^{(m)}(T) |_p \|T^m F\|_p \leqslant \frac{1}{m} + 2\|T^m F\|_p.\end{aligned}$$

Consequently, the operator T satisfies the conditions of Lemma 15. ●

We will now consider the case of an absolute contraction T in an L^p-space. In this case, we can succeed in proving that T admits an M_p-calculus under weaker additional restrictions.

THEOREM 13. *Let T be an absolute contraction in a space $L^p(\mathcal{X}, \mathcal{F}, \mu)$. Suppose that the unitary part of T considered as an operator in the Hilbert space $L^2(\mathcal{X}, \mathcal{F}, \mu)$ has an absolutely continuous spectral measure with respect to Lebesgue measure. Then the conclusion of Lemma* 15 *is true.*

PROOF. Let $\varphi_n \in l_A^1$, $| \varphi_n |_p \leqslant \text{const}$, and $\lim_n \varphi_n(\zeta) = 0$ for all $\zeta \in \mathbb{D}$. Then $\| \varphi_n \|_{H^\infty} \leqslant | \varphi_n |_p \leqslant \text{const}$, and consequently the sequence $\{\varphi_n(T)\}_{n \geqslant 1}$ of operators, considered as operators in the Hilbert space L^2, converges to the zero operator in the weak operator topology (see [2], Chapter III, §2.2), i.e.,

$$\int_{\mathcal{X}} (\varphi_n(T)g)h \, d\mu \to 0 \tag{14}$$

for all g and h in the set $L^1 \cap L^\infty$. Since $| \varphi_n(T) |_p \leqslant | \varphi_n |_p \leqslant \text{const}$ (see (1)), it follows from (14) that the operator sequence $\{\varphi_n(T)\}_{n \geqslant 1}$ converges weakly to the zero operator in L^p, since $L^1 \cap L^\infty$ is dense in L^p and $L^{p'}$. Consequently, T satisfies the conditions of Lemma 15. ●

THEOREM 14. *Let T be an absolute contraction in a space $L^p(\mathcal{X}, \mathcal{F}, \mu)$. Suppose that one of the following two conditions is satisfied*:

A) *If $f \in L^p(\mathcal{X}, \mathcal{F}, \mu)$, $f \neq 0$, then there exists an $m \in \mathbb{N}$ such that $\|T^m f\|_p < \|f\|_p$.*

B) *If $f \in L^{p'}(\mathcal{X}, \mathcal{F}, \mu)$, $f \neq 0$, then there exists an $m \in \mathbb{N}$ such that $\|(T^*)^m f\|_{p'} < \|f\|_{p'}$.*

Then the operator T admits an M_p-calculus satisfying conditions 1)–4).

PROOF. It suffices to consider case A). We will verify that in this case the operator T, regarded as an operator in L^2, is a completely nonunitary contraction. Suppose that this is not so. Then there exists a nonzero function f in L^2 for which

$$\|T^n f\|_{L^2} = \|f\|_{L^2}, \qquad n \in \mathbb{N}.$$

Consider the space $X = L^p + L^2$ that consists of all functions $g = g_1 + g_2$, where $g_1 \in L^p$ and $g_2 \in L^2$. We will consider, on X, the following interpolation norm:

$$\|g\|_X \stackrel{\text{def}}{=} \inf\{\|g_1\|_{L^p} + M\|g_2\|_{L^2} : g = g_1 + g_2, g_1 \in L^p, g_2 \in L^2\},$$

where M is a positive number, whose choice will be made more precise below.

We will show that there exist functions $g \in L^p$ and $h \in L^2$ such that $f = g + h$ and $\|f\|_X = \|g\|_p + M\|h\|_2$. Let $g_n \in L^p$, $h_n \in L^2$, $f = g_n + h_n$, and $\|g_n\|_p + M\|h_n\|_2 \to \|f\|_X$. Choosing a subsequence, if necessary, we may assume that the sequences $\{\|g_n\|_p\}_{n\geq 1}$ and $\{\|h_n\|_2\}_{n\geq 1}$ converge. By virtue of the reflexivity of the spaces L^p and L^2 (we are assuming that $1 < p < +\infty$), we can extract subsequences $\{g_{n_k}\}_{k\geq 1}$ and $\{h_{n_k}\}_{k\geq 1}$ that, respectively, converge weakly to functions g in L^p and h in L^2. Then $\|g\|_p = \lim_n \|g_n\|_p$, $\|h\|_2 = \lim_n \|h_n\|_2$, $f = g + h$, and $\|f\|_X = \|g\|_p + M\|h\|_2$. We will prove that it is possible to choose the number M so that $g \neq \mathbf{0}$.

There exist positive numbers c_1 and c_2 such that the function

$$\tilde{g} = f | \{x \colon c_1 < |f(x)| < c_2\}$$

belongs to L^p, and $\|\tilde{g}\|_p > 0$. Let $\tilde{h} \stackrel{\text{def}}{=} f - \tilde{g}$. Since $\|\tilde{h}\|_2 < \|f\|_2$, the number M can be chosen so large that $\|f\|_X \leq \|\tilde{g}\|_p + M\|\tilde{h}\|_2 < M\|f\|_2$. If $g = \mathbf{0}$ for such a choice of M, then $\|f\|_X = M\|f\|_2$. Consequently, $g \neq \mathbf{0}$.

In view of condition A), there exists an $m \in \mathbb{N}$ such that $\|T^m g\|_p < \|g\|_p$. Then

$$\|T^m f\|_X = \|T^m g + T^m h\|_X \leq \|T^m g\|_p + M\|T^m h\|_2 < \|g\|_p + M\|h\|_2 = \|f\|_X.$$

Let R be the adjoint operator to T considered as an operator in the Hilbert space L^2, i.e. $\int (T\varphi)\bar{\psi} = \int \varphi\overline{(R\psi)}$ for $\varphi, \psi \in L^2$. Then for all $n \in \mathbb{N}$ we have $R^n T^n f = f$, since $\|T^n f\|_2 = \|f\|_2$. The operator R is an absolute contraction, and, consequently, also a contraction in X. We have $\|f\|_X = \|R^m T^m f\|_X \leq \|T^m f\|_X < \|f\|_X$. Contradiction. ●

We now consider the problem of constructing a maximal functional calculus for absolute contractions in a separable symmetric space X of measurable functions. Consider the following class of analytic functions in the unit disc $\mathbb{D}$. We will say that a function, analytic in $\mathbb{D}$, belongs to the space M_X if there exists a sequence $\{f_n\}_{n\geq 0}$ of polynomials such that

$$\lim_n f_n(\zeta) = f(\zeta) \quad \text{for } \zeta \in \mathbb{D}, \qquad \text{and} \quad \sup_n |f_n|_X = c < +\infty.$$

As the norm $|f|_X$ of a function f in M_X, we take the infimum of such numbers c over all choices for the sequence $\{f_n\}_{n\geq 0}$. It is easily verified that the class M_X is a Banach algebra with respect to multiplication, the operator of multiplication by z is an isometry in M_X, $l^1_A \subset M_X \subset H^\infty$, and $M_X = M_p$ if $X = L^p$.

The following assertion permits us to construct an M_X-calculus for absolute contractions of the space X, under additional restrictions of the same type as complete nonunitariness.

THEOREM 13′. *Let X be a separable symmetric space. Suppose that the conjugate space X^* is also separable. Further, let T be an absolute contraction in X such that its unitary part, as an operator in the Hilbert space $L^2[0, 1]$, has an absolutely continuous spectral measure with respect to Lebesgue measure. Then T admits an M_X-calculus that satisfies conditions* 1)–3) *and*

4′) $|\varphi(T)|_X \leq |\varphi|_X$, $\varphi \in M_X$.

COROLLARY. *Let the spaces X and X^* be separable, and let T be an absolute contraction in X. If one of the sequences $\{T^n\}_{n\geq 0}$ or $\{(T^*)^n\}_{n\geq 0}$ converges strongly to the zero operator, then the conclusion of Theorem* 13′ *is true.*

THEOREM 14′. *Let T be an absolute contraction in a reflexive symmetric space X. Suppose that one of the following two conditions is satisfied*:

A′) *If $f \in X, f \neq \mathbf{0}$, then there exists an $n \in \mathbf{N}$ such that $\|T^n f\|_X < \|f\|_X$.*

B′) *If $f \in X^*, f \neq \mathbf{0}$, then there exists an $n \in \mathbf{N}$ such that $\|(T^*)^n f\|_{X^*} < \|f\|_{X^*}$.*

Then T admits an M_X-calculus that satisfies conditions 1)–3) *and* 4′).

Theorem 13′ (14′) is proved in precisely the same way as Theorem 13 (14).

PROOF OF THE COROLLARY TO THEOREM 13′. Let the operator sequence $\{T^n\}_{n\geq 0}$ converge strongly to the zero operator. We will show that in this case the operator T, as an operator in L^2, satisfies the condition $\lim_n T^n = \mathbf{0}$ in the strong operator topology, and consequently satisfies the condition of Theorem 13′. Let $f \in L^\infty$. We have

$$\|T^n f\|_2^2 = \int_0^1 |T^n f| \cdot |T^n f| \leq \|T^n f\|_X \|T^n f\|_{X^*} \leq \|T^n f\|_X \|f\|_{X^*},$$

since T is an absolute contraction and, consequently, a contraction in the space X^*. Thus, $\lim_n \|T^n f\|_2^2 = 0$.

The case where the operator sequence $\{(T^*)^n\}_{n\geq 0}$ converges strongly to the zero operator can be analyzed similarly. ●

§9. Substitutions in multipliers

In this section, estimates of operator polynomials in terms of the multiplier norm will be used to study p-multipliers themselves.

THEOREM 15. *Let $f \in l_A^1$, $|\hat{f}(0)| < 1$, and $\|f\|_{l_A^1} \leq 1$. Then*

$$\varphi \in M_p \Rightarrow \varphi \circ f \in M_p, \qquad |\varphi \circ f|_p \leq |\varphi|_p.$$

PROOF. Consider the operator T in l_A^p defined by $T_g = fg$, $g \in l_A^p$. Then T is an absolute contraction, since $|f|_q \leq \|f\|_{l_A^1}$ for $1 \leq q \leq +\infty$. Consequently, by Theorem 9,

$$|\varphi \circ f|_p = |\varphi(T)|_{l_A^p} \leq |\varphi|_p,$$

if $\varphi \in \mathcal{P}_A$.

Now let φ be an arbitrary p-multiplier. In view of the condition $|\hat{f}(0)| < 1$, the operator T, considered as an operator in the Hilbert space l_A^2, is a completely nonunitary contraction. Consequently, by Theorem 13, T admits an M_p-calculus. It is easily verified that the operator $\varphi(T)$ in l_A^2 is the operator of multiplication by the function $\varphi \circ f$ if $\varphi \in H^\infty$, and hence, for $\varphi \in M_p$, the operator $\varphi(T)$ in l_A^p also is the operator of multiplication by the function $\varphi \circ f$, and $|\varphi(T)|_{l_A^p} = |\varphi \circ f|_p \leq |\varphi|_p$. ●

We remark that a direct, purely analytic proof of these theorems, based on the definition of multiplier, would also be of interest.

COROLLARY. *Let R be an operator in a space $L^p(\mathcal{X}, \mathcal{F}, \mu)$ such that $|\varphi(R)|_p \leq |\varphi|_p$ for every polynomial φ. If $f \in l_A^1$ and $\|f\|_{l_A^1} \leq 1$, then inequality* (1) *holds for the operator $T = f(R)$.*

PROOF. Let $f_n = \sum_{0\leq k\leq n} \hat{f}(k) z^k$. Put $T_n = f_n(R)$. Let $\varphi \in \mathcal{P}_A$. Then

$$|\varphi(T_n)|_p = |(\varphi \circ f_n) R|_p \leq |\varphi \circ f_n|_p \leq |\varphi|_p,$$

by Theorem 15 (the case where $|\hat{f}(0)|=1$ is trivial). It now remains to observe that

$$|T-T_n|_p=|(f-f_n)(R)|_p=\sum_{k>n}|\hat{f}(n)|\to 0.$$

Consequently, $|\varphi(T)|_p\leqslant|\varphi|_p$. ●

We now present similar theorems on substitutions for the case of multipliers for L^p-spaces on locally compact Abelian groups, and multipliers for symmetric spaces.

Let G be a locally compact Abelian group, and f a bounded function on the dual group Γ. Let T_f be the linear transformation given on $L^2(G)\cap L^p(G)$ (with respect to the Haar measure on G) by([10])

$$(T_f g)\hat{}=f\hat{g},\qquad g\in L^2\cap L^p.$$

The function f is called a *multiplier* for the space $L^p(G)$ if the operator T_f is bounded in $L^p(G)$. We denote the set of all multipliers for $L^p(G)$ by $M_p(G)$, and as the norm of a multiplier f, we take the norm of the operator T_f in $L^p(G)$.

THEOREM 16. *Let μ be a regular Borel measure on a locally compact group G. Suppose that $\int_G d|\mu|\leqslant 1$, and $|\hat{\mu}(\gamma)|<1$ for almost all $\gamma\in\Gamma$. Then*

$$\varphi\in M_p\Rightarrow\varphi\circ\hat{\mu}\in M_p(G),\qquad\|\varphi\circ\hat{\mu}\|_{M_p(G)}\leqslant|\varphi|_p.$$

In the case where $G=\mathbb{R}^n$ or $G=\mathbb{Z}^n$, we can obtain a sharper assertion.

THEOREM 17. *Let μ be a regular Borel measure on $\mathbb{R}^n$ (or $\mathbb{Z}^n$). Suppose that $\int_G d|\mu|\leqslant 1$ and $|\mu(\{0\})|<1$, where $G=\mathbb{R}^n$ (or $\mathbb{Z}^n$). Then*

$$\varphi\in M_p\Rightarrow\varphi\circ\hat{\mu}\in M_p(G),\qquad\|f\circ\hat{\mu}\|_{M_p(G)}\leqslant|\varphi|_p.$$

The proofs of these assertions are similar to the proof of Theorem 15, and make use of Theorems 9 and 13.

THEOREM 15′. *Let X be a separable symmetric space, $f\in l^1_A$, $|\hat{f}(0)|<1$, and $\|f\|_{l^1_A}\leqslant 1$. Then*

$$\varphi\in M_X\Rightarrow\varphi\circ f\in M_X,\qquad|\varphi\circ f|_X\leqslant|\varphi|_X.$$

PROOF. First, let $\varphi\in\mathcal{P}_A$. Then $|\varphi\circ f|_X=|(\varphi\circ f)(U_{\Psi_\theta})|_X$. (The notation is borrowed from §7; the operator $(\varphi\circ f)(U_{\Psi_\theta})$ is defined, since $\varphi\circ f\in l^1_A$.) We have

$$|\varphi\circ f|_X=|(\varphi\circ f)(U_{\Psi_\theta})|_X=|\varphi(f(U_{\Psi_\theta}))|_X\leqslant|\varphi|_X$$

by Theorem 10, since $f(U_{\Psi_\theta})$ is an absolute contraction in X.

If φ is an arbitrary function in M_X and $\varphi_n\in\mathcal{P}_A$, $\lim_n\varphi_n(\zeta)=\varphi(\zeta)$ for $\zeta\in\mathbb{D}$, and $\sup_n|\varphi_n|_X\leqslant c$, then $(\varphi_n\circ f)(\zeta)\to f(\zeta)$ for $\zeta\in\mathbb{D}$, and $|\varphi_n\circ f|_X\leqslant|\varphi_n|_X\leqslant c$. Consequently, $\varphi\circ f\in M_X$, and $|\varphi\circ f|_X\leqslant|\varphi|_X$. ●

THEOREM 16′. *Let X be a separable symmetric space with a separable conjugate space, and let μ be a complex Borel measure on $\mathbb{T}$ such that $\int_{\mathbb{T}}|\mu|\leqslant 1$ and $|\hat{\mu}(n)|<1$ for all $n\in\mathbb{Z}$. Then*

$$\varphi\in M_X\Rightarrow\{\varphi(\hat{\mu}(n))\}_{n\in\mathbb{Z}}\in M(X),\qquad\|\{\varphi(\hat{\mu}(n))\}_{n\in\mathbb{Z}}\|_{M(X)}\leqslant|\varphi|_X.$$

The proof is similar to the proof of the corresponding assertions for the case of L^p, with the use of Theorems 10 and 13′.

([10]) $\hat{g}$ denotes the Fourier transform of g. Regarding the concepts used here, see [46].

§10. Weighted shift operators. Multipliers for weighted l^p-spaces

Here, we consider the behavior of the norms of polynomials in weighted shift operators. It turns out that, under natural restrictions on the weight sequence $\{w_n\}$, inequality (1) becomes an equality for such operators, i.e.

$$|\varphi(S)|_p = |\varphi|_p, \qquad \varphi \in \mathcal{P}_A,$$

where S is the weighted shift operator. In passing, without the use of previous results, we obtain a proof of a theorem of A. K. Kitover. For simplicity, we assume that $1 < p < +\infty$.

We denote by S the shift operator in $l^p(w)$ and $l^p(\mathbf{Z}, w)$:

$$S(x_0, x_1, x_2, \ldots) = (0, x_0, x_1, \ldots), \qquad (x_0, x_1, x_2, \ldots) \in l^p(w);$$
$$S\{x_n\}_{n\in\mathbf{Z}} = \{x_{n-1}\}_{n\in\mathbf{Z}}, \qquad \{x_n\}_{n\in\mathbf{Z}} \in l^p(\mathbf{Z}, w).$$

The proof of the following lemma is based on the device used by Kitover in his theorem cited above. It is worth noting that very similar reasoning, involving the variable of integration $(l^p(L^p) = L^p(l^p))$, is encountered in the estimation of operators in the papers of Cotlar [27], Calderón [28], and Coifman and Weiss [14] (see also the survey [15]).

LEMMA 16. *Let U be an isometric operator in a space $L^p(\mathcal{X}, \mathcal{F}, \mu)$, and $\{w_n\}_{n\geq 0}$ a sequence of positive numbers such that*

$$\sup_n \frac{w_{n+1}}{w_n} < +\infty, \qquad \overline{\lim_n} \sqrt[n]{w_n^{-1}} \leq 1. \tag{15}$$

Then $|\varphi(U)|_p = |\varphi(S)|_{l^{p'}(w)}$ for any polynomial φ.

PROOF. Let R be the shift operator in $l^{p'}(w, L^{p'}(\mathcal{X}, \mathcal{F}, \mu))$:

$$R(g_0, g_1, g_2, \ldots) = (\mathbf{0}, g_0, g_1, \ldots).$$

Since $\sup_n(w_{n+1}/w_n) < +\infty$, R is continuous in $l^{p'}(w, L^{p'}(\mathcal{X}, \mathcal{F}, \mu))$.

Let $v_n = w_n^{-p'/p}$. The space conjugate to $l^{p'}(w, L^{p'})$ is $l^p(v, L^p)$, with respect to the duality

$$(f, g) = \sum_{i\geq 0} \int_{\mathcal{X}} f_i g_i \, d\mu, \qquad f = \{f_i\}_{i\geq 0} \in l^p(v, L^p), \quad g = \{g_i\}_{i\geq 0} \in l^{p'}(w, L^{p'}).$$

Then

$$R^*(f_0, f_1, f_2, \ldots) = (f_1, f_2, \ldots), \qquad f = \{f_i\}_{i\geq 0} \in l^p(v, L^p).$$

Let $0 < \lambda < 1$ and $V = \lambda U$. Define an operator W: $L^p \to l^p(v, L^p)$ by

$$Wf = (f, Vf, V^2 f, \ldots).$$

It follows from (15) that

$$\|Wf\|^p_{l^p(v, L^p)} = \|f\|_p^p \sum_{i\geq 0} \lambda^{pi} w_i^{-p'/p} < +\infty.$$

Thus, W is a continuous operator from L^p to $l^p(v, L^p)$. We have $(R^*)^n W = WV^n$ for every natural number n. Consequently, for $\varphi \in \mathcal{P}_A$ we have $\varphi(R^*)W = W\varphi(V)$, and

$$\|W\varphi(V)f\|_{l^p(v,L^p)} \leqslant |\varphi(R^*)|_{l^p(v,L^p)} \|Wf\|_{l^p(v,L^p)}; \tag{16}$$

$$\|W\varphi(V)f\|^p_{l^p(v,L^p)} = \sum_{i\geqslant 0} w_i^{-p'/p} \|V^i\varphi(V)f\|_p^p = \|\varphi(V)f\|_p^p \sum_{i\geqslant 0} \lambda^{pi} w_i^{-p'/p};$$

$$\|Wf\|^p_{l^p(v,L^p)} = \sum_{i\geqslant 0} w_i^{-p'/p} \|V^i f\|_p^p = \|f\|_p^p \sum_{i\geqslant 0} \lambda^{pi} w_i^{-p'/p}.$$

Thus, it follows from (16) that

$$|\varphi(V)|_p \leqslant |\varphi(R^*)|_{l^p(v,L^p)} = |\varphi(R)|_{l^{p'}(w,L^{p'})}.$$

Letting λ approach one, we obtain that $|\varphi(U)|_p \leqslant |\varphi(R)|_{l^{p'}(w,L^{p'})}$. It now remains to show that

$$|\varphi(R)|_{l^{p'}(w,L^{p'})} = |\varphi(S)|_{l^{p'}(w)} \quad .$$

Let $f = \{f_i\}_{i\geqslant 0} \in l^{p'}(w, L^{p'})$ and $\varphi(R)f = \{g_i\}_{i\geqslant 0}$. Then

$$\|\varphi(R)f\|^{p'}_{l^{p'}(w,L^{p'})} = \sum_{i\geqslant 0} w_i \int_{\mathfrak{X}} |g_i|^{p'}\, d\mu = \int_{\mathfrak{X}} \Big(\sum_{i\geqslant 0} w_i |g_i(x)|^{p'} \Big)\, d\mu(x)$$

$$= \int_{\mathfrak{X}} \|\varphi(S)\{f_i(x)\}_{i\geqslant 0}\|^{p'}_{l^{p'}(w)}\, d\mu(x) \leqslant |\varphi(S)|^{p'}_{l^{p'}(w)} \int_{\mathfrak{X}} \Big(\sum_{i\geqslant 0} w_i |f_i(x)|^{p'} \Big)\, d\mu(x)$$

$$= |\varphi(S)|^{p'}_{l^{p'}(w)} \sum_{i\geqslant 0} w_i \int_{\mathfrak{X}} |f_i|^{p'}\, d\mu = |\varphi(S)|^{p'}_{l^{p'}(w)} \|f\|^{p'}_{l^{p'}(w,L^{p'})}.$$

Thus, $|\varphi(R)|_{l^{p'}(w,L^{p'})} \leqslant |\varphi(S)|_{l^{p'}(w)}$. Conversely, let $h \in L^{p'}$, $\|h\|_{p'} = 1$, $c = \{c_i\}_{i\geqslant 0} \in l^{p'}(w)$, $f_i = c_i h$, and $f = \{f_i\}_{i\geqslant 0}$. Then, $\|f\|_{l^{p'}(w,L^{p'})} = \|c\|_{l^{p'}(w)}$ and

$$\|\varphi(R)f\|_{l^{p'}(w,L^{p'})} = \|\varphi(S)c\|_{l^{p'}(w)},$$

whence we obtain that

$$|\varphi(R)|_{l^{p'}(w,L^{p'})} \geqslant |\varphi(S)|_{l^{p'}(w)}. \quad \bullet$$

COROLLARY 1 (A. K. Kitover's theorem). *If U is an isometric operator in a space $L^p(\mathfrak{X}, \mathfrak{F}, \mu)$, then $|\varphi(U)|_p \leqslant |\varphi|_p$ for every polynomial φ.*

PROOF. It is sufficient to apply Lemma 16 with $w_n = 1$, $n \geqslant 0$. Then

$$|\varphi(U)|_p \leqslant |\varphi(S)|_{l^{p'}} = |\varphi|_{p'} = |\varphi|_p. \quad \bullet$$

COROLLARY 2. *Let $\{w_n\}_{n\geqslant 0}$ be a sequence of positive numbers that satisfies condition* (15). *Then $|\varphi(S)|_{l^p(w)} \geqslant |\varphi|_p$ for any polynomial φ.*

PROOF. As the isometric operator U involved in Lemma 16, we take the shift operator in $l^{p'}$. Then $|\varphi|_p = |\varphi(S)|_{l^{p'}} \leqslant |\varphi(S)|_{l^p(w)}$, by Lemma 16. $\bullet$

THEOREM 18. 1) *Let $\{w_n\}_{n\geqslant 0}$ be a nonincreasing sequence of positive numbers such that*

$$\overline{\lim_n} \sqrt[n]{w_n^{-1}} \leqslant 1.$$

Then the operator S in $l^p(w)$ admits an M_p-calculus, and $|\varphi(S)|_{l^p(w)} = |\varphi|_p$ *for any p-multiplier* φ.

2) *Let* $\{w_n\}_{n\in\mathbb{Z}}$ *be a nonincreasing positive function on* $\mathbb{Z}$ *such that*

$$\overline{\lim_{n\to+\infty}} \sqrt[n]{w_n^{-1}} \leqslant 1 \quad \textit{or} \quad \overline{\lim_{n\to+\infty}} \sqrt[n]{w_{-n}} \leqslant 1.$$

Then the operator S in $l^p(\mathbb{Z}, w)$ admits an M_p-calculus, and $|\varphi(S)|_{l^p(\mathbb{Z},w)} = |\varphi|_p$ *for any p-multiplier* φ.

PROOF. 1) First, let $\varphi \in \mathcal{P}_A$. By Theorem 9 we have $|\varphi(S)|_{l^p(w)} \leqslant |\varphi|_p$, and, by Corollary 2 of Lemma 16, the reverse inequality. Consequently, $|\varphi(S)|_{l^p(w)} = |\varphi|_p$.

Now let $\varphi \in M_p$. It is easily verified that the shift operator in the Hilbert space $l^2(w)$ is a completely nonunitary contraction. Consequently, by Theorem 13,

$$|\varphi(S)|_{l^p(w)} \leqslant |\varphi|_p.$$

Consider the Poisson means φ_r of the function φ. Since the Taylor series of φ_r converges absolutely, we have $|\varphi_r(S)|_{l^p(w)} \leqslant |\varphi_r|_p$, $r < 1$.

Since $\overline{\lim}_n \sqrt[n]{w_n^{-1}} \leqslant 1$, it follows that $\varphi(S)$ is unitarily equivalent to the operator of multiplication by φ in the space $l^p_A(w)$ consisting of functions analytic in the disc $\mathbb{D}$. Then $|\varphi_r(S)|_{l^p(w)} \leqslant |\varphi(S)|_{l^p(w)}$. (This is proved in exactly the same way as the fact that $|\varphi_r|_p \leqslant |\varphi|_p$; see [45], Theorem 5.) Consequently, $|\varphi(S)|_{l^p(w)} \geqslant |\varphi_r|_p$. Since $\lim_{r\to 1-0} |\varphi_r|_p = |\varphi|_p$ (see [45]), we have $|\varphi(S)|_{l^p(w)} = |\varphi|_p$.

2) By virtue of Theorems 9 and 13, $|\varphi(S)|_{l^p(\mathbb{Z},w)} \leqslant |\varphi|_p$, $\varphi \in M_p$. First, let $\overline{\lim}_{n\to+\infty} \sqrt[n]{w_n^{-1}} \leqslant 1$, and let $v_n = w_n$ for $n \geqslant 0$. With an element $x = \{x_i\}_{i\geqslant 0}$ of $l^p(v)$, let us associate the element $y = \{y_i\}_{i\in\mathbb{Z}}$ of $l^p(\mathbb{Z}, w)$ such that $y_i = x_i$ for $i \geqslant 0$ and $y_i = 0$ for $i < 0$. Then

$$\|y\|_{l^p(\mathbb{Z},w)} = \|x\|_{l^p(v)} \qquad \text{and} \qquad \|\varphi(S)y\|_{l^p(\mathbb{Z},w)} = \|\varphi(S)x\|_{l^p(v)}.$$

Since, in view of part 1), already proved, we have $|\varphi(S)|_{l^p(v)} = |\varphi|_p$, it follows that $|\varphi(S)|_{l^p(\mathbb{Z},w)} \geqslant |\varphi|_p$.

Now, let $\overline{\lim}_{n\to+\infty} \sqrt[n]{w_{-n}} \leqslant 1$. In this case the adjoint operator S^* is unitarily equivalent to the shift operator in $l^{p'}(\mathbb{Z}, u)$, where $u_n = w_{-n}^{-p'/p}$, $n \in \mathbb{Z}$. Then it is easily seen that $\overline{\lim}_{n\to+\infty} \sqrt[n]{u_n^{-1}} \leqslant 1$, and, consequently,

$$|\varphi(S)|_{l^p(\mathbb{Z},w)} = |\varphi(S)|_{l^p(\mathbb{Z},u)} = |\varphi|_p, \qquad \varphi \in M_p. \quad \bullet$$

We now pass to the characterization of the multipliers for the space $l^p_A(w)$, $\overline{\lim}_n \sqrt[n]{w_n^{-1}} \leqslant 1$. A function g, analytic in $\mathbb{D}$, is called a *multiplier* for $l^p_A(w)$ (we denote by $M_p(w)$ the set of all such multipliers) if

$$f \in l^p_A(w) \Rightarrow fg \in l^p_A(w).$$

The norm $\|g\|_{M_p(w)}$ equals, by definition, the norm of the multiplication operator $f \to fg$ in $l^p_A(w)$.

THEOREM 19. *Let a nonincreasing sequence $\{w_n\}_{n\geqslant 0}$ of positive numbers satisfy the condition $\overline{\lim}_n \sqrt[n]{w_n^{-1}} \leqslant 1$. Then $M_p = M_p(w)$, together with equality of the norms.*

PROOF. By virtue of Theorem 18, the shift operator in $l^p(w)$ admits an M_p-calculus, and $|\varphi(S)|_{l^p(w)} = |\varphi|_p$, $\varphi \in M_p$. But the operator $\varphi(S)$ in $l^p(w)$ is unitarily equivalent to the operator of multiplication by φ in $l^p_A(w)$. Consequently, $M_p \subset M_p(w)$, and $\|\varphi\|_{M_p(w)} = |\varphi|_p$ for $\varphi \in M_p$.

Now let $\varphi \in M_p(w)$. Then $\|\varphi_r\|_{M_p(w)} \leqslant \|\varphi\|_{M_p(w)}$ for $0 < r < 1$, and $\varphi_r \in l^1_A$. Consequently, $\varphi_r \in M_p$ and $|\varphi_r|_p = \|\varphi_r\|_{M_p(w)} \leqslant$ const. But then we can conclude that $\varphi \in M_p$. ●

An analogous assertion can be proved for multipliers for the spaces $L^p_A(\mathbb{R}_+, w)$.

BIBLIOGRAPHY

1. Johann [John] von Neumann, *Eine Spektraltheorie für allgemeine Operatoren eines unitären Raumes*, Math. Nachr. **4** (1951), 258–281.

2. Béla Sz.-Nagy and Ciprian Foiaş, *Analyse harmonique des opérateurs de l'espace de Hilbert*, Masson, Paris, and Akad. Kiadó, Budapest, 1967; new English transl., North-Holland, Amsterdam, Amer. Elsevier, New York, and Akad. Kiadó, Budapest, 1970.

3. Edward Nelson, *The distinguished boundary of the unit operator ball*, Proc. Amer. Math. Soc. **12** (1961), 994–995.

4. Ciprian Foiaş, *Sur certains théorèmes de J. von Neumann concernant les ensembles spectraux*, Acta Sci. Math. (Szeged) **18** (1957), 15–20.

5. V. È. Katsnel'son and V. I. Matsaev, *Spectral sets for operators in a Banach space and estimates of functions of finite-dimensional operators*, Teor. Funktsiĭ, Funktsional. Anal. i Prilozhen. Vyp. 3 (1966), 3–10. (Russian)

6. Vladimir [V. V.] Peller, *L'inégalité de von Neumann, la dilatation isométrique et l'approximation par isométries dans L^p*, C. R. Acad. Sci. Paris Sér. A-B **287** (1978), A311–A314.

7. ______, *Isometric dilation of contractions, approximation by isometries, and von Neumann's inequality in an L^p space*, Preprint No. P-2-78, Leningrad. Otdel. Mat. Inst. Steklov. (LOMI), Leningrad, 1978. (Russian)

8. ______, *An analog of J. von Neumann's inequality for the space L^p*, Dokl. Akad. Nauk SSSR **231** (1976), 539–542; English transl. in Soviet Math. Dokl. **17** (1976).

9. ______, *Estimates of operator polynomials in an L^p-space in terms of the multiplier norm*, Zap. Nauchn. Sem. Leningrad. Otdel. Mat. Inst. Steklov. (LOMI) **65** (1976), 132–148; English transl. in J. Soviet Math. **16** (1981), no. 3.

10. ______, *Approximation by isometries, and V. I. Matsaev's conjecture for absolute contractions of the space L^p*, Funktsional. Anal. i Prilozhen. **12** (1978), no. 1, 38–50; English transl. in Functional Anal. Appl. **12** (1978).

11. ______, *Estimates of operator polynomials in symmetric spaces. A functional calculus for absolute contractions*, Mat. Zametki **25** (1979), 899–912; English transl. in Math. Notes **25** (1979).

12. N. K. Nikol'skiĭ, *Five problems on invariant subspaces*, Zap. Nauchn. Sem. Leningrad. Otdel. Mat. Inst. Steklov. (LOMI) **23** (1971), 115–127; English transl. in J. Soviet Math. **2** (1974), no. 4.

13. Paul R. Halmos, *A Hilbert space problem book*, Van Nostrand, Princeton, N. J., 1967.

14. Ronald R. Coifman and Guido Weiss, *Operators associated with representations of amenable groups, singular integrals induced by ergodic flows, the rotation method and multipliers*, Studia Math. **47** (1973), 285–303.

15. ______, *Transference methods in analysis*, Conf. Board Math. Sci. Regional Conf. Ser. Math., No. 31, Amer. Math. Soc., Providence, R. I., 1976.

16. R. R. Coifman, R. Rochberg and G. Weiss, *Applications of transference: the L^p version of von Neumann's inequality and the Littlewood-Paley-Stein theory*, Preprint.

17. R. G. Douglas, *Canonical models*, Topics in Operator Theory, Math. Surveys, No. 13, Amer. Math. Soc., Providence, R. I., 1974, pp. 161–218.

18. Arnold Lebow, *On von Neumann's theory of spectral sets*, J. Math. Anal. Appl. **7** (1963), 64–90.

19. M. A. Akcoglu, *A pointwise ergodic theorem in L^p-spaces*, Canad. J. Math. **27** (1975), 1075–1082.

20. Elena Stroescu, *Isometric dilations of contractions on Banach spaces*, Pacific J. Math. **47** (1973), 257–262.

21. Mustafa A. Akcoglu and P. Ekkehard Kopp, *Construction of dilations of positive L_p-contractions*, Math. Z. **155** (1977), 119–127.

22. M. A. Akcoglu and L. Sucheston, *Dilations of positive contractions on L_p spaces*, Canad. Math. Bull. **20** (1977), 285–292.

23. D. Dacunha-Castelle and J. L. Krivine, *Applications des ultraproduits à l'étude des espaces et des algèbres de Banach*, Studia Math. **41** (1972), 315–334.

24. Mustafa A. Akcoglu, *Positive contractions of L_1-spaces*, Math. Z. **143** (1975), 5–13.

25. E. M. Semenov, *Imbedding theorems for Banach spaces of measurable functions*, Dokl. Akad. Nauk SSSR **156** (1964), 1292–1295; English transl. in Soviet Math. Dokl. **5** (1964).

26. Didier Dacunha-Castelle and Jean-Louis Krivine, *Sous-espaces de L^1*, C. R. Acad. Sci. Paris Sér. A-B **280** (1975), A645–A648.

27. Mischa Cotlar, *A unified theory of Hilbert transforms and ergodic theorems*, Rev. Mat. Cuyana **1** (1955), 105–167.

28. A. P. Calderón, *Ergodic theory and translation-invariant operators*, Proc. Nat. Acad. Sci. U.S.A. **59** (1968), 349–353.

29. Paul R. Halmos, *Normal dilations and extensions of operators*, Summa Brasil Math. **2** (1950), 125–134.

30. M. A. Akcoglu and L. Sucheston, *Remarks on dilations in L_p-spaces*, Proc. Amer. Math. Soc. **53** (1975), 80–82.

31. Stefan Banach, *Théorie des operations linéaires*, PWN, Warsaw, 1932; reprint, Chelsea, New York, 1955.

32. John Lamperti, *On the isometries of certain function-spaces*, Pacific J. Math. **8** (1958), 459–466.

33. Joram Lindenstrauss and Lior Tzafriri, *Classical Banach spaces*, Lecture Notes in Math., Vol. 338, Springer-Verlag, 1973.

34. N. Bourbaki, *Topologie générale*, Chaps. 1, 2, 4th ed., Actualités Sci. Indust., No. 1142, Hermann, Paris, 1965; English transl., Hermann, Paris, and Addison-Wesley, Reading, Mass., 1966.

35. T. Ando, *Contractive projections in L_p spaces*, Pacific J. Math. **17** (1966), 391–405.

36. V. A. Rohlin, *On the fundamental ideas of measure theory*, Mat. Sb. **25 (67)** (1949), 107–150; English transl. in Amer. Math. Soc. Transl. (1) **10** (1962).

37. A. Ionescu Tulcea, *Ergodic properties of isometries in L^p spaces*, $1 < p < \infty$, Bull. Amer. Math. Soc. **70** (1964), 366–371.

38. Alberto de la Torre, *A simple proof of the maximal ergodic theorem*, Canad. J. Math. **28** (1976), 1073–1075.

39. V. Z. Vulikh, *Introduction to the theory of partially ordered spaces*, Fizmatgiz, Moscow, 1951; English transl., Noordhoff, 1967.

40. S. G. Kreĭn, Ju. I. Petunin and E. M. Semenov, *Interpolation of linear operators*, "Nauka", Moscow, 1978; English transl., Amer. Math. Soc., Providence, R. I., 1982.

41. B. S. Mityagin, *An interpolation theorem for modular spaces*, Mat. Sb. **66 (108)** (1965), 473–482. (Russian)

42. M. G. Zaĭdenberg, *On isometric classification of symmetric spaces*, Dokl. Akad. Nauk SSSR **234** (1977), 283–286; English transl. in Soviet Math. Dokl. **18** (1977).

43. Paul R. Halmos, *Lectures on ergodic theory*, Math. Soc. Japan, Tokyo, 1956; reprint, Chelsea, New York, 1960.

44. S. A. Vinogradov, *Multipliers of power series with sequence of coefficients from l^p*, Zap. Nauchn. Sem. Leningrad. Otdel. Mat. Inst. Steklov. (LOMI) **39** (1974), 30–39; English transl. in J. Soviet Math. **8** (1977), no. 1.

45. N. K. Nikol'skiĭ, *On spaces and algebras of Toeplitz matrices operating in l^p*, Sibirsk. Mat. Zh. **7** (1966), 146–158; English transl. in Siberian Math. J. **7** (1966).

46. Edwin Hewitt and Kenneth A. Ross, *Abstract harmonic analysis*. Vols. I, II, Springer-Verlag, 1963, 1970.

Translated by M. M. ARCHER

Труды Матем. Инст. Стеклов.
155(1981)

Proceedings of the Steklov
Institute of Mathematics
1983, Issue 1

MEN'SHOV'S CORRECTION THEOREM AND GAUSSIAN PROCESSES

UDC 517.948: 513.8 + 519.4

S. V. KHRUSHCHEV

ABSTRACT. In the first part of this article a simple proof is given for the well-known theorem of Men'shov that for any continuous function f on the unit circle $\mathbf{T}$ and any positive number σ there is a function f^* with uniformly convergent Fourier series such that $\operatorname{meas}\{\xi \in \mathbf{T}: f \neq f^*\} < \sigma$. It is shown that this proof goes through also in the case of compact abelian groups of finite dimension with an ergodic epimorphism. The multidimensional tori $\mathbf{T}^n$, $n = 1, 2, \ldots$, and the dyadic group $\{-1, 1\}^N$ are examples of such groups. In the concluding part the problem of "correcting" sample functions of stationary Gaussian processes on $\mathbf{T}$ is considered. Let

$$\mathfrak{F}l^1 \stackrel{\text{def}}{=} \left\{ f \in C(\mathbf{T}): \sum_{n \in \mathbf{Z}} |\hat{f}(n)| < +\infty \right\},$$

and let $(X_n)_{n \geq 1}$ be a sequence of complex-valued independent Gaussian random variables with zero mean and unit variance. A continuous function $f \in C(\mathbf{T})$ is said to be *incorrectable* if $f | E \notin \mathfrak{F}l^1 | E$ for any $E \subset \mathbf{T}$ of positive Lebesgue measure on the circle. It is shown that the property of being incorrectable is typical for functions in the set $C(\mathbf{T}) \setminus \mathfrak{F}l^1$. In particular, it is shown that almost all the sample functions of the process

$$X(\zeta) = \sum_{n \geq 2} \frac{\log \log n}{n} \zeta^n X_n$$

are incorrectable. The relation

$$\hat{f}(n) = O\left(\frac{\sqrt{\log n} \log \log n}{n} \right)$$

holds for the Fourier coefficients $\hat{f}(n)$ of almost all these sample functions. Recent results of Katznelson and Olevskiĭ are used in the proof.

Bibliography: 29 titles.

Contents

1980 *Mathematics Subject Classification*. Primary 42A20, 42A28, 60G15, 60G17; Secondary: 22C05, 22D40, 26A15, 26B15, 42A24, 42C10.

© 1983 American Mathematical Society
0081-5438/83 $1.00 + $.25 per page

§0. Introduction

This paper consists of three parts. In the first part a simple proof is presented for the well-known Men'shov correction theorem (see [1], or [2], Chapter VI, §§1–4), and various applications of this proof are considered (§§2 and 3). In the second part (§4) the Men'shov theorem is carried over to compact abelian groups of finite dimension that have an ergodic epimorphism. Finally, the third part (§§5–7) can be regarded as an illustration of the sharpness of Men'shov's theorem. We describe our main results in more detail.

Let ζ be the identity mapping of the unit circle $\mathbb{T}$ onto itself, and m normalized Lebesgue measure on $\mathbb{T}$. Each summable function f on the circle has a Fourier series:

$$f \sim \sum_{k \in \mathbb{Z}} \hat{f}(k)\zeta^k. \tag{0.1}$$

Here $\mathbb{Z}$ is the ring of all integers, and $\hat{f}(k) \overset{\text{def}}{=} \int_{\mathbb{T}} f\bar{\zeta}^k dm$ are the Fourier coefficients of f.

The problem of harmonic synthesis of a function f is the problem of recovering f from its Fourier series. The problem has various solutions, depending on what is meant by the word "recovering." If, say, f belongs to the space $C(\mathbb{T})$ of all continuous functions on $\mathbb{T}$ with the usual norm $\|f\|_\infty = \sup\{|f(z)|: z \in \mathbb{T}\}$, then the sequence of partial sums of its Fourier series converges to it almost everywhere on $\mathbb{T}$ (the theorem of Carleson [3]), and in this sense the Fourier series (0.1) recovers f. On the other hand, not all the functions in $C(\mathbb{T})$ have the property that $\lim_n \|S_n(f) - f\|_\infty = 0$. Therefore, synthesis of continuous functions by means of uniform convergence turns out to be impossible. Let

$$U(\mathbb{T}) \overset{\text{def}}{=} \left\{ f \in C(\mathbb{T}): \lim_n \|S_n(f) - f\|_\infty = 0 \right\}.$$

The Men'shov theorem, which is discussed in §2, asserts that any continuous function coincides on sets of measure arbitrarily close to 1 with functions which can be synthesized with respect to uniform convergence of Fourier series.

MEN'SHOV'S THEOREM. *Suppose that f is in $C(\mathbb{T})$ and σ is a positive number. Then there exists a function f^* in $U(\mathbb{T})$ such that $m\{f \neq f^*\} < \sigma$.*

The new proof of Men'shov's theorem presented in §2 originated from a study of the proof in [2]. The simplification is achieved due to techniques in harmonic analysis that are now well known. After preparing the preliminary publication [4], I found out that many specialists in orthogonal series knew of the possibility of using these techniques for simplifying the proof in Bari's book. However, as far as I know, no simple proofs of Men'shov's theorem have been published; therefore, the simple proof in this paper is presented in such a way that it can be read completely independently of the remainder of the text.

The problem of linearity of an operator realizing the correction is studied in §3. It is shown that there is no linear continuous operator T: $C(\mathbb{T}) \to U(\mathbb{T})$ such that $m\{Tf \neq f\} < 1$ for any continuous function f. On the other hand, it turns out that such a "correction" operator exists if we barely "diminish" the space $C(\mathbb{T})$. More precisely, let ω be a modulus of continuity, and let Λ_ω be the Banach space of all continuous functions f on the circle whose moduli of continuity ω_f satisfy the relation $\omega_f(\delta) = O(\omega(\delta))$, $\delta \to 0+$. Then for any number $\sigma > 0$ there exists a bounded linear operator $T\colon \Lambda_\omega \to U$ such that $m\{f \neq Tf\} < \sigma, f \in \Lambda_\omega$.

§4 contains a generalization of the Men'shov theorem to the case of compact abelian groups of finite dimension with an ergodic epimorphism. The dyadic group $\{-1, 1\}^\infty$ and the multidimensional tori $\mathbb{T}^n$ $(n = 1, 2, \ldots)$ are such groups. The central role in the approach to the Men'shov theorem taken in §2 is played by the ergodic epimorphism $\tau(\zeta) = \zeta^2$ of the group $\mathbb{T}$. This circumstance is underscored once again by the possibility of extending the Men'shov theorem to compact abelian groups with ergodic epimorphisms. On the other hand, it is easy to use this method to get an analogue of the Men'shov theorem for a certain class of multiplicative orthonormal Vilenkin systems that includes the Walsh system. We mention that in [5] Arutyunyan announced that an analogue of the Men'shov theorem holds for piecewise continuous orthonormal systems on $[0, 1]$ having a certain property (X) introduced in [6]. In particular, the Walsh system and a multiple trigonometric system have this property. Still earlier Kotlyar [7] proved that any measurable function f on $[0, 1]$ can be corrected on a set of arbitrarily small measure in such a way that the series for the resulting function f^* in both the trigonometric system and the Walsh system converge uniformly. Some improvements of the last result and results for the torus were announced in [8].

In §§5–7 we study the possibility of correcting the sample functions of stationary Gaussian processes on the unit circle $\mathbb{T}$. The results obtained in this part are based on a method developed by Katznelson [9] and Olevskiĭ [10].

For $1 \leqslant p < +\infty$ let

$$\mathfrak{F}l^p(\mathbb{T}) \stackrel{\text{def}}{=} \left\{ f \in C(\mathbb{T})\colon \|f\|_p \stackrel{\text{def}}{=} \sum_{k \in \mathbb{Z}} |\hat{f}(k)|^p < +\infty \right\}.$$

It is clear that $\mathfrak{F}l^1(\mathbb{T}) \subset U(\mathbb{T})$. Is it possible to replace the class $U(\mathbb{T})$ by $\mathfrak{F}l^1(\mathbb{T})$ in the Men'shov theorem? This question, which was posed in the well-known survey article [11] of Ul'yanov, received a negative answer in [9].

KATZNELSON'S THEOREM. *There exists a continuous function on the circle that does not coincide on any set of positive Lebesgue measure with any function having an absolutely convergent Fourier series.*

The function f constructed by Katznelson had modulus of continuity $\omega_f(\delta)$ satisfying the condition $\omega_f(\delta) = O(|\log \delta|^{-1})$, $\delta \to 0+$. A sharper result, which, moreover, answered a question posed in [11], was obtained in full scope in [10].

OLEVSKIĬ'S THEOREM. *For every p, $1 \leqslant p < 2$, there exists a function f in $C(\mathbb{T})$ with modulus of continuity $\omega_f(\delta) = O(\delta^{1/p - 1/2})$, $\delta \to 0+$, such that $f|E \notin \mathfrak{F}l^p|E$ for any set E of positive Lebesgue measure on $\mathbb{T}$.*

We remark that, by a theorem of Szász ([2], Chapter IX, §12), a function satisfying the condition $\omega_f(\delta) = O(\delta^{1/p - 1/2 + \varepsilon})$, $\varepsilon > 0$, is automatically in $\mathfrak{F}l^p$.

DEFINITION. A continuous function f is said to be *incorrectable* if for any set $E \subset \mathbb{T}$ with $mE > 0$ the restriction $f|E$ does not belong to the class $\mathfrak{F}l^1|E$.

Let $(X_n)_{n \in \mathbb{Z}}$ be a sequence of complex-valued independent Gaussian random variables on a probability space $(\Omega, \mathfrak{A}, P)$ with zero means $\mathsf{E}X_n = 0$, $n \in \mathbb{Z}$, and unit variances $\mathsf{E}|X_n|^2 = 1$, $n \in \mathbb{Z}$. The algebra $\mathfrak{A}$ of sets is assumed to be complete.

Let $(a_n)_{n\in\mathbf{Z}}$ be a square-summable sequence of nonnegative numbers. Let us consider the stationary Gaussian process

$$X(\zeta) = \sum_{n\in\mathbf{Z}} a_n X_n \zeta^n$$

on the circle $\mathbb{T}$. A *sample function* of the process $X(\zeta)$ is defined to be a function

$$X(\zeta, \Theta) = \sum_{n\in\mathbf{Z}} a_n X_n(\Theta)\zeta^n$$

on $\mathbb{T}$ obtained by fixing an elementary event $\Theta \in \Omega$. We assume that there exist numbers β and γ, $0 < \beta < 1$, such that

$$\left(\sum_{2^j\leqslant|n|<2^{j+1}} a_n^2\right)^{1/2} = O(2^{-\beta j}\cdot j^{\gamma}). \tag{0.2}$$

Then (see [12], p. 66) with probability 1

$$\omega_X(\delta) = O(\delta^{\beta}\log^{1/2+\gamma}(1/\delta)), \qquad \delta\to 0+.$$

An even sharper result is valid, and we shall use it in what follows.

LEMMA A. *There exists a constant $C > 0$ not depending on n such that the probability of the event $\omega_X(2^{-n}) > C\cdot 2^{-\beta_n}\cdot\log^{1/2+\gamma}(2^n)$ is less than 2^{1-2n}.*

A proof can be found in Kahane's book [12] (p. 67).

The main result in the third part gives a criterion for the sample functions of the process X to be incorrectable.

THEOREM A. *Suppose that, in addition to* (0.2), *the sequence $(a_n)_{n\in\mathbf{Z}}$ satisfies the condition*

$$\lim_{n\to+\infty} na_n = +\infty.$$

Then the sample functions of the stationary Gaussian process $X(\zeta)$ are incorrectable with probability 1.

The conditions of this theorem hold for the process

$$X(\zeta) = \sum_{n=2}^{\infty}\frac{\log\log n}{n}X_n\zeta^n. \tag{0.3}$$

Here (0.2) is valid with $\beta = 1/2$ and any $\gamma > 0$. Consequently, with probability 1 the sample functions of the process X have modulus of continuity

$$O\left(\sqrt{\delta(\log(1/\delta))^{1+\varepsilon}}\right), \qquad \varepsilon > 0.$$

Since any sample function of the process (0.3) admits an analytic continuation to the unit disk $\mathbb{D} = \{z\in\mathbb{C}\colon |z|<1\}$, we get a continuous incorrectable function which constitutes the boundary values on $\mathbb{T}$ of a function in the disk algebra. Of course, with respect to estimates of the modulus of continuity the functions constructed in (0.3) are subservient to those constructed in Olevskiĭ's paper. On the other hand, we get that the property of incorrectability is typical for functions not in $\mathfrak{F}l^1$. Moreover, since $X_n = o(\sqrt{\log n})$ with probability 1 (see [12], p. 121), we get an example of an incorrectable function f whose Fourier coefficients satisfy the condition

$$\hat{f}(n) = o\left(\frac{\sqrt{\log n} \cdot \log\log n}{n}\right).$$

In (0.3) we have $a_n = (\log\log n)/n$. Any arbitrarily slowly increasing sequence $\lambda_n = na_n$ can be chosen in place of the sequence $(\log\log n)_{n\geqslant 2}$. However, it unfortunately does not work to set $a_n = 1/n$. The methods used here are too coarse to allow inclusion of this important particular case.

We conclude the Introduction by listing some references dealing with correction theorems. For the functions summable on the circle Men'shov proved a correction theorem (see [2], Chapter VI, §6) asserting that any summable function f can be corrected outside a given perfect nowhere dense set in such a way that the Fourier series of the resulting function converges almost everywhere. Kheladze [13] improved this theorem in the following direction. For any summable function f there exists a function s, $m\{s \neq 1\} < \varepsilon$, taking only the values ± 1 and such that the Fourier series of the functions sf and $\widetilde{sf}$ converge almost everywhere. Correction theorems for functions satisfying a Lipschitz condition in the mean were considered by Oskolkov in [14]. Finally, an analogue of Olevskiĭ's theorem for the Haar system was obtained in [15].

I use this opportunity to express my appreciation to S. B. Stechkin, who made me aware of Arutyunyan's work. I thank V. N. Sudakov, whose advice led to an improved presentation of the material in §§5–7, and N. K. Nikol'skiĭ, who carefully read through the manuscript and eliminated many errors.

§1. Main notation

To each function f on the unit circle $\mathbb{T}$ there corresponds a 2π-periodic function f_* on the line $\mathbb{R}$: $f_*(t) \stackrel{\text{def}}{=} f(e^{it})$, $t \in \mathbb{R}$. A function f on $\mathbb{T}$ is said to be *piecewise linear* if the function f_* has this property. A simple example of a piecewise linear function is the function Δ_ε, $0 < \varepsilon \leqslant \pi$, determined by the formula

$$(\Delta_\varepsilon)_*(t) = \max(0, 1 - |t|/\varepsilon), \qquad |t| \leqslant \pi.$$

It is easy to verify that

$$\hat{\Delta}_\varepsilon(n) = (2/\pi)\left(\sin^2(n\varepsilon/2)/\varepsilon n^2\right),$$

and, therefore, the Fourier series of Δ_ε converges absolutely. Clearly

$$\|\Delta_\varepsilon\|_1 = \sum_{n\in\mathbb{Z}} |\hat{\Delta}_\varepsilon(n)| = \Delta_\varepsilon(1) = 1.$$

For every $\theta \in \mathbb{T}$ the symbol τ_θ denotes the corresponding translation operator on the group: $\mathbb{T}$: $\tau_\theta f(\zeta) = f(\zeta\bar{\theta})$. Let $\varepsilon = 2^{-n}\pi$, where n is a positive integer, and let F_n be the group of 2^nth roots of unity. Then the family of functions $(\tau_\theta\Delta_\varepsilon)_{\theta\in F_{n+1}}$ forms a partition of unity on $\mathbb{T}$. Figure 1 illustrates this fact for $n = 2$ in terms of the functions $(\Delta_\varepsilon)_*$. Thus, $\sum_{\theta\in F_{n+1}} \tau_\theta\Delta_{2^{-n}\pi} \equiv 1$ for any $n = 1, 2, \ldots$, and the multiplicity of the covering of $\mathbb{T}$ by the interiors of the supports of the functions $\tau_\theta\Delta_{2^{-n}\pi}$ does not exceed two.

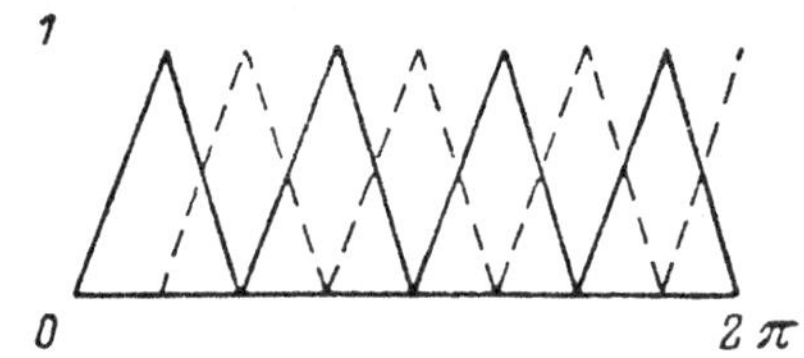

FIGURE 1

We mention a useful property of the family $(\tau_\theta\Delta_{2^{-n}\pi})_{\theta\in F_{n+1}}$. If

$$x=\sum_{\theta\in F_{n+1}}\lambda_\theta\tau_\theta\Delta_{2^{-n}\pi},$$

then $|\lambda_\theta|=|x(\theta)|\leqslant\|x\|_\infty$. Indeed, if $\theta\in F_{n+1}$, then among the functions of our family only $\Delta_{2^{-n}\pi}(\zeta\bar{\theta})$ is nonzero at the point θ, and it equals 1 there. Therefore, $\lambda_\theta=x(\theta)$.

If $f\in C(\mathbb{T})$, let $\tilde{f}$ be the *harmonic conjugate* of f:

$$\tilde{f}\sim\sum_{k\in\mathbb{Z}}\hat{f}(k)\operatorname{sign}(k)\zeta^k.$$

In the space $U(\mathbb{T})$ we single out the subspace U of functions f for which $\tilde{f}\in U(\mathbb{T})$. The inclusion $U\subset U(\mathbb{T})$ is strict. The space U is a Banach space in the norm

$$\|f\|_U=\sup_{n\geqslant0}\left\|\sum_{k=0}^{n}\hat{f}(k)\zeta^k\right\|_\infty+\sup_{n\geqslant1}\left\|\sum_{k=-n}^{-1}\hat{f}(k)\zeta^k\right\|_\infty.$$

A function ω mapping the interval $[0,+\infty)$ into itself is called a *modulus of continuity* if it is continuous, $\omega(0)=0$, and $\omega(x)\leqslant\omega(x+y)\leqslant\omega(x)+\omega(y)$ for all $x, y\geqslant0$. The modulus of continuity of a continuous function f in $C(\mathbb{T})$ is defined to be the modulus of continuity

$$\omega_f(\delta)\overset{\text{def}}{=}\sup_{|x-y|\leqslant\delta}\left|f_*(x)-f_*(y)\right|.$$

Finally, $V(e_i\colon i\in I)$ denotes the closed linear span of a family $(e_i)_{i\in I}$ of vectors in a topological vector space.

§2. A proof of Men'shov's theorem

We show that the correction f^* can be chosen from the smaller class U (this result was announced in [5]). Given a sequence $(\varepsilon_k)_{k\geqslant1}$ of positive numbers, it is not hard to construct a sequence $(f_k)_{k\geqslant0}$ of piecewise linear functions such that $\|f_k\|_\infty\leqslant\varepsilon_k$, $k=1,2,\dots$, and $f=\sum_{k\geqslant0}f_k$, where the series converges absolutely in $C(\mathbb{T})$. After this, the correction of f can be found in the form $f^*=\sum_{k\geqslant0}f_k^*$.

The sequence $(f_k)_{k\geqslant0}$ can be constructed by induction, with the following argument as a basis. Let n be a nonnegative integer, and let

$$X_n=V(\tau_\theta\Delta_{2^{-n}\pi}\colon\theta\in F_{n+1}).$$

For brevity set $\varphi_\theta=\tau_\theta\Delta_{2^{-n}\pi}$, $\theta\in F_{n+1}$. It is clear that $(\varphi_\theta)_{\theta\in F_{n+1}}$ is a partition of unity of $\mathbb{T}$. Therefore, for any $f\in C(\mathbb{T})$

$$f=\sum_{\theta\in F_{n+1}}f\varphi_\theta=\sum_{\theta\in F_{n+1}}[f-f(\theta)]\varphi_\theta+\sum_{\theta\in F_{n+1}}f(\theta)\varphi_\theta. \tag{2.1}$$

From this it follows that

$$\operatorname{dist}(f, X_n) \stackrel{\text{def}}{=} \inf_{x \in X_n} \|f - x\|_\infty \leqslant \sum_{\theta \in F_{n-1}} \omega_f(2^{-n}\pi)\varphi_\theta = \omega_f(2^{-n}\pi)$$

and, consequently, $\lim_{n \to +\infty} \operatorname{dist}(f, X_n) = 0$.

After constructing an expansion of a function f in a series of piecewise linear functions, we are justified in studying the finite-dimensional correction problem.

LEMMA 2.1. *For any positive integer n and any number $r \geqslant 1$ there exists a linear mapping $T: X_n \to U$ such that*

(a) $\sum_{\theta \in F_{n+1}} m(\varphi_\theta \neq T\varphi_\theta) < 4r^{-1}$, *and*

(b) $\|Tx\|_U \leqslant 8r\|x\|_\infty$ *for any* $x \in X_n$.

We first show how to finish the proof of Men'shov's theorem, and then we prove the lemma.

Let $(r_k)_{k \geqslant 0}$ be a sequence of positive integers such that $\sum_{k \geqslant 0} r_k^{-1} < \sigma/4$, and suppose that $\sum_{k \geqslant 0} \varepsilon_k r_k < +\infty$. Let us choose a sequence of piecewise linear functions $(f_k)_{k \geqslant 0}$, $f_k \in X_{n_k}$, such that

$$f = \sum_{k \geqslant 0} f_k, \quad \|f_k\|_\infty \leqslant \varepsilon_k, \qquad k = 0, 1, \ldots.$$

For each k we construct for the number r_k an operator T_k having properties (a) and (b) of Lemma 2.1. Let $f^* = \sum_{k \geqslant 0} T_k f_k$. Since $\|T_k f_k\|_U \leqslant 8r_k\|f_k\|_\infty$, it follows that

$$\|f^*\|_U \leqslant 8 \sum_{k \geqslant 0} r_k \varepsilon_k < +\infty.$$

Finally,

$$m\{f \neq f^*\} \leqslant \sum_{k \geqslant 0} m\{f_k \neq T_k f_k\} < 4 \sum_{k \geqslant 0} r_k^{-1} < \sigma. \quad \bullet$$

REMARK. The set $\{f \neq f^*\}$ lies in a set A with $mA < \sigma$ that depends only on the modulus of continuity of f. This set is completely determined by the sequence $(X_{n_k})_{k \geqslant 0}$ of subspaces chosen for approximation and by the sequence of numbers $(r_k)_{k \geqslant 0}$.

PROOF OF LEMMA 2.1. Let n and r be fixed, and let $h = \mathbf{1} - 2r\Delta_{\pi/r}$. It is clear that $\hat{h}(0) = 0$ and $\hat{h}(n) = -2r\hat{\Delta}_{\pi/r}(n)$, $n \neq 0$. Therefore, $\|h\|_1 \leqslant 2r$. The elements φ_θ, $\theta \in F_{n+1}$, form a basis of the finite-dimensional space X_n, and, consequently, it suffices to define the values of the linear operator T on the elements φ_θ. We set $T\varphi_\theta = \varphi_\theta h(\zeta^{N_\theta})$, where the choice of the large positive integers $(N_\theta)_{\theta \in F_{n+1}}$ will be settled later.

LEMMA 2.2. *For any sufficiently large positive integer N the inequality*

$$m\{\varphi_\theta h(\zeta^N) \neq \varphi_\theta\} < (2/r)m(\operatorname{supp} \varphi_\theta)$$

holds for all $\theta \in F_{n+1}$.

PROOF. It is clear that

$$\{\varphi_\theta h(\zeta^N) \neq \varphi_\theta\} \subset \{h(\zeta^N) \neq \mathbf{1}\} \cap \operatorname{supp}(\varphi_\theta).$$

The set $\{h \neq \mathbf{1}\}$ is a circular arc with measure equal to r^{-1}. The preimage of this arc under the mapping ζ^N consists of N equal uniformly separated disjoint arcs of common measure r^{-1}. Therefore, if N is large, then a number of small arcs proportional to the measure $m(\mathrm{supp}(\varphi_\theta))$ falls into each arc of $\mathrm{supp}(\varphi_\theta)$. It remains to note that each point of the circle lies interior to the supports of no more than two functions of the family $(\varphi_\theta)_{\theta\in F_{n+1}}$, and that all the measures $m(\mathrm{supp}\,\varphi_\theta)$ are equal. ●

The reader familiar with ergodic theory will observe that this lemma amounts, in essence, to a proof that the endomorphism ζ^k, $k\geqslant 2$, of $\mathbb{T}$ is strongly mixing (see [16], Chapter 1, §1).

Let us now estimate the norm $\|Tx\|_U$. Since

$$Tx=\sum_{k\neq 0}\hat{h}(k)\sum_{\theta\in F_{n+1}}x(\theta)\varphi_\theta\zeta^{N_\theta k},$$

it follows that

$$\|Tx\|_U\leqslant 2r\sup_{k\neq 0}\left\|\sum_{\theta\in F_{n+1}}x(\theta)\varphi_\theta\zeta^{N_\theta k}\right\|_U.$$

Lemma 2.2 gives us that $m\{x\neq Tx\}<4r^{-1}$. The proof of Lemma 2.1 is concluded by Lemma 2.3.

LEMMA 2.3. *For any positive integer N_0 there exist positive integers N_θ, $\theta\in F_{n+1}$, greater than N_0 such that*

$$\sup_{k\neq 0}\left\|\sum_{\theta\in F_{n+1}}x(\theta)\varphi_\theta\zeta^{N_\theta k}\right\|_U\leqslant 4\|x\|_\infty.$$

PROOF. Since φ_θ coincides with the function $\tau_\theta\Delta_{2^{-n}\pi}$, it follows that $\|\varphi_\theta\|_1=\|\varphi_\theta\|_\infty=1$. Therefore, there exist trigonometric polynomials P_θ, $\theta\in F_{n+1}$, satisfying the inequalities

$$\|\varphi_\theta-P_\theta\|_1\leqslant 2^{-(n+1)},\qquad \|P_\theta\|_1\leqslant 1,\qquad \theta\in F_{n+1}. \tag{2.2}$$

The *support of a trigonometric polynomial* is defined to be the smallest interval $\{a,a+1,\dots,b\}$ of the set $\mathbb{Z}$ outside which the Fourier coefficients of the polynomial vanish. Let J_θ be the support of P_θ. If the numbers N_θ and their nonnegative pairwise differences are sufficiently large, say $N_\theta-N_\xi\notin\bigcup_{s\in F_{n+1}}(J_s-J_s)$, then the supports $J_\theta+kN_\theta$ of the polynomials $p_\theta\zeta^{N_\theta k}$ are disjoint and all lie either in the set $\{n\in\mathbb{Z}: n\geqslant 0\}$ $(k>0)$ or in the set $\{n\in\mathbb{Z}: n\leqslant -1\}$.

Suppose for definiteness that $k>0$. Let us estimate the norm in U of the trigonometric polynomial

$$P=\sum_{\theta\in F_{n+1}}x(\theta)P_\theta\zeta^{N_\theta k}.$$

Since the supports of the polynomials $P_\theta\zeta^{N_\theta k}$ are disjoint, for any positive integer m there exist a subset I_m of F_{n+1} and an index $j=j(m)\in F_{n+1}$ such that

$$S_m(P)=\sum_{\theta\in I_m}x(\theta)P_\theta\zeta^{N_\theta k}+S_m\big(x(j)P_j\zeta^{N_j k}\big).$$

It is now useful to recall that $\|P_\theta\|_1\leqslant 1$ (see (2.2)). Hence,

$$\left\|S_m\big(x(j)P_j\zeta^{N_j k}\big)\right\|\leqslant\|x\|_\infty\|P_j\|\leqslant\|x\|_\infty.$$

Further,

$$\left|\sum_{\theta\in I_m} x(\theta)P_\theta\zeta^{N_\theta k}\right| \leqslant \sum_{\theta\in I_m} |x(\theta)|\|P_\theta-\varphi_\theta\|_1 + \left|\sum_{\theta\in I_m} x(\theta)\varphi_\theta\zeta^{N_\theta k}\right|$$
$$\leqslant \|x\|_\infty + \sum_{\theta\in F_{n+1}} |x(\theta)|\varphi_\theta \leqslant 2\|x\|_\infty$$

(see (2.2); the identity $\sum_{\theta\in F_{n+1}}\varphi_0 \equiv \mathbf{1}$ is used in the last inequality). Therefore, $\|S_m(P)\|_\infty \leqslant 3\|x\|_\infty$. It remains to observe that

$$\|S_m(x-P)\|_\infty \leqslant \|x\|_\infty \sum_{\theta\in F_{n+1}} \|\varphi_\theta - P_\theta\|_1 \leqslant \|x\|_\infty. \quad \bullet$$

REMARK 1. It is interesting to note the connection between the Men'shov theorem and the problems discussed in S. V. Kislyakov's article in this issue [17]. Let $\alpha(x) = \exp(-x/B)$, where B is a positive constant. Then for any sufficiently large number η and any continuous function f, $\|f\|_\infty = 1$, there exist functions g and h such that

$$f = g + h, \quad \|g\|_U \leqslant \eta, \quad \|h\|_2 \stackrel{\text{def}}{=} \left(\int_{\mathbb{T}} |h|^2\,dm\right)^{1/2} \leqslant \alpha(\eta).$$

In other words, any continuous function can be represented as the sum of a function in U and a function with a small L^2-norm. The Men'shov theorem allows us to draw an analogous conclusion, the only difference being that the concept of "smallness" is taken in another sense. We choose the special sequence $(r_k)_{k\geqslant 0}$: $r_k = 4(k^2+1)/\sigma$, $k \geqslant 0$. Let $f \in C(\mathbb{T})$, and let $\varepsilon_k = 2\|f\|_\infty(1+k^4)^{-1}$, $k \geqslant 0$.

The arguments of this section can be used to construct a function f^* in U such that

$$m\{f \neq f^*\} < \sigma \quad \text{and} \quad \|f^*\|_U \leqslant 8\sum_{k\geqslant 0} r_k\varepsilon_k \leqslant \frac{\text{const}\|f\|_\infty}{\sigma}.$$

Thus, any function f, $\|f\|_\infty \leqslant 1$, can be represented in the form $f = g + h$, where $\|g\|_U \leqslant \text{const}\cdot\eta$ and the function h equals zero outside an open set of measure not more than $1/\eta$.

REMARK 2. The method of proof of the Men'shov theorem enables us to show that a set of type H (see [2], Chapter XII, §5) is an interpolation set for the space U. Recall that E is a set of type H if there exist a number $\delta > 0$, an increasing sequence $(n_k)_{k\geqslant 1}$ of positive integers, and a sequence $(J_k)_{k\geqslant 1}$ of arcs of length δ such that none of the sets $G(n_k, \delta) \stackrel{\text{def}}{=} \{\xi \in \mathbb{T}: \xi^{n_k} \in J_k\}$ intersect E. To prove that $C(\mathbb{T})|E = U|E$ we need to choose the auxiliary functions h used in proving Lemma 2.2 with supports in the arcs J_k, $k \geqslant 1$. Oberlin recently proved [29] that every closed set E of Lebesgue measure zero is an interpolation set for U.

§3. Properties of the Men'shov correcting operators

Let $0 < \sigma < 1$. We consider the set $\mathfrak{M}_\sigma$ of all mappings T: $C(\mathbb{T}) \to U(\mathbb{T})$ such that $m\{f \neq Tf\} < \sigma$ for any $f \in C(\mathbb{T})$. Men'shov's theorem implies that $\mathfrak{M}_\sigma \neq \varnothing$ for any $\sigma > 0$. The correction operator T constructed in §2 corrects continuous functions in such a way that the sets $\{f \neq Tf\}$, though they have a small measure, are distributed over the whole circle $\mathbb{T}$. It turns out that for any operator $T \in \mathfrak{M}_\sigma$,

$0 < \sigma < 1$, it is impossible to include all the sets $\{f \neq Tf\}$, $f \in C(\mathbb{T})$, in a single measurable set with measure less than 1. This was established by Men'shov in [18]. The simple proof given below for this theorem uses the same ideas as Men'shov's original proof.

THEOREM (MEN'SHOV [18]). *Let E be a closed subset of $\mathbb{T}$ with $mE > 0$. Then $U(\mathbb{T})|E \neq C(\mathbb{T})|E$.*

PROOF. It is well known (see [19], Chapter VIII, §1, Theorem 1.2) that for any sequence of positive numbers $(\lambda_n)_{n\geq 1}$ satisfying the condition $\lambda_n = o(\log n)$ there is a function f in $C(\mathbb{T})$ such that $\lambda_n = o(S_n(f, 1))$ on some subsequence of the set of all positive integers. We show that $f|E \notin U(\mathbb{T})|E$, if the sequence $(\lambda_n)_{n\geq 1}$ is suitably chosen.

Let $E^* = \{t \in [-\pi, \pi)\colon e^{it} \in E\}$. Without loss of generality it can be assumed that 0 is a point of density of E^*. Since the Lebesgue constants $\int_{-\pi}^{\pi} |t^{-1} \sin nt|\, dt$ increase as $\log n$ and since 0 is a point of density of E^*, the following lemma is no surprise.

LEMMA 3.1.

$$a_n \overset{\text{def}}{=} \int_{[-\pi,\pi)\setminus E^*} \left|\frac{\sin nt}{t}\right| dt = o(\log n), \qquad n \to +\infty.$$

PROOF. Let $G = [-\pi, \pi)\setminus E^*$, and suppose that $\mathbf{1}_G$ is the characteristic function of G. It is clear that

$$a_n \leqslant \frac{m(G \cap (-1/n, 1/n))}{1/n} + \int_{1/n\leqslant|t|\leqslant\pi} \frac{\mathbf{1}_G(t)}{|t|}\, dt.$$

Let $\delta > 0$ and $F(t) = \int_{-t}^{t} \mathbf{1}_G\, ds$. Then

$$\int_{\delta\leqslant|t|\leqslant\pi} \frac{\mathbf{1}_G(t)}{|t|}\, dt = \int_\delta^\pi \frac{1}{t}\, dF(t) = \left.\frac{F(t)}{t}\right|_\delta^\pi + \int_\delta^\pi \frac{F(t)}{t^2}\, dt.$$

Since G has density zero at the point 0, it follows that $F(t) = o(t)$, $t \to 0+$. ●

We choose the sequence $(\lambda_n)_{n\geq 1}$ so that $a_n \leqslant \lambda_n$, $\lambda_n \uparrow +\infty$, and $\lambda_n = o(\log n)$. Let f be a corresponding function. If $g \in C(\mathbb{T})$ and $g|E = f|E$, then $(g_* - f_*)|E_* = 0$. Therefore,

$$|S_n(f - g, 1)| \leqslant \text{const}\|f - g\|_\infty \int_G \left|\frac{\sin nt}{t}\right| dt = O(\lambda_n).$$

Consequently, for infinitely many indices n

$$S_n(f, 1) - S_n(g, 1) = o(S_n(f, 1)).$$

This implies that $\overline{\lim}_{n\to+\infty} |S_n(g, 1)| = +\infty$. ●

This theorem enables us to extract some information about the mappings in $\mathfrak{M}_\sigma$. Let

$$\mathfrak{M} = \{T\colon C(\mathbb{T}) \to U(\mathbb{T})\colon m\{f \neq Tf\} < 1, f \in C(\mathbb{T})\}.$$

THEOREM 3.1. *Let T be a continuous linear operator acting from the Banach space $C(\mathbb{T})$ to the Banach space $U(\mathbb{T})$. Then $T \notin \mathfrak{M}$.*

PROOF. We suppose that $T \in \mathfrak{M}$ and show first that $T \in \mathfrak{M}_\sigma$ for some σ, $0 < \sigma < 1$. Let

$$E_f \overset{\text{def}}{=} \{\xi \in \mathbb{T}: f(\xi) = Tf(\xi), f \in C(\mathbb{T})\}.$$

Since $T \in \mathfrak{M}$, $mE_f > 0$ for any f. Let us consider the set $L_n \overset{\text{def}}{=} \{f \in C(\mathbb{T}): mE_f \geqslant 1/n\}$. Then $C(\mathbb{T}) = \bigcup_{n \geqslant 1} L_n$. It is not hard to see that L_n is closed. Indeed, if $\lim_m f_m = f$ in $C(\mathbb{T})$ and $f_m \in L_n$ for $m = 1, 2, \ldots$, then $\lim_m Tf_m = Tf$ pointwise, since T is continuous. Therefore, if $x \in E_{f_m}$ for infinitely many values of m, it follows that $x \in E_f$. Hence,

$$\bigcap_{m \geqslant 1} \bigcup_{k \geqslant m} E_{fk} \subset E_f$$

and, consequently, $mE_f \geqslant 1/n$. By the Baire category theorem, one of the sets L_n contains a ball $B(f_0, \delta)$ of the space $C(\mathbb{T})$ of radius $\delta > 0$ about some f_0. The next lemma allows us to conclude from this that $mE_f \geqslant 1/n$ for any f in $C(\mathbb{T})$.

LEMMA 3.2. *Suppose that f and g are nonzero functions in $C(\mathbb{T})$. Then $mE_{f+\lambda g} = m(E_f \cap E_g)$ for all complex numbers λ with the possible exception of a not more than countable subset of $\mathbb{C}$.*

PROOF. The linearity of T implies that $E_{f+\lambda g} \supset E_f \cap E_g$ and

$$E_{f+\lambda g} \cap \big((E_f \backslash E_g) \cup (E_g \backslash E_f)\big) = \varnothing,$$

if $\lambda \neq 0$. For each $\lambda \in \mathbb{C}$ we consider the set S_λ of points x, $x \notin E_f \cup E_g$, such that $T(f + \lambda g)(x) = f(x) + \lambda g(x)$. It is clear that

$$S_\lambda = \{x \in \mathbb{T} \backslash (E_f \cup E_g): T_f(x) - f(x) = \lambda(g(x) - Tg(x))\}.$$

Since $Tf - f$ and $g - Tg$ do not vanish on $\mathbb{T} \backslash (E_f \cup E_g)$, the sets S_λ, $\lambda \neq 0$, are disjoint. Consequently, $mS_\lambda > 0$ for no more than countably many values of λ. But if $mS_\lambda = 0$ and $\lambda \neq 0$, then, obviously,

$$mF_{f+\lambda g} = m(E_f \cap E_g). \quad \bullet$$

If $f_0 = 0$, then $B(0, \delta) \subset L_n$, and the linearity of T implies that $C(\mathbb{T}) \subset L_n$. But then $T \in \mathfrak{M}_\sigma$ for $\sigma = 1 - 1/2n$. Suppose that $f_0 \not\equiv 0$, and let $g \in B(0, \delta)$. We apply Lemma 3.2 to $f = f_0$ and g. There exists a λ, $0 < \lambda < 1$, such that

$$mE_{f_0+\lambda g} = m(F_{f_0} \cap E_g).$$

Since $f_0 + \lambda g \in L_n$, it follows that $mE_g \geqslant m(E_{f_0} \cap E_g) \geqslant 1/n$ for any function g in $B(0, \delta)$. Again from the linearity of T it follows that $mE_g \geqslant 1/n$ for any $g \in C(\mathbb{T})$. Consequently, $T \in \mathfrak{M}_{1-1/2n}$ also in this case.

Thus, $mE_f \geqslant 1/n$ for every f in $C(\mathbb{T})$. Applying Lemma 3.2 again, we get that $m(E_f \cap E_g) \geqslant 1/n$ for any f and g. From this it is easy to prove by induction that $m(\bigcap_{f \in I} E_f) \geqslant 1/n$ for any finite subset I of $C(\mathbb{T})$.

Let us now consider the closed set $E = \bigcap_{f \in C(\mathbb{T})} E_f$. It is clear that $mE \geqslant 1/n$ (see [20], Chapter IV, §4.6). On the other hand, $f|E = Tf|E$ for any function $f \in C(\mathbb{T})$. This contradicts the theorem of Men'shov in [18]. $\bullet$

Despite its nonlinearity, the operator $f \to f^*$ constructed in §2 was obtained with the aid of a sequence of linear correction operators defined on the finite-dimensional spaces X_n, $n \geqslant 0$. If the subspaces X_n were linearly independent, then on the linear span of the X_n we could construct a linear operator inducing the given operators T_n: $X_n \to U$ on the subspaces X_n. We could then try to extend this operator to the closure of the span in some topology. Of course, by Theorem 3.1, this closure cannot coincide with the whole of $C(\mathbb{T})$, but it may nevertheless be very "close" to $C(\mathbb{T})$. Let ω be a modulus of continuity, and

$$\Lambda_\omega \stackrel{\text{def}}{=} \{f \in C(\mathbb{T})\colon \omega_f(\delta) \leqslant \text{const}\, \omega(\delta), \delta > 0\}.$$

THEOREM 3.2. *For any modulus of continuity ω and any number $\sigma > 0$ there exist a bounded linear operator T: $\Lambda_\omega \to U$ and an open set G, $mG < \sigma$, such that $\{f \neq Tf\} \subset G$ for any f in Λ_ω.*

This theorem is a slight improvement of the corresponding theorem of Men'shov (see [18]). In the original formulation it was not required that the operator T be linear. The proof of the theorem follows the scheme presented in §2. The program of the proof sketched above requires that the spaces X_n, $n \geqslant 0$, be linearly independent. However, it is not hard to see that $\{\lambda \cdot \mathbf{1}\colon \lambda \in \mathbb{C}\} \subset X_n \subset X_{n+1}$, $n \geqslant 0$. Therefore, before proceeding to the proof of Theorem 3.2, we modify the linear spaces X_n in such a way that they become linearly independent.

LEMMA 3.3. *Let $(\theta_n)_{n \geqslant 0}$ be a sequence of points of $\mathbb{T}$ such that the sets $\theta_n F_{2^{n+1}}$ are disjoint. If*

$$\mathbf{0} = \sum_{k=0}^{n} f_k,$$

where $f_k \in \tau_{\theta_k} X_k$, then all the functions f_k are constant.

PROOF. Let A_k be the preimage of the set $\theta_k F_{2^{k+1}}$ under the mapping $t \to e^{it}$ of $\mathbb{R}$ onto $\mathbb{T}$. The function $(f_n)_*$ is piecewise linear, and its linearity can be violated only at the points of A_n. But since $(f_n)_* = -\sum_{k=0}^{n-1}(f_k)_*$ and since the sets A_k, $k = 0, 1, \ldots, n$, are pairwise disjoint, $(f_n)_*$ is linear also at the points of A_n. Therefore, $(f_n)_*$ is a bounded linear function on the line. Consequently, $f_n \equiv \text{const}$. The successive application of this argument obviously shows that all the functions f_k, $k = 0, 1, \ldots, n$, are constant. ●

Let Y_n be the linear subspace of $\tau_{\theta_n} X_n$ consisting of the functions f orthogonal to the constants: $\hat{f}(0) = \int_{\mathbb{T}} f\,dm = 0$. It follows immediately from Lemma 3.3 that the linear spaces $\{\lambda \mathbf{1}\}$, Y_0, $Y_1, \ldots$ are linearly independent.

PROOF OF THEOREM 3.2. For the time being let $(n_k)_{k \geqslant 0}$ be an arbitrary sequence of positive integers, and $(r_k)_{k \geqslant 0}$ a sequence such that $\sum_{k \geqslant 0} r_k^{-1} < \sigma/4$. It follows from Lemma 2.1 that for each positive integer k there exist a linear operator T_k: $\tau_{\theta_k} X_{n_k} \to U$ and an open set G_k such that $mG_k < 4r_k^{-1}$, and $\{T_k x \neq x\} \subset G_k$ for any $x \in \tau_{\theta_k} X_{n_k}$, and $\|T_k x\|_U \leqslant 8r_k \|x\|_\infty$.

We now define a linear operator T on the linear span $\mathfrak{L}$ of the spaces $\{\lambda \mathbf{1}\}$, Y_{n_0}, $Y_{n_1}, \ldots$ as follows: $T\mathbf{1} = \mathbf{1}$, and $T | Y_{n_k} = T_k | Y_{n_k}$. Then $\{Tx \neq x\} \subset \bigcup_{k \geqslant 0} G_k \stackrel{\text{def}}{=} G$, and $mG < \sigma$. If on $\mathfrak{L}$ we define the norm

$$\|x\| = |\lambda| + \sum_{k \geqslant 0} r_k \|x_k\|_\infty,$$

where $\lambda \in \mathbf{C}$, $x_k \in Y_{n_k}$, and $x = \lambda \mathbf{1} + \sum_{k \geqslant 0} x_k$ (such a representation of x is unique), then T turns out to be a bounded linear operator from the normed space $(\mathfrak{L}, \|\cdot\|)$ to the Banach space U. Let $Y(n_k)$ be the Banach space obtained by completing $\mathfrak{L}$ in this norm. It is clear that the elements of $Y(n_k)$ are precisely the continuous functions x for which there exist a complex number λ and an infinite sequence $(x_k)_{k \geqslant 0}$, $x_k \in Y_{n_k}$, such that

$$x = \lambda \cdot \mathbf{1} + \sum_{k \geqslant 0} x_k \quad \text{and} \quad \sum_{k \geqslant 0} r_k \|x_k\|_\infty < +\infty.$$

Obviously, T can be extended to a bounded operator from $Y(n_k)$ to U, and $\{x \neq Tx\} \subset G$ for every $x \in Y(n_k)$.

To conclude the proof we establish the following "imbedding theorem."

LEMMA 3.4. *For any modulus of continuity ω there exists a sequence $(n_k)_{k \geqslant 0}$ such that $\Lambda_\omega \subset Y(n_k)$.*

The proof of Lemma 3.4 amounts to a refinement of the arguments at the beginning of §2. It is not hard to see that it suffices to establish the inclusion $\Lambda_\omega \subset Y(n_k)$ for real functions. Therefore, from here to the end of this section all arguments will be carried out over the field of real numbers. For brevity we let ω_n denote the modulus of continuity of the function $\Delta_{2^{-n}\pi}$.

LEMMA 3.5. *Suppose that $x \in C(\mathbf{T})$, $\hat{x}(0) = 0$, and n is a nonnegative integer. Then there exists an element x_n in Y_n such that*:

(a) $\|x_n\|_\infty \leqslant \|x\|_\infty$;

(b) $\|x - x_n\|_\infty \leqslant \omega_x(2^{-n+1}\pi)$;

(c) $\omega_{x - x_n}(\delta) \leqslant \omega_x(\delta) + 4\|x\|_\infty \omega_n(\delta)$.

PROOF. Recall that $Y_n = \{x \in \tau_{\theta_n} X_n \colon \hat{x}(0) = 0\}$. Since the inequalities (a)–(c) are invariant with respect to the group of transformations $(\tau_\theta)_{\theta \in \mathbf{T}}$, we can assume without loss of generality that $\theta_n = 1$. Let $(\varphi_\theta)_{\theta \in F_{n+1}}$ be the standard basis of X_n generated by the translates of the function $\Delta_{2^{-n}\pi}$ by elements of the group F_{n+1}. It is not hard to see that $x_n = \sum_{\theta \in F_{n+1}} \lambda_\theta \varphi_\theta$ belongs to the space Y_n if and only if $\sum_{\theta \in F_{n+1}} \lambda_\theta = 0$. Let us show first that for some choice of points $(z_\theta)_{\theta \in F_{n+1}}$, $z_\theta \in \operatorname{supp}(\varphi_\theta)$, $\theta \in F_{n+1}$, the element

$$x_n = \sum_{\theta \in F_{n+1}} x(z_\theta) \varphi_\theta$$

belongs to Y_n. It will be shown below that it also satisfies conditions (a), (b), and (c). To do this we consider the auxiliary function

$$y(\zeta) = \sum_{\theta \in F_{n+1}} x(\theta\zeta).$$

Recall that $\varphi_\theta(\theta) = 1$, $\theta \in F_{n+1}$. It is clear that

$$\int_{\mathbf{T}} y \, dm = 2^{n+1} \int_{\mathbf{T}} x \, dm = 0.$$

Moreover, $y(\theta\zeta) = y(\zeta)$ for $\theta \in F_{n+1}$, since the set F_{n+1} is a group. A continuous real function with integral zero vanishes at at least one point of the circle. Since it is also periodic, there is a point ξ on the small arc joining 1 and $\exp(2\pi i/2^{n+1})$, such that $y(\xi) = 0$. It remains to set $z_\theta \stackrel{\text{def}}{=} \theta\xi$, $\theta \in F_{n+1}$.

The inequality (a) obviously follows from the formula defining x_n.

To verify (b) we need to observe that

$$x - x_n = \sum_{\theta \in F_{n+1}} (x - x(z_\theta))\varphi_\theta$$

and that $|x - x(z_\theta)| \leqslant \omega_x(2^{-n+1}\pi)$ on $\operatorname{supp}\varphi_\theta$.

Let us verify (c). Let s and t be two arbitrary points on $\mathbb{T}$. Then

$$\begin{aligned}(x - x_n)(s) - (x - x_n)(t) &= x(s) - x(t) - (x_n(s) - x_n(t)) \\ &= x(s) - x(t) - \sum_{\theta \in F_{n+1}} x(z_\theta)(\varphi_\theta(s) - \varphi_\theta(t)).\end{aligned}$$

Since $\varphi_\theta(s) - \varphi_\theta(t)$ is nonzero for at most four indices in F_{n+1}, we have

$$\omega_{x - x_n}(\delta) \leqslant \omega_x(\delta) + 4\|x\|_\infty \omega_n(\delta). \quad \bullet$$

We make the choice of a sequence $(n_k)_{k \geqslant 0}$. Let us first choose a monotone positive sequence $(\varepsilon_k)_{k \geqslant 0}$ such that

$$\sum_{k \geqslant 0} \varepsilon_k r_k < +\infty \quad \text{and} \quad \varepsilon_0 = 4\omega(2\pi).$$

We then determine n_0 from the inequality $\omega(2^{-n_0+1}\pi) < \varepsilon_1$. If $n_0, n_1, \ldots, n_k$ have already been constructed, then n_{k+1} is chosen in such a way that $n_{k+1} > n_k$ and the inequality

$$\omega(\delta) + \varepsilon_0 \omega_{n_0}(\delta) + \varepsilon_1 \omega_{n_1}(\delta) + \cdots + \varepsilon_k \omega_{n_k}(\delta) < \varepsilon_{k+2}$$

holds for $\delta = 2^{-n_k+1} \cdot 2\pi$.

Let x be a continuous function such that $\hat{x}(0) = 0$ and $\omega_x(\delta) \leqslant \omega(\delta)$. It is clear that $\|x\|_\infty \leqslant \omega(2\pi)$. With the help of Lemma 3.5 we choose an element $x_{n_0} \in Y_{n_0}$ such that

$$\|x_{n_0}\|_\infty \leqslant \|x\|_\infty < \varepsilon_0.$$

Then $\|x - x_{n_0}\|_\infty \leqslant \omega(2^{-n_0+1}\pi) < \varepsilon_1$. We apply Lemma 3.5 to the element $x := x - x_{n_0}$ and to $n = n_1$. Then there is an element $x_{n_1} \in Y_{n_1}$ such that

$$\|x_{n_1}\|_\infty \leqslant \|x - x_{n_0}\|_\infty < \varepsilon_1,$$

and

$$\|x - x_{n_0} - x_{n_1}\|_\infty \leqslant \omega_{x - x_{n_0}}(2^{-n_1+1}\pi) < \omega(2^{-n_1+1}\pi) + \varepsilon_0 \omega_{n_0}(2^{-n_1+1}\pi) < \varepsilon_2.$$

Continuing the construction by induction, we get a sequence $(x_{n_k})_{k \geqslant 0}$ such that

$$x = \sum_{k \geqslant 0} x_{n_k} \quad \text{and} \quad \sum_{k \geqslant 0} \|x_{n_k}\| r_k < +\infty.$$

Consequently, $x \in Y(n_k)$. Since every element f in Λ_ω can be represented in the form $f = \lambda \mathbf{1} + x$, where $\lambda \in \mathbb{R}$, $\hat{x}(0) = 0$, and $\omega_x \leqslant \omega$, it is clear that $\Lambda_\omega \subset Y(n_k)$.
$\bullet$

§4. A correction theorem for compact abelian groups

Let G be a compact abelian group, Γ the group of all continuous characters on G, and $C(G)$ the space of all continuous functions on G, equipped with the usual norm

$$\|f\|_\infty = \sup_{x\in G} |f(x)|.$$

Let m be Haar measure on G, $mG = 1$. The Fourier coefficients of an f in $C(G)$ are determined by

$$\hat{f}(\gamma) = \int_G f\bar{\gamma}\, dm, \qquad \gamma \in \Gamma.$$

We restrict ourselves to metrizable groups G. In the language of characters this means that the group Γ is at most countable.

DEFINITION. A family $\mathcal{A} = (A_n)_{n\geq 0}$ of finite subsets of Γ is called a *summation family* if $\Gamma = \bigcup_{n\geq 0} A_n$ and $A_n \subset A_{n+1}$, $n = 0, 1, \ldots$.

Let $\mathcal{A}$ be a summation family and let $f \in C(G)$. Then

$$S_n(f, \mathcal{A}) \stackrel{\text{def}}{=} \sum_{\gamma\in A_n} \hat{f}(\gamma)\gamma.$$

The space of functions on G whose Fourier series are uniformly summable with respect to the summation family $\mathcal{A}$ is denoted by $U(G, \mathcal{A})$. More precisely,

$$U(G, \mathcal{A}) \stackrel{\text{def}}{=} \left\{ f \in C(G) \colon \lim_{n\to+\infty} \|f - S_n(f, \mathcal{A})\|_\infty = 0 \right\}.$$

The group G with the σ-algebra of Borel sets and Haar measure m becomes a measure space. Every continuous epimorphism τ of G is a (Haar) measure-preserving transformation of G. By duality, an epimorphism τ induces a monomorphism of the character group: $\tau^*\gamma \stackrel{\text{def}}{=} \gamma \circ \tau$, $\gamma \in \Gamma$. By a theorem of Rokhlin (see [21], or [22], p. 76), an epimorphism τ is ergodic if and only if the orbit $O^+_{\tau^*}(\gamma) \stackrel{\text{def}}{=} \{\gamma, \tau^*\gamma, \tau^{*2}\gamma, \ldots\}$ is infinite for each character $\gamma \not\equiv \mathbf{1}$.

Let $\dim G$ be the topological dimension of the group G. It is well known that $\dim G$ is equal to the free rank of the group Γ of characters (see [23], Theorem 24.28).

THEOREM 4.1. *Let G be a compact abelian group of finite dimension* ($\dim G < +\infty$) *with an ergodic epimorphism τ, and let $\mathcal{A}$ be any summation family in Γ. Then for any continuous function f on G and any number $\varepsilon > 0$ there is a function $f^* \in U(G, \mathcal{A})$ such that $m\{f \neq f^*\} < \varepsilon$.*

DISCUSSION. When is there an ergodic epimorphism τ on a compact abelian group G? It is clear that such an epimorphism exists if there are no elements of finite order in Γ. In this case we can set $\tau^*\gamma = \gamma^2$ (the group operation in the set Γ of characters is written multiplicatively). A structure theorem in the theory of topological groups (see [23], §9) shows that the finite-dimensional groups under consideration are exhausted by the tori $\mathbb{T}, \mathbb{T}^2, \ldots$. If any of these groups is multiplied by some finite group, then we get a group without ergodic epimorphisms, since the periodic part of the corresponding group of characters is finite. However, Men'shov's theorem can nevertheless be extended to such groups. Let us now consider the ring Δ_p of p-adic integers (see [23]), p a prime number. In the natural topology Δ_p is a compact topological group. Its character group is isomorphic to the group $Z(p^\infty)$ (see [23]) formed by the roots of unity whose degrees are powers of p. Since the number of elements of any given order in $Z(p^\infty)$ is finite, the group Δ_p does not have an ergodic epimorphism. It would be interesting to find out whether the correction theorem is valid when $G = \Delta_p$.

On the other hand, there are fairly many zero-dimensional groups with an ergodic epimorphism. For example, the dyadic group $\mathbf{D} \stackrel{\text{def}}{=} \{(x_n)_{n\geqslant 1}: x_n = 0, 1\}$, is such a group, where addition is defined coordinatewise modulo two. The shift $(\tau x)_n = x_{n+1}$, $n \geqslant 1$, is an ergodic epimorphism. Since G is a metrizable group, the measure space $(G, \mathfrak{B}, m)$ is a Lebesgue space. If, in addition, m is a continuous measure, then in the category of measure spaces there is an isomorphism T of the interval $[0, 1]$ onto G. The family $(\gamma \circ T)_{\gamma\in\hat{G}}$ of functions forms a multiplicative system (see [24], Russian pp. 457–493).* In this way we can extend Men'shov's theorem to various multiplicative systems, including the Walsh system.

The proof of Theorem 4.1 repeats in concept that for $\mathbb{T}$ presented in §2.

Definition. A neighborhood V of zero in a compact abelian group G is said to be *covering* if there is a finite family $(x_i)_{i\in I}$ of elements in G such that the sets $x_i + V$, $i \in I$, are disjoint and

$$m\left(G \setminus \bigcup_{i\in I} (x_i + V)\right) = 0.$$

It is not hard to see that Card $I = (mV)^{-1}$. Let $\mathfrak{F} = \mathfrak{F}(G)$ be the family of all covering neighborhoods in G.

Lemma 4.1. *Let G be a compact abelian group. Then there exists a subfamily $\mathfrak{F}_0$ of $\mathfrak{F}$ that forms a base for the topology of G at the point* 0 *and is such that $m(V + V)/mV \leqslant 3^{\dim G}$ for every neighborhood $V \in \mathfrak{F}_0$.*

Proof. Let $\gamma_1, \ldots, \gamma_n$ be an arbitrary finite family of characters on G, and $\varepsilon > 0$. The neighborhoods of zero

$$U(\gamma_1, \ldots, \gamma_n; \varepsilon) = \left\{x \in G: |\gamma_k(x) - 1| < \varepsilon, k = 1, \ldots, n\right\}$$

form a base of the topology of G at 0. The mapping

$$\gamma: G \to \mathbb{T}^n, \qquad \gamma(x) \stackrel{\text{def}}{=} (\gamma_1(x), \ldots, \gamma_n(x))$$

is an epimorphism of G onto the closed subgroup $\gamma(G)$ of the group $\mathbb{T}^n$. Let V be a covering neighborhood of the group $\gamma(G)$ contained in the neighborhood $\{\xi \in \gamma(G): |\xi_k - 1| < \varepsilon, \xi = (\xi_1, \ldots, \xi_k)\}$, and $(x_i)_{i\in I}$ a corresponding family of elements of $\gamma(G)$. Let $x_i = \gamma(y_i)$, $i \in I$. Then $\gamma^{-1}(x_i + V) = y_i + \gamma^{-1}(V)$. Every homomorphism of compact groups preserves normalized Haar measure, and, therefore, the covering neighborhood $\gamma^{-1}(V)$ is contained in the neighborhood $U(\gamma_1, \ldots, \gamma_n; \varepsilon)$. Further, the closed subgroups of $\mathbb{T}^n$ are topologically isomorphic to groups of the form $\mathbb{T}^d \times F$,

* *Editor's note.* The reference is to the appendix of the Russian translation, prepared by N. Ya. Vilenkin.

where $0 \leqslant d \leqslant n$ and F is a finite group. Consequently, $\gamma(G)$ is isomorphic to a group $\mathbb{T}^d \times F$. Observe that $d \leqslant \dim G$, since the free rank of $\gamma(G)$ is equal to d. But the group $\mathbb{T}^d$ always contains arbitrarily small neighborhoods V (say, in the form of cubes, if $\mathbb{T}^d$ is represented as $\mathbb{R}^d/\mathbb{Z}^d$) that are covering. Since V is a cube, $m(V+V)/mV = 3^d$. ●

Let $\mathfrak{F}l^1(G)$ be the Banach space of all absolutely convergent Fourier series on G with the usual norm:

$$\|f\|_1 \stackrel{\text{def}}{=} \sum_{\gamma\in\Gamma} |\hat{f}(\gamma)|.$$

Fix a neighborhood $V \in \mathfrak{F}$, and let $(x_i)_{i\in I}$ be a family of elements of G corresponding to it. For $i \in I$ let

$$\varphi_i(t) \stackrel{\text{def}}{=} \frac{1}{mV}\cdot \mathbf{1}_V * \mathbf{1}_V(t - x_i), \qquad t \in G.$$

LEMMA 4.2. *For every $i \in I$, $\varphi_i \in \mathfrak{F}l^1(G)$,*

$$\|\varphi_i\|_\infty = \|\varphi_i\|_1 = 1 \qquad \text{and} \qquad \sum_{i\in I} \varphi_i(t) \equiv \mathbf{1}.$$

PROOF. The equality $\|\varphi_i\|_\infty = 1$ follows from the formula for the convolution, and the Parseval equality shows that

$$\frac{1}{mV}\|\mathbf{1}_V*\mathbf{1}_V\|_1 = \frac{1}{mV}\sum_{\gamma\in\Gamma}\left|\hat{\mathbf{1}}_V(\gamma)\right|^2 = \frac{1}{mV}\int_G \mathbf{1}_V^2\, dm = 1.$$

Let us check the last identity of the lemma. We have

$$\sum_{i\in I}\varphi_i(t) = \sum_{i\in I}\frac{1}{mV}\int_G \mathbf{1}_V(t - x_i - s)\mathbf{1}_V(s)\,dm(s)$$
$$= \frac{1}{mV}\int_G \mathbf{1}_V(s)\sum_{i\in I}\mathbf{1}_V(t - x_i - s)\,dm(s).$$

But the sum $\sum_{i\in I}\mathbf{1}_V(t - x_i)$ is the characteristic function of the set $\bigcup_{i\in I}(x_i + V)$. ●

The functions in the family $(\varphi_i)_{i\in I}$ will play the role of the triangular functions in §2 in our proof. Note that $(\varphi_i)_{i\in I}$ is a linearly independent family of functions. Indeed, φ_i is equal to 1 at the point x_i, and $\varphi_j(x_i) = 0$ for $j \neq i$, because $\sum_{j\in I}\varphi_j \equiv \mathbf{1}$. Therefore,

$$\max_{i\in I}|c_i| \leqslant \left\|\sum_{i\in I} c_i\varphi_i\right\|_\infty.$$

To each neighborhood $V \in \mathfrak{F}_0$ we assign the finite-dimensional vector space $E_V = \mathbf{V}(\varphi_i\colon i \in I)$. By following the arguments in §2, it is not difficult to find for any sequence $(\varepsilon_k)_{k\geqslant 0}$ of positive numbers a family $(V_k)_{k\geqslant 0}$ of neighborhoods and a sequence of functions $f_k \in E_{V_k}$ such that $f = \sum_{k\geqslant 0} f_k$ and $\|f_k\|_\infty \leqslant \varepsilon_k$, $k \geqslant 1$. To conclude the proof of Theorem 4.1 we must now prove the analogue of Lemma 2.1.

LEMMA 4.3. *For any open set $V \in \mathfrak{F}_0$ and any number $r \geqslant 1$ there exist a linear operator $T\colon E_V \to U(G, \mathfrak{A})$ and a number $R \geqslant r$ such that*

$$m\{f \neq Tf\} < \frac{2}{R}\cdot 3^{\dim G}; \qquad \|Tf\|_U \leqslant 8R\|f\|_\infty, \qquad f \in E_V.$$

PROOF. Let W be a neighborhood of zero such that $mW \leqslant 1/r$ and $W \in \mathfrak{F}_0$. We set $R = 1/mW$ and

$$h = \mathbf{1} - \frac{1}{(mW)^2} \cdot \mathbf{1}_W * \mathbf{1}_W.$$

Then $\hat{h}(\mathbf{1}) = 0$ (here $\mathbf{1}$ is the character on G identically equal to 1) and $\|h\|_1 \leqslant R$. The correction operator $T\colon E_V \to U$ is found in the following form:

$$\text{if } f = \sum_{i\in I} \lambda_i \varphi_i, \quad \text{then } Tf = \sum_{i \in I} \lambda_i \varphi_i h \circ \tau^{N_i},$$

where we shall choose the positive integers N_i, $i \in I$, a little later.

An ergodic epimorphism τ on G is strongly mixing (see [16], p. 12):

$$\lim_{n\to+\infty} m(\tau^{-n}A \cap B) = mA \cdot mB$$

for any measurable subsets A and B of G. Therefore,

$$\lim_{n\to+\infty} m\{\varphi_i h \circ \tau^n \neq \varphi_i\} = m(\operatorname{supp} \varphi_i) m\{h \neq \mathbf{1}\}.$$

Since V and W are open sets,

$$m(\operatorname{supp} \varphi_i) = m(V + V) \quad \text{and} \quad m\{h \neq \mathbf{1}\} = m(W + W).$$

This implies that there exists a positive integer N_0 such that

$$m\{\varphi_i \cdot h \circ \tau^n \neq \varphi_i\} < 2 \cdot 3^{2\dim G} \cdot mV \cdot mW$$

for all $n > N_0$ uniformly in $i \in I$. Since, obviously, for any $f \in E_V$ and since card $I = 1/mW$, it follows that $m\{f \neq Tf\} < 2R^{-1} \cdot 3^{2\dim G}$.

We now verify the second assertion of Lemma 4.3. It is clear that

$$Tf = \sum_{1\neq\gamma\in\Gamma} \hat{h}(\gamma) \sum_{i \in I} \lambda_i \varphi_i \tau^{*N_i}\gamma.$$

Let B be a finite set of characters such that

$$\sum_{\gamma\notin B} |\hat{h}(\gamma)| \sum_{i\in I} \|\varphi_i\|_1 = \operatorname{card} I \cdot \sum_{\gamma \notin B} |\hat{h}(\gamma)| < 1.$$

Then

$$\|Tf\|_U \leqslant \|h\|_1 \sup_{1\neq\gamma\in B} \left\| \sum_{i\in I} \lambda_i \varphi_i \tau^{*N_i}\gamma \right\|_U + \|f\|_\infty$$

(we use the inequality $|\lambda_i| \leqslant \|f\|_\infty$).

LEMMA 4.4. *For any positive integer N_0 and any finite set of characters B there exist positive integers N_i, $i \in I$, greater than N_0 such that*

$$\sup_{1\neq\gamma\in B} \left\| \sum_{i\in I} \lambda_i \varphi_i \tau^{*N_i}\gamma \right\|_U \leqslant 4\|f\|_\infty.$$

The proof follows the scheme of the proof of Lemma 2.4. We concentrate our attention on the part of this proof which was trivial in the case of the group $\mathbb{T}$. It is convenient to assume that $I = \{1, \ldots, n\}$. We approximate each function φ_i in the metric of $\mathfrak{F}l^1$ by a trigonometric polynomial T_i: $\|\varphi_i - T_i\|_1 \leqslant (\operatorname{card} I)^{-1}$. Let $B_i \overset{\text{def}}{=} \{\gamma \in \Gamma\colon \hat{T}_i(\gamma) \neq 0\}$ be the spectrum of T_i. We obviously need the following lemma in order to be able to carry over the proof for the circle to the general case.

LEMMA 4.5. *For any number N_0 there are numbers $N_1, \dots, N_m$ such that:*

(a) *for any character $\gamma \in B$, $\gamma \neq \mathbf{1}$, the sets $B_i + \tau^{*N_i}\gamma$, $i = 1, 2, \dots, m$, are disjoint; and*

(b) *for every positive integer n and every $\gamma \in B$ the set A_n, $A_n \in \mathcal{A}$, intersects at most one of the sets $B_i + \tau^{*N_i}\gamma$ in a proper subset.*

The proof is by induction on m. We shall use the fact that card $O^+_{\tau^*}(\gamma) = +\infty$, if $\gamma \neq \mathbf{1}$. The assertion of the lemma is obvious if $m = 1$. Assuming that it holds for $m = k$, we prove it for $m = k + 1$. Let

$$S = \bigcup_{\substack{i=1 \\ \gamma \in B,\, \gamma \neq \mathbf{1}}}^{m} (B_i + \tau^{*N_i}\gamma).$$

It is clear that card $S < +\infty$. The set $B_{k+1} + \tau^{*N_{k+1}}\gamma$ intersects S if $\tau^{*N_{k+1}}\gamma \in S - B_{k+1} \overset{\text{def}}{=} \{s - b: s \in S, b \in B_{k+1}\}$. But card$(S - B_{k+1}) < +\infty$. Therefore, for sufficiently large N_{k+1} the inclusion $\tau^{*N_{k+1}}\gamma \in S - B_{k+1}$ does not hold for all $\gamma \in B$ (use the facts that card $B < +\infty$ and card $O^+_{\tau^*}(\gamma) = +\infty$). Consequently, for such large values of N_{k+1} the set $B_{k+1} + \tau^{*N_{k+1}}\gamma$ does not intersect any of the sets $B_i + \tau^{*N_i}\gamma$, $i = 1, \dots, k$.

Suppose now that n is the smallest index such that $S \subset A_n$. Since card $A_n < +\infty$, we can choose an N_{k+1} large enough so that $(B_{k+1} + \tau^{*N_{k+1}}\gamma) \cap A_n = \varnothing$ by applying the preceding arguments to A_n instead of S. ●

§5. A criterion for incorrectability

In this section a necessary condition is given for a continuous function f on $\mathbb{T}$ to be correctable. We verify in the subsequent sections that with probability 1 this condition fails to hold for the sample functions of the process X considered in the theorem. The scheme for the finite-dimensional approximations and the statement of the correction problem in terms of two orthonormal systems are due to Olevskiĭ [10].

Let $C > 0$ and $\delta > 0$.

DEFINITION 1. A continuous function f is said to be *(C, δ)-correctable* if there exist a closed subset E, $mE > \delta$, and a function $f^* \in \mathfrak{F}l^1$ such that $f|E = f^*|E$ and $\|f^*\|_1 \leqslant C$.

On the circle $\mathbb{T}$ we consider the group F_n of all nth roots of unity, and define

$$\mu_n \overset{\text{def}}{=} \sum_{\zeta \in F_n} \delta(\zeta),$$

where $\delta(\zeta)$ is the Dirac measure at the point ζ. In the space $L^2(\mu_n)$, endowed with the natural inner product

$$(x, y) \overset{\text{def}}{=} \int x\bar{y}\, d\mu_n,$$

we can single out two complete orthonormal systems: $(e_k)_1^n$ and $(e_k^*)_1^n$. The first consists of the characteristic functions e_k of the singleton subsets of F_n, and the second of the functions $e_k^* = n^{-1/2}\zeta^k | F_n$.

DEFINITION 2. Let $E \subset F_n$. An element x of $L^2(\mu_n)$ is said to be *C-correctable outside the set E* if there exists an element $x^* \in L^2(\mu_n)$ such that

$$x|E = x^*|E \quad \text{and} \quad \sum_{k=1}^{n} |(x^*, e_k^*)| < C\sqrt{n}.$$

An element x is said to be (C, δ)-*correctable* if it is C-correctable outside some set E such that card $E > n\delta$.

It is easy to grasp the approximation meaning of Definition 2 in relation to Definition 1 if we observe that $(x, e_k^*)/\sqrt{n}$ is a Riemann sum for the integral $\hat{f}(k) = \int_{\mathbb{T}} f\bar{\zeta}^k dm$ under the condition that $x = f | F_n$.

Let us now consider the restriction operator $\pi_n \colon C(\mathbb{T}) \to L^2(\mu_n)$. The next lemma shows how the coefficients of $\pi_n f$ in the basis $(e_k^*)_1^n$ are connected with the Fourier coefficients of f.

LEMMA 5.1. *If $f \in C(\mathbb{T})$ and $1 \leqslant k \leqslant n$, then*

$$\frac{1}{\sqrt{n}}(\pi_n f, e_k^*) = \sum_{m \equiv k (\mathrm{mod}\ u)} \hat{f}(m). \tag{5.1}$$

REMARK 1. The sum in (5.1) is understood in the sense of Cesàro summation.

PROOF. Since

$$\frac{1}{\sqrt{n}}(\pi_n f, e_k^*) = \frac{1}{n}\int_{\mathbb{T}} f\bar{\zeta}^k \, d\mu_n = f * \left(\frac{\zeta^k d\mu_n}{n}\right)(1),$$

it remains to observe that

$$f * \left(\frac{\zeta^k d\mu_n}{n}\right) \sim \sum_{m \equiv k (\mathrm{mod}\ n)} \hat{f}(m)\zeta^m$$

and that the Fourier series of a continuous function is Cesàro-summable to this function. ●

REMARK 2. If $f \in \Lambda_\beta$ for some $\beta > 0$, then the Fourier series of this function and, consequently, the Fourier series of the convolution $f*(\zeta^k d\mu_n/n)$ converge uniformly on $\mathbb{T}$. Hence, in this case the summation on the right-hand side of (5.1) can be understood in the usual sense.

LEMMA 5.2. *Let $f \in C(\mathbb{T})$ and suppose that it is a (C, δ)-correctable function. Then for any positive integer n there is a number $\theta \in \mathbb{T}$ such that the element $\pi_n(\tau_\theta f)$ is (C, δ)-correctable.*

PROOF. Since f is (C, δ)-correctable, there exist a closed set E, $mE > \delta$, and a function f^*, $\|f^*\|_1 < C$, such that $f^* | E = f | E$. We fix the positive integer n and consider on $\mathbb{T}$ the auxiliary function

$$\varphi_n(\xi) = \frac{1}{n}\int_{\mathbb{T}} \mathbf{1}_E(\zeta\bar{\xi})\, d\mu_n(\zeta) = \frac{\operatorname{card}(F_n \cap \xi E)}{n}.$$

It is clear that $0 \leqslant \varphi_n \leqslant 1$. Moreover, $\int_{\mathbb{T}} \varphi_n \, dm = mE > \delta$. Therefore, there is a number $\theta \in \mathbb{T}$ such that

$$\delta < \varphi_n(\theta) = \operatorname{card}(F_n \cap \theta E)/n.$$

It is not hard to see that $\tau_\theta f | \theta E = \tau_\theta f^* | \theta E$. Let $E_n = F_n \cap \theta E$ and $x^* = \pi_n(\tau_\theta f^*)$. Then $x^* | E_n = \tau_\theta f | E_n$. Using Lemma 5.1, we get that

$$\sum_{k=1}^{n} |(x^*, e_k^*)| = \sqrt{n}\sum_{k=1}^{n}\left|\sum_{m \equiv k (\mathrm{mod}\ n)} (\tau_\theta f^*)\hat{\ }(m)\right| \leqslant \sqrt{n} \cdot \|f^*\|_1 < C\sqrt{n}.$$

Consequently, $\pi_n(\tau_\theta f)$ is (C, δ)-correctable. ●

§6. Reduction to an estimate of the Gaussian measure of a convex body

Let

$$b_{kn} = \left(\sum_{m \equiv k (\mathrm{mod}\ n)} a_m^2 \right)^{1/2}.$$

Let us consider in $L^2(\mu_n)$ the convex set (the sequence (a_m) is defined in the Introduction)

$$K = \left\{ x \in L^2(\mu_n) \colon \sum_{k=1}^{n} b_{kn} |(x, e_k^*)| \leqslant C \right\}.$$

Complex-valued independent Gaussian random variables $(Z_k)_1^n$ with zero mean and unit variance induce a Gaussian measure γ_n in $L^2(\mu_n)$:

$$\gamma_n(G) = P\left\{ \sum_{k=1}^{n} Z_k e_k^* \in G \right\}. \tag{6.1}$$

Let B_n be the unit ball of $L^2(\mu_n)$. The codimension of a linear subspace M of $L^2(\mu_n)$ over the field of complex numbers is denoted by $\mathrm{codim}_{\mathbf{C}}\, M$. In this section we show that the proof of Theorem A can be reduced to the proof of the following theorem.

THEOREM B. *Let $(\varepsilon_n)_{n \geqslant 1}$ be an arbitrary sequence of positive numbers converging to zero, and let $\delta > 0$. Then there exist a constant* const *depending only on $(\varepsilon_n)_{n \geqslant 1}$ and δ and a sequence $(\alpha_n)_{n \geqslant 1}$, $\lim_n \alpha_n = 0$, such that*

$$\sup_{\mathrm{codim}_{\mathbf{C}} M \geqslant n\delta} \gamma_n(K + M + \varepsilon_n B_n) \leqslant \mathrm{const}\, \alpha_n^n. \tag{6.2}$$

It is clear that the sample functions of the process X are incorrectable with probability 1 if for any pair (C, δ) of positive numbers the probability is zero that a sample function of X is (C, δ)-correctable. According to Lemma 5.2, the possibility of correcting the sample functions depends on the possibility of correcting the random elements $\pi_n(\tau_\theta X)$ of the space $L^2(\mu_n)$. From here on the pair (C, δ) is assumed to be fixed. Let 2^Y be the collection of all subsets of a set Y. We introduce the family

$$\mathcal{E}_n = \{ E \in 2^{F_n} \colon \mathrm{card}\, E \geqslant n\delta \},$$

and let $E \in \mathcal{E}_n$. The set of all elements that are C-correctable outside E admits a simple geometric description. Let

$$K_0 \overset{\mathrm{def}}{=} \left\{ x \in L^2(\mu_n) \colon \sum_{k=1}^{n} |(x, e_k^*)| \leqslant C\sqrt{n} \right\},$$

$$M = M(E) \overset{\mathrm{def}}{=} \{ x \in L^2(\mu_n) \colon \mathrm{supp}\, x \cap E = \varnothing \}.$$

Then the desired set is obviously equal to $K_0 + M(E)$. Let

$$\sigma_n(E, \theta) = \{ \omega \in \Omega \colon \pi_n(\tau_\theta X) \in K_0 + M(E) \},$$

and let J be the set of $\omega \in \Omega$ for which $X(\zeta, \omega)$ is (C, δ)-correctable. It follows from Lemma 5.2 that for any n

$$J \subset \bigcup_{E \in \mathcal{E}_n} \bigcup_{\theta \in \mathbf{T}} \sigma_n(E, \theta).$$

We show below that the set on the right-hand side can be included in a measurable set J_n such that $\lim_n P(J_n) = 0$. Since the σ-algebra $\mathcal{Q}$ is complete with respect to the probability measure P, it follows from this that $J \in \mathcal{Q}$ and $P(J) = 0$.

The next lemma shows that the sets $\sigma_n(E, \theta)$ cannot be uniformly distributed in the probability space Ω as θ runs through the set of points of the circle.

LEMMA 6.1. *Let $\theta \in \mathbb{T}$, and suppose that θ' is the point of the finite set F_{2^n} closest to θ. Suppose that $\sigma_n \in \mathcal{Q}$ and*

$$\omega_X(2^{-n}) \leqslant C \cdot 2^{-\beta n} \cdot \log^{\gamma+1/2}(2^n)$$

for $\omega \notin \sigma_n$. Then $P(\sigma_n) \leqslant 2^{1-2n}$, and

$$\|\pi_n(\tau_\theta X) - \pi_n(\tau_{\theta'} X)\|_2 \leqslant Cn^{\gamma+1/2} \cdot 2^{-\beta n}$$

whenever $\omega \notin \sigma_n$.

PROOF. The inequality $P(\sigma_n) \leqslant 2^{1-2n}$ follows from Lemma A, stated in the Introduction. If $\omega \notin \sigma_n$, then

$$\|\pi_n(\tau_\theta X) - \pi_n(\tau_{\theta'} X)\|_2^2 = \int_{\mathbb{T}} \left|X(\zeta\bar{\theta}) - X(\varepsilon\bar{\theta}')\right|^2 d\mu_n$$

$$\leqslant \omega_X^2(2^{-n}) \operatorname{card} F_n \leqslant nC^2 \cdot 2^{-2\beta n} \cdot n^{2\gamma+1}. \quad \bullet$$

This lemma shows that for any $\theta \in \mathbb{T}$ there is a $\theta' \in F_{2^n}$ such that

$$\pi_n(\tau_{\theta'} X) \in \pi_n(\tau_\theta X) + \beta_n B_n, \tag{6.3}$$

where $\beta_n = Cn^{\gamma+1} \cdot 2^{-\beta n}$, under the assumption, of course, that $\omega \notin \sigma_n$. We introduce the auxiliary measurable sets:

$$\sigma_n(E, \theta, \varepsilon) \stackrel{\text{def}}{=} \{\omega \in \Omega : \pi_n(\tau_\theta X) \in K_0 + M(E) + \varepsilon B_n\}.$$

The inclusion (6.3) implies at once the inclusion

$$\bigcup_{E \in \mathcal{E}_n} \bigcup_{\theta \in \mathbb{T}} \sigma_n(E, \theta) \subset \bigcup_{E \in \mathcal{E}_n} \bigcup_{\theta \in F_{2^n}} \sigma_n(E, \theta, \beta_n) \cup \sigma_n. \tag{6.4}$$

Let us now calculate the probability of the event $\sigma_n(E, \theta, \beta_n)$. For this it is useful to find the coefficients of the random vector $\pi_n(\tau_\theta X)$ in the basis $(e_k^*)_1^n$. It follows from Lemma 5.1 that

$$(\pi_n(\tau_\theta X), e_k^*) = \sqrt{n} \sum_{m \equiv k \pmod{n}} a_m X_m \bar{\theta}^m = \sqrt{n}\, b_{kn} Z_k^\theta,$$

where $b_{kn} = (\sum_{m \equiv k \pmod{n}} a_m^2)^{1/2}$, and the complex-valued Gaussian variables $(Z_k^\theta)_1^n$ are independent and have zero mean and unit variance. Therefore,

$$\pi_n(\tau_\theta X) = \sum_{k=1}^{n} \sqrt{n}\, b_{kn} Z_k^\theta e_k^*.$$

To compute the desired probability it is convenient to subject the space $L^2(\mu_n)$ to the auxiliary linear transformation T determined by the equalities

$$Te_k^* = n^{-1/2} b_{kn}^{-1} e_k^*, \qquad k = 1, 2, \ldots, n.$$

It is then clear that $\pi_n(\tau_\theta X) \in K_0 + \varepsilon B_n + M$ if and only if

$$\sum_{k=1}^{n} Z_k^\theta e_k^* \in TK_0 + \varepsilon TB_n + TM.$$

It is not hard to show that $TK_0 = K$ and that

$$TB_n \subset n^{-1/2}\Big(\max_k b_{kn}^{-1}\Big)B_n.$$

Since T is invertible, codim TM = codim $M > n\delta$. Hence,

$$P(\sigma_n(E,\theta,\beta_n)) \leqslant \gamma_n\Big(K + \beta_n n^{-1/2}\Big(\max_k b_{kn}^{-1}\Big)B_n + TM\Big).$$

Let $\varepsilon_n \overset{\text{def}}{=} \beta_n n^{-1/2}\max_k b_{kn}^{-1}$. It is obvious that $b_{kn} \geqslant a_k$. By assumption,

$$\lim_{n\to+\infty} na_n = +\infty.$$

Therefore, there is a constant $c > 0$ such that $\inf_k b_{kn} \geqslant c/n$. Consequently,

$$\varepsilon_n \leqslant Cn^{\gamma+1}\cdot 2^{-\beta_n}n^{-1/2}c^{-1}n = \left(\frac{C}{c}\right)n^{\gamma+3/2}\cdot 2^{-\beta_n} \xrightarrow[n\to+\infty]{} 0.$$

Applying Theorem B to the chosen sequence $(\varepsilon_n)_{n\geqslant 1}$ and the fixed number $\delta > 0$, we get

$$P(\sigma_n(E,\theta,\beta_n)) \leqslant \text{const}\ \alpha_n^n,$$

where $\lim_n \alpha_n = 0$. It now follows from (6.4) that

$$P(J_n) \leqslant \sum_{E\in\mathcal{E}_n}\sum_{\theta\in F_{2^n}} P(\sigma_n(E,\theta,\beta_n)) + P(\sigma_n)$$
$$\leqslant 2^{2n}\cdot\text{const}\ \alpha_n^n + 2^{1-2n} \to 0, \qquad n\to+\infty.$$

§7. Proof of Theorem B

Let L be an arbitrary subspace of $L^2(\mu_n)$ such that $\dim_{\mathbf{C}} L \geqslant n\delta$, and M its orthogonal complement. We consider the orthogonal projection P_L of $L^2(\mu_n)$ onto L. It is well known that the image of the Gaussian measure γ_n under P_L is the standard Gaussian measure γ_L in L. Therefore,

$$\gamma_n(K + M + \varepsilon_n B_n) = \gamma_L(P_L K + \varepsilon_n P_L B_n).$$

The next lemma allows us to replace the computation of the Gaussian measure of the convex body $P_L K + \varepsilon_n P_L B_n$ in L by the computation of its volume in L.

LEMMA 7.1. *Among all the convex bodies in $\mathbf{C}^n$ with a given volume v the ball with center at zero has the largest Gaussian measure.*

This lemma is undoubtedly well known. For lack of a precise reference we give a proof here.

PROOF. Let us identify $\mathbf{C}^n$ with $\mathbb{R}^{2n}$. Then

$$d\gamma_n(x) = (1/\pi^n)e^{-\|x\|^2}dV_{2n}.$$

Let K be a convex body in $\mathbb{R}^{2n}$ with $\mathrm{vol}_{2n}(K) = v$. Integration in spherical coordinates shows that

$$\gamma_n(K) = \frac{1}{\pi^n}\int_0^{+\infty} e^{-r^2}\cdot I(r)\,dr, \qquad v = \int_0^{+\infty} I(r)\,dr,$$

where $I(r)$ is the area of the surface cut out by K on the sphere of radius r with center at zero. Let $\sigma_{2n}(r)$ be the total surface of this sphere. It is clear that $0 \leqslant I(r) \leqslant \sigma_{2n}(r)$. We remark that if $\int_0^{+\infty}\varphi_1 dr = \int_0^{+\infty}\varphi_2 dr$ and supp φ_2 lies to the right of supp φ_1, then

$$\int_0^{+\infty} e^{-r^2}\varphi_1\,dr \geqslant \int_0^{+\infty} e^{-r^2}\varphi_2\,dr.$$

Let r_v be the radius of a ball whose volume equals v. In the preceding inequality we set $\varphi_1 = \sigma_{2n}(r) - I(r)$ for $0 < r \leqslant r_v$, $\varphi_1 \equiv 0$ for $r > r_v$, and we set $\varphi_2(r) = I(r)$ for $r > r_v$, $\varphi_2 \equiv 0$ for $r < r_v$. Then

$$\int_0^{+\infty} \varphi_1\,dr = v - \int_0^{r_v} I(r)\,dr = \int_{r_v}^{\infty} I(r)\,dr = \int_0^{+\infty} \varphi_2\,dr.$$

Hence,

$$\int_0^{r_v} \{\sigma_{2n}(r) - I(r)\}e^{-r^2}\,dr \geqslant \int_{r_v}^{\infty} e^{-r^2} I(r)\,dr. \quad \bullet$$

LEMMA 7.2. *Let K be a convex body in $\mathbb{C}^n$ whose volume v is less than $(n-1)^n\pi^n/n!$. Then $\gamma_n(K) \leqslant v$.*

PROOF. Let r_v be the radius of a ball in $\mathbb{C}^n$ with volume v. By the preceding lemma,

$$\gamma_n(K) \leqslant \gamma_n(r_v B_n) = \frac{1}{\pi^n} 2n\omega_{2n}\int_0^{r_v} e^{-r^2} r^{2n-1}\,dr.$$

Here $\omega_{2n} = \pi^n/n!$ is the volume of the unit ball in $\mathbb{C}^n$. Then the surface area of the unit sphere is equal to $2n\omega_{2n}$, and so

$$\gamma_n(r_v B_n) = \frac{n}{n!}\int_0^{r_v^2} r^{n-1} e^{-r}\,dr.$$

The numbers r_v and v are connected by the relation $v = r_v^{2n}\pi^n/n!$, and since, by assumption, $v < (n-1)^n\pi^n/n!$, it follows that $r_v^2 < n-1$. On the other hand, the function $r \to r^{n-1}e^{-2}$ increases monotonically on $[0, n-1)$. Therefore,

$$\gamma_n(r_v B_n) \leqslant (n/n!)e^{-r_v^2}\cdot r_v^{2n} \leqslant (n/n!)r_v^{2n} = (n/\pi^n)v < v. \quad \bullet$$

This lemma enables us to get rid of the Gaussian measure in (6.2) and to reduce the proof of Theorem B to the proof of the inequality

$$\sup_{\dim_{\mathbb{C}} L \geqslant n\delta} \operatorname{vol}_{\dim L}(P_L K + \varepsilon_n P_L B_n) \leqslant \operatorname{const} \alpha_n^n. \tag{7.1}$$

Here vol_m denotes the volume in the m-dimensional space L.

So far we have considered the body K as a subset of $L^2(\mu_n)$. Since our problem has been reduced to an estimate of the volume of a certain body connected in the specified way with the compact set K, it is appropriate to forget the presence of the complex structure in $L^2(\mu_n)$ and to regard $L^2(\mu_n)$ as a vector space over $\mathbb{R}$. Let

$$e_{k1}^* = e_k^*, \qquad e_{k2}^* = ie_k^*, \qquad k = 1, 2, \ldots, n.$$

Then $(e_{kj}^*)_{k=1,\ldots,n;\, j=1,2}$ is a complete orthonormal system in the real Hilbert space $L^2(\mu_n)$ with inner product $(x, y)_{\mathbb{R}} \overset{\text{def}}{=} \operatorname{Re}(x, y)_{\mathbb{C}}$. It is clear that for any $x \in L^2(\mu_n)$ and $k = 1, \ldots, n$

$$(x, e_k^*)_{\mathbb{C}} = (x, e_{k1}^*)_{\mathbb{R}} + i(x, e_{k2}^*)_{\mathbb{R}}.$$

Therefore,

$$|(x, e_{k1}^*)_{\mathbb{R}}| + |(x, e_{k2}^*)_{\mathbb{R}}| \leqslant \sqrt{2}\,|(x, e_k^*)_{\mathbb{C}}|,$$

which implies that the compact set K is contained in the convex polyhedron

$$K_1 \stackrel{\text{def}}{=} \left\{ x \in L^2(\mu_n) \colon \sum_{k=1}^{n} b_{kn}\left(|(x, e_{k1}^*)_{\mathbb{R}}| + |(x, e_{k2}^*)_{\mathbb{R}}|\right) \leqslant C\sqrt{2} \right\}.$$

Any isometry J of $L^2(\mu_n)$ onto $\mathbb{R}^{2n}$ carrying the basis (e_{kj}^*) into the standard basis $(e_s)_1^{2n}$ of $\mathbb{R}^{2n}$ induces a bijection α of the set of pairs (k, j), $k = 1, \dots, n$, $j = 1, 2$, onto the interval $[1, 2n]$ of the natural numbers such that

$$Je_{kj}^* = e_s \Leftrightarrow \alpha(k, j) = s.$$

We set $d_s = b_{kn}$ if $s = \alpha(k, j)$, and we choose the bijection α (and with it the isometry J) in such a way that the sequence $(d_s)_1^{2n}$ thus obtained is nonincreasing.

LEMMA 7.3. *There exists a sequence $(c_k)_1^\infty$ that increases monotonically to $+\infty$ and is such that $d_s \geqslant c_s/s$ for any positive integer n and for $s = 1, \dots, n$.*

PROOF. For any n we have the obvious inequality

$$b_{kn} = \left(\sum_{m \equiv k (\operatorname{mod} n)} a_m^2 \right)^{1/2} \geqslant a_k.$$

The sequence $(d_s)_1^{2n}$ is obtained from $(b_{kn})_{k=1}^n$ as follows. Let us consider the nonincreasing rearrangement $(b_{kn}^*)_{k=1}^n$ of $(b_{kn})_{k=1}^n$ and write it with repetitions. The resulting sequence of length $2n$ coincides with $(d_s)_{s=1}^{2n}$, so that $d_s = b_{[(s+1)/2]^n}^*$. Let us now estimate b_{kn}^* from below. It is clear that $b_{kn}^* \geqslant a_k^*$, where $(a_k^*)_{k \geqslant 1}$ is the nonincreasing rearrangement of $(a_k)_{k \geqslant 1}$. We prove that $\lim_k k a_k^* = +\infty$. Indeed, for any $C > 0$ the inequality $a_k \geqslant Ck^{-1}$ holds for all $k \geqslant k(C)$. Since the sequence $1/k$ decreases, $a_k^* \geqslant Ck^{-1}$ for sufficiently large k. Now let

$$c_k^0 = \inf\{n a_n^* \colon n \geqslant k\}.$$

Then $c_n^0 \uparrow +\infty$ and $b_{kn}^* \geqslant c_k^0/k$. Consequently,

$$d_s = b_{[(s+1)/n]n}^* \geqslant \frac{c_{[(s+1)/2]}^0}{[(s+1)/2]} \geqslant \frac{2c_{[(s+1)/2]}^0}{s}$$

and we can set $c_k = 2c_{[(k+1)/2]}^0$. ●

For simplicity of notation let $K = JK_1$ again. Then

$$K = \left\{ x \in \mathbb{R}^{2n} \colon \sum_{s=1}^{2n} d_s |x_s| \leqslant C\sqrt{2} \right\}.$$

We now need some facts from integral geometry. Let L be a linear subspace of $\mathbb{R}^{2n}$, $m = \dim_{\mathbb{R}} L$, Q an arbitrary convex body in L, and B_m the unit ball in L. How do we compute the volume $\operatorname{vol}_m(Q + \varepsilon B_m)$? We have the Steiner formula (see [25], §6.1.8)

$$\operatorname{vol}_m(Q + \varepsilon B_m) = \sum_{j=0}^{m} W_j^{(m)}(Q) \varepsilon^j,$$

where the $W_j^{(m)}(Q)$ are the so-called Minkowski integrals of cross-sectional measures (Quermassintegrale). There is an integral representation for them that justifies this name. Let G_j^{m-j} be the Grassmann manifold of all $(m-j)$-dimensional (over $\mathbb{R}$, of course) subspaces of L, $d\tau$ a probability measure on G_m^{m-j} that is invariant under the rotation group of L acting on the elements of G_m^{m-j}, and ω_k the volume of the unit ball in $\mathbb{R}^k$. It is well known that $\omega_k = \pi^{k/2}/\Gamma(k/2+1)$. Then (see [25], §6.2.5)

$$W_j^{(m)}(Q) = \frac{\omega_m}{\omega_{m-j}} \int_{G_m^{m-j}} \operatorname{vol}_{m-j} P_M(Q)\, d\tau(M).$$

This leads to the obvious estimate

$$W_j^{(m)}(Q) \leqslant \frac{\omega_m}{\omega_{m-j}} \sup_{M \in G_m^{m-j}} \operatorname{vol}_{m-j} P_M(Q). \tag{7.2}$$

The following lemma enables us to get an estimate of $\operatorname{vol}_m P_L(Q)$ that is sufficiently good for our purposes.

LEMMA 7.4. *Let Q be a convex polyhedron in $\mathbb{R}^n$, v_m the largest volume of the m-dimensional faces of Q, and N_m the number of m-dimensional faces ($m \leqslant n$). Then for any subspace $L \subset \mathbb{R}^n$ of dimension m*

$$\operatorname{vol}_m P_L(Q) \leqslant N_m v_m.$$

PROOF. Let Γ_m be the family of all m-dimensional faces of Q. We show below that

$$P_L(Q) = \bigcup_{S \in \Gamma_m} P_L(S). \tag{7.3}$$

Since the volume can only be decreased under an orthogonal projection, the lemma follows from this. To prove (7.3) we introduce the orthogonal complement M of L and show that for any $x \in \mathbb{R}^n$ either $(x+M) \cap Q = \varnothing$ or $x+M$ intersects some m-dimensional face of Q. This follows from the next lemma.

LEMMA 7.5. *Let Q be a convex proper polyhedron in n-dimensional Euclidean space $\mathbb{R}^n$, and M a linear subspace of $\mathbb{R}^n$ with codimension m. If $(x+M) \cap Q \neq \varnothing$, then $x+M$ intersects some m-dimensional face of Q.*

We prove this by induction on n. If $n = 1$, then Q is a closed interval on the line, and there are two possibilities for M: $M = \{\mathbf{0}\}$ and $M = \mathbb{R}$. In the first case $m = 1$ and the lemma is obvious, and in the second case $m = 0$ and, therefore, $x + M = \mathbb{R}$ for every x.

Suppose that the lemma has been proved for all dimensions less than n. Let $(x+M) \cap Q \neq \varnothing$. Then $x+M$ intersects a face of Q, and, consequently, there is an $(n-1)$-dimensional face $Q^* \in \Gamma_{n-1}$ such that $(x+M) \cap Q^* \neq \varnothing$. Let H be the hyperplane in $\mathbb{R}^n$ containing the face Q^*. If it happens that $x + M \subset H$, then the codimension of $x+M$ in H is equal to $n-1-(n-m) = m-1$. The induction hypothesis is applicable to the polyhedron Q^* and the affine subspace $x+M$. Therefore, $x+M$ intersects an $(m-1)$-dimensional face of Q^* and, consequently, of Q. It remains to note that every $(m-1)$-dimensional face is a face of an m-dimensional face.

Let us now consider the case where $x + M \not\subset H$. Then the dimension of $(x+M) \cap H$ is obviously equal to $n-m-1$. Therefore, the codimension of $x+M$ in H is $n-1-(n-m-1) = m$. We again use the induction hypothesis and get that $x+M$ intersects an m-dimensional face of Q. ●

The next lemma enables us to get a sufficiently good estimate[1] of $\mathrm{vol}_m P_L(K)$ for any subspace L of dimension m.

LEMMA 7.6. *Let $(c_k)_{k\geqslant 1}$ be the monotonically increasing sequence whose existence is proved in Lemma* 7.3. *There exists a constant* const > 0 *such that for any subspace L of dimension m*

$$\mathrm{vol}_m P_L(K) \leqslant (\mathrm{const})^n (c_{2n-m} \cdots c_{2n})^{-1}.$$

PROOF. By Lemma 7.4, $\mathrm{vol}_m P_L(K) \leqslant N_m v_m$. The polyhedron K is made up of 2^{2n} simplexes obtained by reflections of the simplex

$$\left\{x \in \mathbb{R}^{2n} \colon \sum_{s=1}^{2n} d_s x_s \leqslant C\sqrt{2}\,,\, x_s \geqslant 0\right\}$$

with respect to the coordinate hyperplanes. Therefore,

$$N_m \leqslant 2^{2n} C_{2n}^{m+1} \leqslant 2^{4n}.$$

To estimate v_m we compute the m-dimensional volume of the part of the hypersurface $p_1 x_1 + \cdots + p_{m+1} x_{m+1} = C$, $x_j \geqslant 0$, that is a face of the simplex $\sum p_s |x_s| \leqslant C$. The surface can be given by the explicit formula

$$x_{m+1} = C/p_{m+1} - (p_1/p_{m+1}) x_1 - \cdots - (p_m/p_{m+1}) x_m,$$

where $p_1 x_1 + \cdots + p_m x_m \leqslant C$, $x_j \geqslant 0$, $j = 1, \ldots, m$. Then the desired volume is

$$\iiint \cdots \int_{p_1 x_1 + \cdots + p_m x_m \leqslant C} \left(1 + \frac{p_1^2}{p_{m+1}^2} + \cdots + \frac{p_m^2}{p_{m+1}^2}\right)^{1/2} dx_1 \cdots dx_m$$
$$= \frac{(p_1^2 + \cdots + p_{m+1}^2)^{1/2}}{m!\, p_{m+1}} \frac{C^m}{p_1 \cdots p_m} = \frac{C^m}{m!} \frac{(p_1^2 + \cdots + p_{m+1}^2)^{1/2}}{p_1 \cdots p_{m+1}}.$$

It is not hard to see that the function $x \to \sqrt{p + x^2}/x$ decreases on the half-line $(0, +\infty)$ for any $p > 0$. The volume of any m-dimensional face of our polyhedron K can be computed by the preceding formula if we identify the set $\{p_1, \ldots, p_{m+1}\}$ with a corresponding subset of $\{d_1, \ldots, d_{2n}\}$. The largest value for the volume is obtained by choosing

$$p_{m+1} = d_{2n}, \qquad p_m = d_{2n-1}, \ldots, p_2 = d_{2n-(m-1)}, \qquad p_1 = d_{2n-m}.$$

Consequently,

$$v_m \leqslant \frac{(C\sqrt{2})^m}{m!} \frac{(d_{2n-m}^2 + \cdots + d_{2n}^2)^{1/2}}{d_{2n-m} \cdots d_{2n}}.$$

We remark that

$$\sum_{s=1}^{2n} d_s^2 = 2 \sum_{k=1}^{n} b_{kn}^2 = 2 \sum_{j \in \mathbf{Z}} a_j^2 < +\infty$$

[1] Such an estimate is given in Lemma 1 of [10] for the case $d_1 = \cdots = d_n = 1$. Related arguments were used even earlier in [26] and [27].

and that, by Lemma 7.3, $d_s \geqslant c_s/s$. Hence,

$$v_m \leqslant \frac{(C\sqrt{2})^m}{m!} \frac{2\Sigma a_j^2}{c_{2n-m}/(2n-m)\cdots c_{2n}/2n}\left(2\sum_{j\in\mathbf{Z}} a_j^2\right)(C\sqrt{2})^m$$
$$\times \frac{(2n)!}{m!\,(2n-m)!}(c_{2n-m}\cdots c_{2n})^{-1}$$
$$\leqslant (\text{const})^n (c_{2n-m}\cdots c_{2n})^{-1}. \quad \bullet$$

We can now conclude the proof of Theorem B. To prove (7.1) it obviously suffices to verify that

$$\sup_{\dim_{\mathbb{R}} L \geqslant 2n\delta} \operatorname{vol}_{\dim L}(P_L K + \varepsilon_n P_L B_{2n}) \leqslant \text{const}\, \alpha_n^n.$$

Let us fix an arbitrary subspace L of $\mathbb{R}^{2n}$ such that $m \overset{\text{def}}{=} \dim_{\mathbb{R}} L \geqslant 2n\delta$. From Steiner's formula and (7.2) it follows that

$$\operatorname{vol}_m(P_L K + \varepsilon_n P_L B_{2n}) = \sum_{j=0}^{m} W_j^{(m)}(P_L K)\varepsilon_n^j$$
$$\leqslant \sum_{j=0}^{m} \frac{\omega_m}{\omega_{m-j}} \varepsilon_n^j \sup_{M\in G_m^{m-j}(L)} \operatorname{vol}_{m-j} P_M P_L K.$$

Since $M \subset L$, $P_M P_L = P_M$. Moreover,

$$\frac{\omega_m}{\omega_{m-j}} = \frac{\pi^{m/2}}{\Gamma(m/2+1)} \frac{\Gamma((m-j)/2+1)}{\pi^{m-j/2}} \leqslant \pi^{j/2}.$$

Therefore,

$$\operatorname{vol}_m(P_L K + \varepsilon_n P_L B_{2n}) \leqslant \sum_{j=0}^{m} (\sqrt{\pi}\,\varepsilon_n)^j \sup_{\dim M = m-j} \operatorname{vol}_{m-j} P_M K.$$

We now use Lemma 7.3. Then

$$\sup_{\dim M = m-j} \operatorname{vol}_{m-j} P_M K \leqslant \text{const}(c_{2n-(m-j)}\cdots c_{2n})^{-1}.$$

Collecting everything together, we have

$$\operatorname{vol}_m(P_L K + \varepsilon_n P_L B_{2n}) \leqslant (\text{const})^n \sum_{j=0}^{m} \varepsilon_n^j (c_{2n-(m-j)}\cdots c_{2n})^{-1}$$
$$\leqslant (\text{const})^n \left\{ \sum_{j=0}^{[n\delta]} \varepsilon_n^j (c_{2n-(m-j)}\cdots c_{2n})^{-1} + \sum_{j=[n\delta]+1}^{m} \varepsilon_n^j (c_{2n-(m-j)}\cdots c_{2n})^{-1} \right\}.$$

Since $m \geqslant 2n\delta$, $(m-j) \geqslant n\delta$ for $j = 0, \ldots, [n\delta]$. Therefore,

$$\operatorname{vol}_m(P_L K + \varepsilon_n P_L B_{2n}) \leqslant (\text{const})^n n\delta (c_{2n}\cdots c_{2n-[n\delta]})^{-1} + (\text{const})^n n\varepsilon_n^{n\delta}.$$

It follows from Lemma 7.3 that $\lim_k c_k = +\infty$. Moreover, the product $c_{2n}\cdots c_{2n-[n\delta]}$ contains $[n\delta]+1$ factors. It is not hard now to construct a sequence α_n, $\lim_n \alpha_n = 0$, such that

$$(\text{const})^n n\delta (c_{2n}\cdots c_{2n-[n\delta]})^{-1} + (\text{const})^n n\varepsilon_n^{n\delta} = o(\alpha_n^n). \quad \bullet$$

BIBLIOGRAPHY

1. D. Menchoff [Men'shov], *Sur la convergence uniforme des séries de Fourier*, Mat. Sb. **11** (**53**) (1942), 67–96.

2. N. K. Bari, *Trigonometric series*, Fizmatgiz, Moscow, 1961; English transl., Vols. I, II, Macmillan, New York, and Pergamon Press, Oxford, 1964.

3. Lennart Carleson, *On convergence and growth of partial sums of Fourier series*, Acta Math. **116** (1966), 135–157.

4. S. V. Khrushchev, *A simple proof of Men'shov's correction theorem*, Preprint no. R-1-78, Leningrad. Otdel. Mat. Inst. Steklov. (LOMI), Leningrad, 1978. (Russian)

5. F. G. Arutyunyan, *Representation of functions by multiple series*, Akad. Nauk Armyan. SSR Dokl. **64** (1977), 72–76. (Russian)

6. ______, *Representation of measurable functions by almost everywhere convergent series*, Mat. Sb. **90** (**132**) (1973), 483–520; English transl. in Math. USSR Sb. **19** (1973).

7. B. D. Kotlyar, *Walsh series and the theorem of Men'shov on "correction" of functions*. Izv. Akad. Nauk SSSR Ser. Mat. **30** (1966), 1193–1200. (Russian)

8. G. M. Chernomashentsev, *On the "correction" of functions of several variables defined on the torus*, Dokl. Akad. Nauk SSSR **176** (1967), 277–278; English transl. in Soviet Math. Dokl. **8** (1967).

9. Y. Katznelson, *On a theorem of Menchoff*, Proc. Amer. Math. Soc. **53** (1975), 396–398.

10. A. M. Olevskiĭ, *The existence of functions with unremovable Carleman singularities*, Dokl. Akad. Nauk SSSR **238** (1978), 796–799; English transl. in Soviet Math. Dokl. **19** (1978).

11. P. L. Ul'yanov, *Solved and unsolved problems in the theory of trigonometrical and orthogonal series*, Uspekhi Mat. Nauk **19** (1964), no. 1 (115), 3–69; English transl. in Russian Math. Surveys **19** (1964).

12. Jean-Pierre Kahane, *Some random series of functions*, Heath, Lexington, Mass., 1968.

13. Sh. V. Kheladze, *On the convergence of Fourier series almost everywhere and in L*, Dokl. Akad. Nauk SSSR **232** (1977), 535–537; English transl. in Soviet Math. Dokl. **18** (1977).

14. K. I. Oskolkov, *Uniform modulus of continuity of summable functions on sets of positive measure*, Dokl. Akad. Nauk SSSR **229** (1976), 304–306; English transl. in Soviet Math. Dokl. **17** (1976).

15. Yu. S. Fridlyand, *The nonremovable Carleman singularity for the Haar system*, Mat. Zametki **14** (1973), 799–807; English transl. in Math. Notes **14** (1973).

16. Patrick Billingsley, *Ergodic theory and information*, Wiley, 1965.

17. S. V. Koslyakov, *The Fourier coefficients of the boundary values of functions analytic in the disk and in the bidisk*, this issue.

18. D. E. Men'shov, *On Fourier series of continuous functions*, Moskov. Gos. Univ. Uchen. Zap. **148** (1951), Mat. 4, 108–132. (Russian)

19. A. Zygmund, *Trigonometric series*, 2nd rev. ed., Vol. I, Cambridge Univ. Press, 1959.

20. N. Bourbaki, *Intégration*, Chaps. 1–5, Actualités Sci. Indust. nos. 1175, 1244, Hermann, Paris, 1952, 1956.

21. V. A. Rokhlin, *On endomorphisms of compact commutative groups*, Izv. Akad. Nauk SSSR Ser. Mat. **13** (1949), 329–340. (Russian)

22. James R. Brown, *Ergodic theory and topological dynamics*, Academic Press, 1976.

23. Edwin Hewitt and Kenneth A. Ross, *Abstract harmonic analysis*. Vol. I, Springer-Verlag, 1963.

24. Stefan Kacmarz and Hugo Steinhaus, *Theorie der Orthogonalreihen*, PWN, Warsaw, 1935; reprint, Chelsea, New York, 1951; Russian transl., Fizmatgiz, Moscow, 1958.

25. H. Hadwiger, *Vorlesungen über Inhalt, Oberfläche und Isoperimetrie*, Springer-Verlag, 1957.

26. B. S. Kashin, *General orthonormal systems and some questions from approximation theory*, Dissertation, Moscow, 1977; abstract published in Mat. Zametki **26** (1979), 299–315; English transl. in Math. Notes **26** (1979).

27. ______, *Widths of some finite-dimensional sets and classes of smooth functions*, Izv. Akad. Nauk SSSR Ser. Mat. **41** (1977), 334–351; English transl. in Math. USSR Izv. **11** (1977).

28. K. I. Oskolkov, *Approximation properties of summable functions on sets of full measure*, Mat. Sb. **103** (**145**) (1977), 563–589; English transl. in Math. USSR Sb. **32** (1977).

29. Daniel M. Oberlin, *Uniformly convergent Fourier series and sets of measure zero*, preprint, 1978.

Translated by H. H. McFADEN

ABCDEFGHIJ–AMS–89876543

COPYING AND REPRINTING

Individual readers of this publication, and nonprofit libraries acting for them, are permitted to make fair use of the material, such as to copy an article for use in teaching or research. Permission is granted to quote brief passages from this publication in reviews provided the customary acknowledgement of the source is given.

Republication, systematic copying, or multiple reproduction of any material in this publication (including abstracts) is permitted only under license from the American Mathematical Society. Requests for such permission should be addressed to the Executive Director, American Mathematical Society, P. O. Box 6248, Providence, Rhode Island 02940.

The appearance of the code on the first page of an article in this journal indicates the copyright owner's consent for copying beyond that permitted by Sections 107 or 108 of the U. S. Copyright Law, provided that the fee of $1.00 plus $.25 per page for each copy be paid directly to Copyright Clearance Center, Inc., 21 Congress Street, Salem, Massachusetts 01970. This consent does not extend to other kinds of copying, such as copying for general distribution, for advertising or promotion purposes, for creating new collective works or for resale.